UCLA Symposia on Molecular and Cellular Biology, New Series

Series Editor, C. Fred Fox

RECENT TITLES

Volume 74
Growth Regulation of Cancer,
Marc E. Lippman, *Editor*

Volume 75
Steroid Hormone Action, Gordon Ringold, *Editor*

Volume 76
Molecular Biology of Intracellular Protein Sorting and Organelle Assembly,
Ralph A. Bradshaw, Lee McAlister-Henn, and Michael G. Douglas, *Editors*

Volume 77
Signal Transduction in Cytoplasmic Organization and Cell Motility, Peter Satir, John S. Condeelis, and Elias Lazarides, *Editors*

Volume 78
Tumor Progression and Metastasis,
Garth L. Nicolson and Isaiah J. Fidler, *Editors*

Volume 79
Altered Glycosylation in Tumor Cells,
Christopher L. Reading, Senitiroh Hakomori, and Donald M. Marcus, *Editors*

Volume 80
Protein Recognition of Immobilized Ligands,
T.W. Hutchens, *Editor*

Volume 81
Biological and Molecular Aspects of Atrial Factors, Philip Needleman, *Editor*

Volume 82
Oxy-Radicals in Molecular Biology and Pathology, Peter A. Cerutti, Irwin Fridovich, and Joe M. McCord, *Editors*

Volume 83
Mechanisms and Consequences of DNA Damage Processing, Errol C. Friedberg and Philip C. Hanawalt, *Editors*

Volume 84
Technological Advances in Vaccine Development, Laurence Lasky, *Editor*

Volume 85
B Cell Development, Owen N. Witte, Norman R. Klinman, and Maureen C. Howard, *Editors*

Volume 86
Synthetic Peptides: Approaches to Biological Problems, Emil Thomas Kaiser and James P. Tam, *Editors*

Volume 87
Gene Transfer and Gene Therapy, I. Verma, R. Mulligan, and A. Beaudet, *Editors*

Volume 88
Molecular Biology of the Eye: Genes, Vision, and Ocular Disease, Joram Piatigorsky, Toshimichi Shinohara, and Peggy S. Zelenka, *Editors*

Volume 89
Liposomes in the Therapy of Infectious Diseases and Cancer, Gabriel Lopez-Berestein and Isaiah J. Fidler, *Editors*

Volume 90
Cell Biology of Virus Entry, Replication, and Pathogenesis, Richard W. Compans, Ari Helenius, and Michael B.A. Oldstone, *Editors*

Volume 91
Bone Marrow Transplantation: Current Controversies, Richard E. Champlin and Robert Peter Gale, *Editors*

Volume 92
The Molecular Basis of Plant Development,
R. Goldberg, *Editor*

Volume 93
Cellular and Molecular Biology of Muscle Development, Laurence H. Kedes and Frank E. Stockdale, *Editors*

Volume 94
Molecular Biology of RNA, Thomas R. Cech, *Editor*

Volume 95
DNA-Protein Interactions in Transcription,
Jay D. Gralla, *Editor*

Volume 96
Stress-Induced Proteins, Mary Lou Pardue, James R. Feramisco, and Susan Lindquist, *Editors*

Volume 97
Molecular Biology of Stress, Shlomo Breznitz and Oren Zinder, *Editors*

Volume 98
Metal Ion Homeostasis: Molecular Biology and Chemistry, Dean H. Hamer and Dennis R. Winge, *Editors*

Volume 99
Human Tumor Antigens and Specific Tumor Therapy, Richard S. Metzgar and Malcolm S. Mitchell, *Editors*

Please contact the publisher for information about previous titles in this series.

UCLA Symposia Board

C. Fred Fox, Ph.D., Director
Professor of Microbiology, University of California, Los Angeles

Charles J. Arntzen, Ph.D.
Director, Plant Science and Microbiology
E.I. du Pont de Nemours and Company

Floyd E. Bloom, M.D.
Director, Preclinical Neurosciences/
Endocrinology
Scripps Clinic and Research Institute

Ralph A. Bradshaw, Ph.D.
Chairman, Department of Biological
Chemistry
University of California, Irvine

Francis J. Bullock, M.D.
Vice President, Research
Schering Corporation

Ronald E. Cape, Ph.D., M.B.A.
Chairman
Cetus Corporation

Ralph E. Christoffersen, Ph.D.
Executive Director of Biotechnology
Upjohn Company

John Cole, Ph.D.
Vice President of Research
and Development
Triton Biosciences

Pedro Cuatrecasas, M.D.
Vice President of Research
Glaxo, Inc.

Mark M. Davis, Ph.D.
Department of Medical Microbiology
Stanford University

J. Eugene Fox, Ph.D.
Vice President, Research
and Development
Miles Laboratories

J. Lawrence Fox, Ph.D.
Vice President, Biotechnology Research
Abbott Laboratories

L. Patrick Gage, Ph.D.
Director of Exploratory Research
Hoffmann-La Roche, Inc.

Gideon Goldstein, M.D., Ph.D.
Vice President, Immunology
Ortho Pharmaceutical Corp.

Ernest G. Jaworski, Ph.D.
Director of Biological Sciences
Monsanto Corp.

Irving S. Johnson, Ph.D.
Vice President of Research
Lilly Research Laboratories

Paul A. Marks, M.D.
President
Sloan-Kettering Memorial Institute

David W. Martin, Jr., M.D.
Vice President of Research
Genentech, Inc.

Hugh O. McDevitt, M.D.
Professor of Medical Microbiology
Stanford University School of Medicine

Dale L. Oxender, Ph.D.
Director, Center for Molecular Genetics
University of Michigan

Mark L. Pearson, Ph.D.
Director of Molecular Biology
E.I. du Pont de Nemours and Company

George Poste, Ph.D.
Vice President and Director of Research
and Development
Smith, Kline and French Laboratories

William Rutter, Ph.D.
Director, Hormone Research Institute
University of California, San Francisco

George A. Somkuti, Ph.D.
Eastern Regional Research Center
USDA-ARS

Donald F. Steiner, M.D.
Professor of Biochemistry
University of Chicago

UCLA Symposia Board membership at the time of the meeting is indicated on the above list.

Molecular Biology
of RNA

Molecular Biology of RNA

Proceedings of a Director's Sponsors-UCLA Symposium
Held at Keystone, Colorado
April 4–10, 1988

Editor

Thomas R. Cech
Howard Hughes Medical Institute
Department of Chemistry and Biochemistry
University of Colorado
Boulder, Colorado

Alan R. Liss, Inc. • New York

**Address all Inquiries to the Publisher
Alan R. Liss, Inc., 41 East 11th Street, New York, NY 10003**

Copyright © 1989 Alan R. Liss, Inc.

Printed in the United States of America.

Under the conditions stated below the owner of copyright for this book hereby grants permission to users to make photocopy reproductions of any part or all of its contents for personal or internal organizational use, or for personal or internal use of specific clients. This consent is given on the condition that the copier pay the stated per-copy fee through the Copyright Clearance Center, Incorporated, 27 Congress Street, Salem, MA 01970, as listed in the most current issue of "Permissions to Photocopy" (Publisher's Fee List, distributed by CCC, Inc.), for copying beyond that permitted by sections 107 or 108 of the US Copyright Law. This consent does not extend to other kinds of copying, such as copying for general distribution, for advertising or promotional purposes, for creating new collective works, or for resale.

Library of Congress Cataloging-in-Publication Data

UCLA Symposium on Molecular Biology of RNA (1988 : Keystone, Colo.)
   Molecular biology of RNA : proceedings of a Director's Sponsors -UCLA Symposium, held at Keystone, Colorado, April 4–10, 1988 / editor, Thomas R. Cech.
     p. cm. — (UCLA symposia on molecular and cellular biology ; new ser., v. 94)
   "The 1988 UCLA Symposium on Molecular Biology of RNA"—Pref.
   Includes bibliographies and index.
   ISBN 0-8451-2693-8
   1. RNA—Congresses. I. Cech, Thomas. II. University of California, Los Angeles. III. Title. IV. Series.
   [DNLM: 1. Molecular Biology—congresses. 2. RNA—congresses. W3 U17N new ser. v. 94 / QU 58 U175m 1988]
QP623.U25   1988
574.87'3283—dc 19
DNLM/DLC
for Library of Congress   88-37911
                                                                                CIP

Contents

Contributors xi
Preface
    Thomas R. Cech xvii

I. STRUCTURE OF RNA

Crystal Structure of a Kinked RNA
    A.C. Dock-Bregeon, B. Chevrier, A. Podjarny, D. Moras, J.S. de
    Bear, G.R. Gough, P.T. Gilham, and J.J. Johnson 1
Phylogenetic Comparisons of U2 Small Nuclear RNA Sequences Suggest a Pseudoknotted Structure
    Manuel Ares, Jr. and A. Haller Igel 13
Pseudoknotted RNA Oligonucleotides
    Jacqueline R. Wyatt, Joseph D. Puglisi, and Ignacio Tinoco, Jr. 25
Dimers of M1 RNA as Determined by Psoralen Crosslinking
    Samuel E. Lipson and John E. Hearst 33

II. RNA CATALYSIS

Towards Defining the Minimal Structural Requirements for Self-Splicing of the Phage T4 *td* Intron
    Marlene Belfort, Jill L. Galloway Salvo, Karen Ehrenman, and
    Timothy Coetzee 49
A Group I Intron in *Bacillus subtilis* Bacteriophage SPO1
    Heidi A. Goodrich, Jonatha M. Gott, Ming-Qun Xu,
    Vincenzo Scarlato, and David A. Shub 59
A Site-Directed Mutation in the *Bacillus subtilis* Ribonuclease P RNA That Selectively Influences Its Protein-Dependent Reaction
    David S. Waugh, Michael W. Sganga, Michael Raff, and
    Norman R. Pace 67
Processing of a Synthetic tRNA Precursor Model by *E. coli* RNase P and M1 RNA
    Christopher K. Surratt, Barbara J. Carter, and Sidney M. Hecht 79
Cleavage of the tRNA-Like RNA of the Turnip Yellow Mosaic Virus by the Catalytic RNA Component of RNase P
    Christopher J. Green and Barbara S. Vold 89
The Self-Cleaving RNAs of Human Hepatitis Delta Virus
    John Taylor, Lamia Sharmeen, Mark Kuo, and Gail Dinter-Gottlieb . . . 99

III. RNA-PROTEIN INTERACTIONS

Discussion Summary: RNA–Protein Interactions
David E. Draper 109

Ribosomal Protein S4 Recognizes Complex Messenger and Ribosomal RNA Structures
David E. Draper, Careen Tang, and Jailaxmi Vartikar 113

The Specificity of the RNA Binding Activity of *Xenopus* Transcription Factor IIIA
Paul J. Romaniuk, Isabel L. de Stevenson, and Qimin You 123

Ultrastructural Analysis of RNA Splicing *In Vivo*
Ann Beyer, Milan Jamrich, and Yvonne Osheim 133

IV. mRNA SPLICING AND EDITING

Structural Characterization of the Intron Containing S25 Ribosomal Protein Genes From Two *Tetrahymena* Species
Jan Engberg, Kirsten Bojsen, and Henrik Nielsen 145

Intron Recognition in Plants
Greg Goodall, Karin Wiebauer, and Witold Filipowicz 155

Differential Accumulation of U4 Small Nuclear RNAs During Chicken Development
Gina M. Korf and William E. Stumph 165

Developmental Stage Specific RNA-Editing of a Mitochondrial Transcript of *Trypanosoma brucei*
Paul Sloof, Hans van der Spek, Janny van den Burg, and Rob Benne . . . 177

Transcript Alteration by mRNA Editing in Kinetoplastid Mitochondria
Jean E. Feagin and Kenneth Stuart 187

The Generation of Poly(A) Heads on Vaccinia Late mRNA: Proposal of a Slippage Mechanism
Hendrik G. Stunnenberg, Luisa de Magistris, and Beate Schwer 199

V. RNA IN TRANSLATION

The Role of 16S RNA in Ribosome Function: *In Vitro* Synthesis, Assembly and Function of 30S Ribosomes Containing Single Base Alterations in the 16S RNA
D. Negre, P.R. Cunningham, C. Weitzmann, R. Denman, J. Colgan, K. Nurse, and J. Ofengand 209

Ribosomal RNA and UGA-Dependent Peptide Chain Termination
Emanuel J. Murgola, H. Ulrich Göringer, Albert E. Dahlberg, and Kathryn A. Hijazi 221

Translation Control of the S10 Ribosomal Protein Operon of *Escherichia coli*
L. Lindahl, P. Shen, and J.M. Zengel 231

Eukaryotic Viral 5'-Leader Sequences Act as Translational Enhancers in Eukaryotes and Prokaryotes
Daniel R. Gallie, Clarence I. Kado, John W.B. Hershey, Michael A. Wilson, and Virginia Walbot 237

Involvement of 3'-Poly(A) Tracts in Protein Synthesis: A Role in mRNP Assembly
David Munroe and Allan Jacobson 257

VI. NOVEL REACTIONS OF tRNA

Glutamyl-tRNAs From Chloroplasts as Cofactors in Non-Ribosomal Enzymatic Reactions
Astrid Schön, Gary O'Neill, David Peterson, and Dieter Söll 271

Host tRNA Reprocessing: A Phage T4-System of RNA-Splicing Mediated by Protein Enzymes
G. Kaufmann, M. Amitsur, D. Chapman, M.J. Gait, R. Levitz, L. Jorissen, I. Morad, and L. Snyder 281

VII. REGULATION OF GENE EXPRESSION AT THE RNA LEVEL

Regulation of Gene Expression by a Small Antisense RNA in Bacteriophage P22
Sha-Mei Liao and William R. McClure 289

Antisense RNA-Mediated Inhibition of Viral Infection in Tissue Culture and Transgenic Mice
Kathy M. Takayama, Shigeki Kuriyama, Susan Weiss, Kiran Chada, Sumiko Inouye, and Masayori Inouye 299

Site-Specific Endonucleolytic Cleavages in the 5' Region of the *E. coli* *ompA* and *bla* mRNA Seem to Initiate Degradation
A. von Gabain, U. Lundberg, Ö. Melefors, and G. Nilsson 311

VIII. RNA 3' END FORMATION

Discussion Summary: Transcription Termination and RNA 3' End Formation
Claire L. Moore 321

Mutational and Chemical Modifications of Rho Factor Affect Its Domain Conformations and Interactions
T. Platt, C.A. Brennan, A.J. Dombroski, and P. Spear 325

Two Separable Activities Are Required for Pre-mRNA 3' End Formation: A Poly(A) Polymerase and a Cleavage/Specificity Factor
Lisa C. Ryner, Yoshio Takagaki, Justina Voulgaris, and James L. Manley 335

Proteins That Interact With the SV40 Late Polyadenylation Signal
Jeffrey Wilusz and Thomas Shenk 351

IX. RNA IN EVOLUTION

Building the RNA World: Evolution of Catalytic RNA in the Laboratory
Gerald F. Joyce 361

A Stereoselective Binding Site For Arginine on the *Tetrahymena* Self-Splicing Intron
Michael Yarus 373

Index 375

Contributors

M. Amitsur, Tel-Aviv University, Tel-Aviv, Israel 69978 [281]

Manuel Ares, Jr., Department of Biology, University of California, Santa Cruz, CA 95064 [13]

Marlene Belfort, Wadsworth Center for Laboratories and Research, New York State Department of Health, Albany, NY 12201-0509 and School of Public Health Sciences, The State University of New York, New York State Department of Health, Albany, NY 12201 [49]

Rob Benne, Department of Biochemistry, University of Amsterdam, Academic Medical Center, 1105 AZ Amsterdam, The Netherlands [177]

Ann Beyer, Department of Microbiology, Cancer Center and Department of Biology, University of Virginia, Charlottesville, VA 22908 [133]

Kirsten Bojsen, Department of Biochemistry B, University of Copenhagen, DK-2200 Copenhagen, Denmark [145]

Catherine A. Brennan, Department of Biochemistry, University of Rochester Medical Center, Rochester, NY 14642 [325]

Barbara J. Carter, Departments of Chemistry and Biology, University of Virginia, Charlottesville, VA 22901 [79]

Thomas R. Cech, Howard Hughes Medical Institute, Department of Chemistry and Biochemistry, University of Colorado, Boulder, CO 80309 [xvii]

Kiran Chada, Department of Biochemistry, UMDNJ-Robert Wood Johnson Medical School, Piscataway, NJ 08854 [299]

D. Chapman, Tel-Aviv University, Tel-Aviv, Israel 69978 [281]

Bernard Chevrier, Laboratoire de Cristallographie Biologique, I.B.M.C. du C.N.R.S., 67084 Strasbourg Cedex, France [1]

Timothy Coetzee, Wadsworth Center for Laboratories and Research, New York State Department of Health, Albany, NY 12201-0509 and Department of Microbiology and Immunology, Albany Medical College, Albany, NY 12208 [49]

John Colgan, Roche Institute of Molecular Biology, Roche Research Center, Nutley, NJ 07110 [209]

The numbers in brackets are the opening page numbers of the contributors' articles.

Contributors

Phillip R. Cunningham, Roche Institute of Molecular Biology, Roche Research Center, Nutley, NJ 07110 [209]

Albert E. Dahlberg, Section of Biochemistry, Brown University, Providence, RI 02912 [221]

J.S. de Bear, Department of Biological Sciences, Purdue University, West Lafayette, IN 47907 [1]

Luisa de Magistris, European Molecular Biology Laboratory, 6900 Heidelberg, Federal Republic of Germany [199]

Robert Denman, Roche Institute of Molecular Biology, Roche Research Center, Nutley, NJ 07110; present address: Institute for Basic Research in Developmental Disabilities, Staten Island, NY 10314 [209]

Isabel Leal de Stevenson, Department of Biochemistry and Microbiology, University of Victoria, Victoria, British Columbia, Canada V8W 2Y2 [123]

Gail Dinter-Gottlieb, Department of Bioscience and Biotechnology, Drexel University, Philadelphia, PA 19104 [99]

Anne-Catherine Dock-Bregeon, Laboratoire de Cristallographie Biologique, I.B.M.C. du C.N.R.S., 67084 Strasbourg Cedex, France [1]

Alicia J. Dombroski, Department of Biochemistry, University of Rochester Medical Center, Rochester, NY 14642 [325]

David E. Draper, Department of Chemistry, Johns Hopkins University, Baltimore, MD 21218 [109, 113]

Karen Ehrenman, Wadsworth Center for Laboratories and Research, New York State Department of Health, Albany, NY 12201-0509 and Department of Microbiology and Immunology, Albany Medical College, Albany, NY 12208 [49]

Jan Engberg, Department of Biochemistry B, University of Copenhagen, DK-2200 Copenhagen, Denmark [145]

Jean E. Feagin, Seattle Biomedical Research Institute, Seattle, WA 98109-1651 [187]

Witold Filipowicz, Friedrich Miescher Institute, Basel CH4002, Switzerland [155]

Michael J. Gait, Department of Molecular Biology, Medical Research Council, Cambridge CB2 2QH, England [281]

Daniel R. Gallie, Department of Biological Sciences, Stanford University, Stanford, CA 93405-5020 [237]

Jill L. Galloway Salvo, Wadsworth Center for Laboratories and Research, New York State Department of Health, Albany, NY 12201-0509 [49]

P.T. Gilham, Department of Biological Sciences, Purdue University, West Lafayette, IN 47907 [1]

Greg Goodall, Friedrich Miescher Institute, Basel CH4002, Switzerland [155]

Heidi A. Goodrich, Department of Biological Sciences, State University of New York at Albany, NY 12222 [59]

H. Ulrich Göringer, Section of Biochemistry, Brown University, Providence, RI 02912 [221]

Jonatha M. Gott, Department of Biological Sciences, State University of New York at Albany, NY 12222; present address: Department of Chemistry and Biochemistry, University of Colorado, Boulder, CO 80309 [59]

Contributors

G.R. Gough, Department of Biological Sciences, Purdue University, West Lafayette, IN 47907 **[1]**

Christopher J. Green, Department of Molecular Biology, SRI International, Menlo Park CA 94025 **[89]**

John E. Hearst, Department of Chemistry, University of California, Berkeley, CA 94720 **[33]**

Sidney M. Hecht, Departments of Chemistry and Biology, University of Virginia, Charlottesville, VA 22901 **[79]**

John W.B. Hershey, Department of Biological Chemistry, University of California, Davis, CA 95616 **[237]**

Kathryn A. Hijazi, Department of Molecular Genetics, The University of Texas, M.D. Anderson Center, Houston, TX 77030 **[221]**

A. Haller Igel, Department of Biology, University of California, Santa Cruz, CA 95064 **[13]**

Masayori Inouye, Department of Biochemistry, UMDNJ-Robert Wood Johnson Medical School, Piscataway, NJ 08854 **[299]**

Sumiko Inouye, Department of Biochemistry, UMDNJ-Robert Wood Johnson Medical School, Piscataway, NJ 08854 **[299]**

Allan Jacobson, Department of Molecular Genetics and Microbiology, University of Massachusetts Medical School, Worcester, MA 01655 **[257]**

Milan Jamrich, Department of Microbiology, Cancer Center and Department of Biology, University of Virginia, Charlottesville, VA 22908; present address: Laboratory of Molecular Genetics, NICHD, National Institutes of Health, Bethesda, MD 20892 **[133]**

John J. Johnson, Department of Biological Sciences, Purdue University, West Lafayette, IN 47907 **[1]**

L. Jorissen, Michigan State University, Lansing, MI 48824 **[281]**

Gerald F. Joyce, The Salk Institute for Biological Studies, La Jolla, CA 92037 **[361]**

Clarence I. Kado, Department of Plant Pathology, University of California, Davis, CA 95616 **[237]**

Gabriel Kaufmann, Department of Biochemistry, Tel-Aviv University, Israel 69978 **[281]**

Gina M. Korf, Department of Chemistry and Molecular Biology Institute, San Diego State University, San Diego, CA 92182 **[165]**

Mark Kuo, Fox Chase Cancer Center, Philadelphia, PA 19111 **[99]**

Shigeki Kuriyama, Department of Biochemistry, UMDNJ-Robert Wood Johnson Medical School, Piscataway, NJ 08854 **[299]**

R. Levitz, Tel-Aviv University, Tel-Aviv, Israel 69978 **[281]**

Sha-Mei Liao, Department of Biological Sciences, Carnegie Mellon University, Pittsburgh, PA 15213 **[289]**

Lasse Lindahl, Department of Biology, University of Rochester, Rochester, NY 14627 **[231]**

Samuel E. Lipson, Department of Chemistry, University of California, Berkeley, CA 94720; present address: CODON, South San Francisco, CA 94080 **[33]**

Urban Lundberg, Department of Bacteriology, Karolinska Institute, S-104 01 Stockholm, Sweden **[311]**

Contributors

James L. Manley, Department of Biological Sciences, Columbia University, New York, NY 10027 **[335]**

William McClure, Department of Biological Sciences, Carnegie Mellon University, Pittsburgh, PA 15213 **[289]**

Öjar Melefors, Department of Bacteriology, Karolinska Institute, S-104 01 Stockholm, Sweden **[311]**

Claire L. Moore, Department of Molecular Biology and Microbiology, Tufts University School of Medicine, Boston, MA 02111 **[321]**

I. Morad, Tel-Aviv University, Tel-Aviv, Israel 69978 **[281]**

Dino Moras, Laboratoire de Cristallographie Biologique, I.B.M.C. du C.N.R.S., 67084 Strasbourg Cedex, France **[1]**

David Munroe, Department of Molecular Genetics and Microbiology, University of Massachusetts Medical School, Worcester, MA 01655 **[257]**

Emanuel J. Murgola, Department of Molecular Genetics, The University of Texas, M.D. Anderson Cancer Center, Houston, TX 77030 **[221]**

Didier Negre, Roche Institute of Molecular Biology, Roche Research Center, Nutley, NJ 07110 **[209]**

Henrik Nielsen, Department of Biochemistry B, University of Copenhagen, DK-2200 Copenhagen, Denmark **[145]**

Gisela Nilsson, Department of Bacteriology, Karolinska Institute, S-104 01 Stockholm, Sweden; present address: Department of Cell Biology, University of Basel, CH-4056 Basel, Switzerland **[311]**

Kelvin Nurse, Roche Institute of Molecular Biology, Roche Research Center, Nutley, NJ 07110 **[209]**

James Ofengand, Roche Institute of Molecular Biology, Roche Research Center, Nutley, NJ 07110 **[209]**

Gary O'Neill, Department of Molecular Biophysics and Biochemistry, Yale University, New Haven, CT 06511 **[271]**

Yvonne Osheim, Department of Microbiology, Cancer Center and Department of Biology, University of Virginia, Charlottesville, VA 22908 **[133]**

Norman R. Pace, Department of Biology and Institute for Molecular and Cellular Biology, Indiana University, Bloomington, IN 47405 **[67]**

David Peterson, Department of Molecular Biophysics and Biochemistry, Yale University, New Haven, CT 06511 **[271]**

Terry Platt, Department of Biochemistry, University of Rochester Medical Center, Rochester, NY 14642 **[325]**

Alberto Podjarny, Laboratoire de Cristallographie Biologique, I.B.M.C. du C.N.R.S., 67084 Strasbourg Cedex, France **[1]**

Joseph D. Puglisi, Department of Chemistry and Laboratory of Chemical Biodynamics, University of California, Berkeley, CA 94720 **[25]**

Michael Raff, Department of Biology and Institute for Molecular and Cellular Biology, Indiana University, Bloomington, IN 47405 **[67]**

Paul J. Romaniuk, Department of Biochemistry and Microbiology, University of Victoria, Victoria, British Columbia, Canada V8W 2Y2 **[123]**

Lisa C. Ryner, Department of Biological Sciences, Columbia University, New York, NY 10027 [335]

Vincenzo Scarlato, Department of Biology, University of California, San Diego, La Jolla, CA 92093; present address: IIGB, C.P 3061, Naples, Italy [59]

Astrid Schön, Department of Molecular Biophysics and Biochemistry, Yale University, New Haven, CT 06511 [271]

Beate Schwer, European Molecular Biology Laboratory, 6900 Heidelberg, Federal Republic of Germany [199]

Michael W. Sganga, Department of Biology and Institute for Molecular and Cellular Biology, Indiana University, Bloomington, IN 47405 [67]

Lamia Sharmeen, Fox Chase Cancer Center, Philadelphia, PA 19111 [99]

Ping Shen, Department of Biology, University of Rochester, Rochester, NY 14627 [231]

Thomas Shenk, Department of Molecular Biology, Princeton University, Princeton, NJ 08544 [351]

David Shub, Department of Biological Sciences, State University of New York, Albany, Albany, NY 12222 [59]

Paul Sloof, Laboratory of Biochemistry, University of Amsterdam, Academic Medical Center, 1105 AZ Amsterdam, The Netherlands [177]

Larry Snyder, Department of Microbiology, Michigan State University, Lansing, MI 48824 [281]

Dieter Söll, Department of Molecular Biophysics and Biochemistry, Yale University, New Haven, CT 06511 [271]

Peggy Spear, Department of Biochemistry, University of Rochester Medical Center, Rochester, NY 14642 [325]

Kenneth Stuart, Seattle Biomedical Research Institute, Seattle, WA 98109-1651 [187]

William E. Stumph, Department of Chemistry and Molecular Biology Institute, San Diego State University, San Diego, CA 92182 [165]

Hendrik G. Stunnenberg, European Molecular Biology Laboratory, 6900 Heidelberg, Federal Republic of Germany [199]

Christopher K. Surratt, Departments of Chemistry and Biology, University of Virginia, Charlottesville, VA 22901; present address: Department of Biochemistry, University of California, Berkeley, CA 94720 [79]

Yoshio Takagaki, Department of Biological Sciences, Columbia University, New York, NY 10027 [335]

Kathy M. Takayama, Department of Biochemistry, UMDNJ-Robert Wood Johnson Medical School, Piscataway, NJ 08854 [299]

Careen Tang, Department of Chemistry, Johns Hopkins University, Baltimore, MD 21218 [113]

John Taylor, Fox Chase Cancer Center, Philadelphia, PA 19111 [99]

Ignacio Tinoco, Jr., Department of Chemistry and Laboratory of Chemical Biodynamics, University of California, Berkeley, CA 94720 [25]

Janny van den Berg, Laboratory of Biochemistry, University of Amsterdam, Academic Medical Center, 1105 AZ Amsterdam, The Netherlands [177]

Hans van der Spek, Laboratory of Biochemistry, University of Amsterdam, Academic Medical Center, 1105 AZ Amsterdam, The Netherlands **[177]**

Jailaxmi Vartikar, Department of Chemistry, Johns Hopkins University, Baltimore, MD 21218 **[113]**

Barbara S. Vold, Department of Molecular Biology, SRI International, Menlo Park, CA 94025 **[89]**

Alexander von Gabain, Department of Bacteriology, Karolinska Institute, S-104 01 Stockholm, Sweden **[311]**

Justina Voulgaris, Department of Biological Sciences, Columbia University, New York, NY 10027 **[335]**

Virginia Walbot, Department of Biological Sciences, Stanford University, Stanford, CA 93405-5020 **[237]**

David S. Waugh, Department of Biology and Institute for Molecular and Cellular Biology, Indiana University, Bloomington, IN 47405 **[67]**

Susan Weiss, Department of Biochemistry, UMDNJ-Robert Wood Johnson Medical School, Piscataway, NJ 08854; present address: Department of Microbiology, University of Pennsylvania, Philadelphia, PA 19104 **[299]**

Carl Weitzmann, Roche Institute of Molecular Biology, Roche Research Center, Nutley, NJ 07110 **[209]**

Karin Wiebauer, Friedrich Miescher Institute, Basel CH4002, Switzerland **[155]**

Michael A. Wilson, John Innes Institute, Norwich, England NR4 7UH **[237]**

Jeffrey Wilusz, Department of Molecular Biology, Princeton University, Princeton, NJ 08544 **[351]**

Jacqueline R. Wyatt, Department of Chemistry and Laboratory of Chemical Biodynamics, University of California, Berkeley, CA 94720 **[25]**

Ming-Qun Xu, Department of Biological Sciences, State University of New York at Albany, NY 12222 **[59]**

Michael Yarus, Department of Molecular Cellular and Developmental Biology, University of Colorado at Boulder, CO 80309-0347 **[373]**

Qimin You, Department of Biochemistry and Microbiology, University of Victoria, Victoria, British Columbia, Canada V8W 2Y2 **[123]**

Janice M. Zengel, Department of Biology, University of Rochester, Rochester, NY 14627 **[231]**

Preface

The 1988 UCLA Symposium on **Molecular Biology of RNA** was held at Keystone, Colorado, April 4–10, 1988. The conference brought together some 300 scientists with interests ranging from RNA chemistry through biochemistry to developmental biology. Methods for predicting and determining RNA structure were presented first, followed by a session on RNA catalysis—the self-splicing and enzymatic activities that some folded RNA molecules have in the absence of proteins. Attention then turned to RNA–protein interactions and the activities of a variety of ribonucleoprotein systems. These systems include the splicesomes and snRNPs involved in mRNA splicing, the ribosome, the signal recognition particle, the enzyme responsible for telomere elongation in *Tetrahymena*, and enzymes involved in mitochondrial DNA synthesis in mammals. On the more biological side, there were presentations of alternative RNA processing in differentiation, antisense RNA, and regulation at the levels of translation and mRNA stability. The final session of the symposium focused on RNA in early evolution: prebiotic RNA synthesis and the origin of protein synthesis. In addition, sessions on interactions with nuclear superstructures and on transcription termination and 3' end formation were held jointly with the concurrent UCLA Symposium **DNA-Protein Interactions in Transcription**.

The articles in this volume are organized along the same lines as the conference, and they exemplify the very broad range of topics and approaches that were presented at the meeting. Meeting topics that are not represented here include nuclear magnetic resonance approaches to RNA structure, the activity of group II self-splicing introns, mechanistic studies of mRNA splicing, tRNA-synthetase interactions, certain enzymes with RNA subunits, and self-cleaving plant infectious RNAs. Thus, while this volume conveys the flavor of the symposium, it does not provide a comprehensive summary of the proceedings.

Special thanks are due to the 1988 UCLA Symposia Director's Sponsors—Cetus Corporation, ICI Pharmaceuticals Group, Monsanto, Schering Corporation, and The Upjohn Company—for financial support of the symposium. Additional funds were generously provided by Synergen, Inc. The success of the meeting was due in large part to the UCLA Symposia staff, and in particular to Robin Yeaton and Betty Handy; their efforts are gratefully acknowledged.

Thomas R. Cech

CRYSTAL STRUCTURE OF A KINKED RNA

A.C. Dock-Bregeon, B. Chevrier, A. Podjarny,
D. Moras

Laboratoire de Cristallographie Biologique
I.B.M.C. du C.N.R.S. 15, rue R. Descartes
67084 Strasbourg Cedex France

J.S. de Bear, G.R. Gough, P.T. Gilham
and J.J. Johnson

Department of Biological Sciences
Purdue University
West Lafayette, Indiana 47907 USA

ABSTRACT The single-crystal X-ray structure of the synthetic oligoribonucleotide $U(UA)_6A$ has been solved at 2.3 Å resolution. The molecule shows the typical features of an A-RNA helix. Two kinks are observed, stabilized by a few H-bonds.

INTRODUCTION

The A and B forms of duplex DNA were established from early X-ray diffraction studies of oriented fibres. Fibre methods were extended to permit least squares refinement of duplex models

This work was supported by the CNRS and a grant from the U.S. National Science Foundation (DMB 8511084 to JEJ). Funds for the synthesis and characterization of the RNA were provided by grants (GM19395, GM 11518 to PTG) from the National Institutes of Health.

which showed that oriented DNA helices were remarkably flexible and dependant on sequence and the conditions of fibre preparation[1]. During the last decade high resolution single crystal X-ray diffraction studies of synthetic DNA oligomers have provided a wealth of information on local effects of the sequence and have suggested the roles of higher order DNA structures in protein-DNA recognition and interaction[2]. In contrast, fibre diffraction studies of duplex RNA revealed only the A helical configuration[3], a result that was readily rationalized by the steric restrictions imposed on the ribose ring by the 2' hydroxyl group.

Structures of tRNA[4-6] revealed a wealth of information on nucleotide stereochemistry but only at 3.0 Å resolution and for molecules with a specific role in protein synthesis. The variety of functional roles now known to be associated with RNA[7] can not be understood in a structural context until single crystal X-ray studies, analogous to the DNA investigations, are performed. Such crystallographic analysis had not been carried out because of problems in synthesizing large quantities of specific oligoribonucleotides. Recently these problems have been overcome and crystals of a tetradecanucleotide were grown. In this communication we report the first crystallographic analysis of a synthetic RNA duplex, $U(UA)_6A$, at 2.3 Å resolution.

DESCRIPTION OF THE STRUCTURE

The $U(UA)_6A$ molecule presented in figure 1 forms a typical right handed A-RNA helix. Table 1 compares the helical parameters observed with those obtained in the fibre structures of RNA[3] and DNA[1] and in the crystal structure[6] of yeast tRNA[Asp].

The average twist angle is 33.1°, corresponding to 10.9 residues per turn in the helix, a value very close to the 11 residues per turn deduced from the fibre-studies. The tilt angle averages 16.7°, corresponding exactly to the

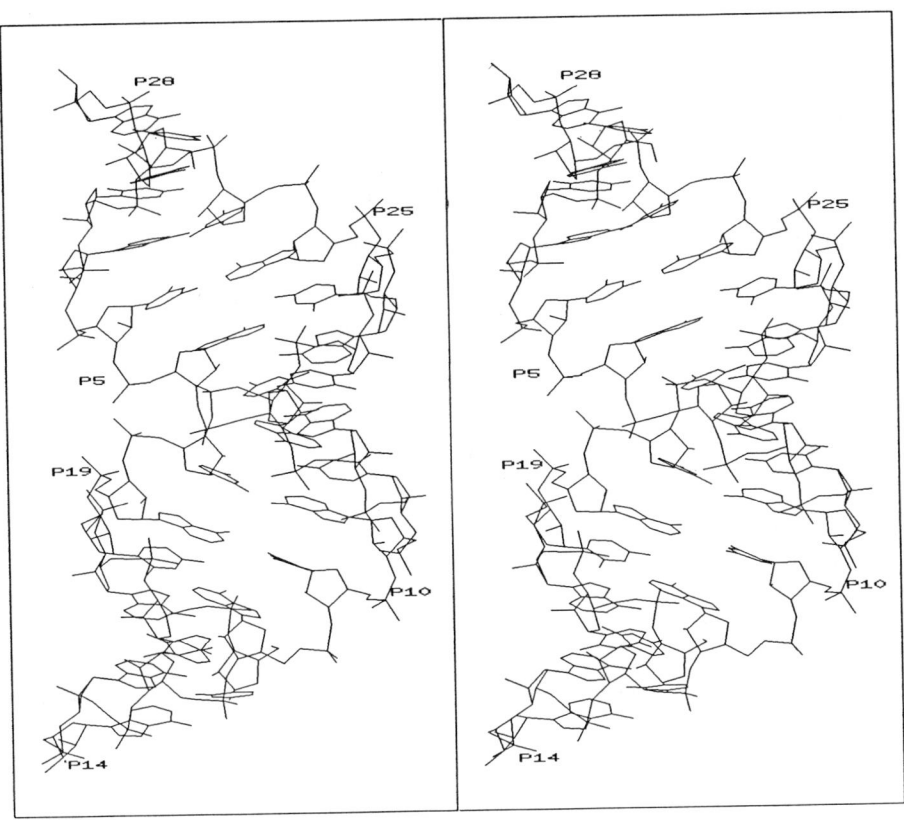

FIGURE 1. Refined model of the tetradecamer (R =13.4 %). The pinching effect due to the two kinks narrows the major groove between P5 and P18.

values deduced from fibre diffraction. This is different from the case of DNA where crystal structures showed tilt angles much smaller than those deduced from fibre -studies[2] (9.0° to 14.0°). However the value of the propeller twist is larger than predicted but quite similar to the one observed in the oligo(dA).oligo(dT) decamer structure[8] (20°). Therefore this value is clearly related to the sequence.

TABLE 1
HELICAL PARAMETERS

	Twist (°)	Rise /res. (Å)	Res. /turn	Groove width Minor (Å)	Groove width Major (Å)	Tilt (°)	Prop. twist (°)
mean s.d.	33.1 (3.0) (3.5)	2.8 (0.2)	10.9	10.2 (0.5)	3.7 (1.5)	16.7 (3.9)	18.6
tRNA^{Asp} (6)	32.7	2.4	11.0	9.8	4.1	16.6	2.8 to 12.4
RNA* fibre	32.7	2.75	11.0	11.3	4.1	16.7	13.8
DNA*	32.7	2.6	11.0	11.0	2.3	22.0	6.0

*helical parameters measured or calculated from a model of sequence U(UA)$_6$A constructed with values from (1) for DNA and (3) for RNA.

The minor groove, flat and wide, and the major groove, deep and narrow, are typical of A-helices. An interesting feature is the variation of the major groove width between 2.2 and 6.2 Å. An opening of this major groove can be observed at the extremity of the helix, between P2 and P21. The smallest width is found between P5 and P18 and reflects a pinching effect, coming from the presence of kinks, i.e. deviations of the helix axis from a straight line, in two points. Therefore the molecule consists of three helical blocks of 5, 5 and 4 base-pairs. For the sake of clarity the kinks have been localized in the sequence in figure 2.

```
  1  2  3  4  5    6  7  8  9    10 11  12 13 14
  U  U  A  U  A    U  A  U  A  U *A  U  A  A
  |  |  |  |  |    |  |  |  |   |   |  |  |  |
  A  A  U  A  U * A  U  A  U  A     U  A  U  U
  28 27 26 25 24   23 22 21 20  19  18 17 16 15
```

FIGURE 2. Sequence of the tetradecamer with asterisks at the kink points.

The angular amplitudes of these kinks (11° and 13°) are giving rise to large translational differences between this structure and the fibre-model or a tRNA stem. The present paper will focus on these kinks, their origin and their stabilization.

DESCRIPTION OF THE KINKS

When fitting the $U(UA)_6A$ structure to the fibre-model RNA a very good aggreement can be observed in the central part of the molecule. The divergences become perceptible at pairs A5-U24 and A11-U18 towards the extremities of the helix. This is not a continuous deformation of the helix. In table 2 the torsion angles describing the structure are reported. The average values are close to those determined in the fibre structure or in the tRNA^{Asp} stems, except for A11 and U24 where two torsion angles differ significantly : α, about P-O5', and γ, about C5'-C4'. These change from the gauche⁻ , gauche⁺ conformation to the less common trans,trans one, the divergence being larger for residue A11.

The modified conformation of residues A11 and U24 leads to an extended conformation of the backbone. The distance between adjacent phosphates at these steps increases significantly from an average value of 5.8 Å to 6.6 Å for P10-P11 and 6.7 Å for P23-P24. Such a conformational change has already been observed in tRNA stems : i.e. G30, G51 and G53 in tRNA^{Asp}(6), and also in two A-DNA oligonucleotides structures[9,10].

TABLE 2
TORSION ANGLES

	α	β	γ	δ	ε	ζ	χ
mean*	-61	169	53	80	-148	-79	-169
s.d.	(14)	(14)	(13)	(10)	(19)	(13)	(9)
A11	159	177	177	88	-114	-106	-178
U24	107	-169	-142	105	-134	-78	-165
tRNA^{Asp} AA stem	-81 (27)	-171 (19)	78 (69)	82 (2)	-168 (19)	66 (21)	-157 (15)
AC stem	-78 (45)	174 (20)	66 (36)	82 (2)	-148 (23)	81 (19)	-167 (13)
RNA fibre polyA.polyU	-62	180	48	83	-151	-74	-164

*without A11 and U24. A28 was also excluded from the calculation as it is affected by end-effects.

Beside the increase of the P-P distance the main effect of the conformational change is an unstacking of the bases of the concerned residues.

The observed kink results from the trend to recover the stacking interaction. This is achieved by small variations in torsion angles, especially χ, about the glycosyl bond. The overall changes result in a re-winding of the helix. This is expressed in the variation of the twist angle[11] plotted in figure 3. The deviation from the mean value is small in the center of the molecule and increases significantly near the kinks.

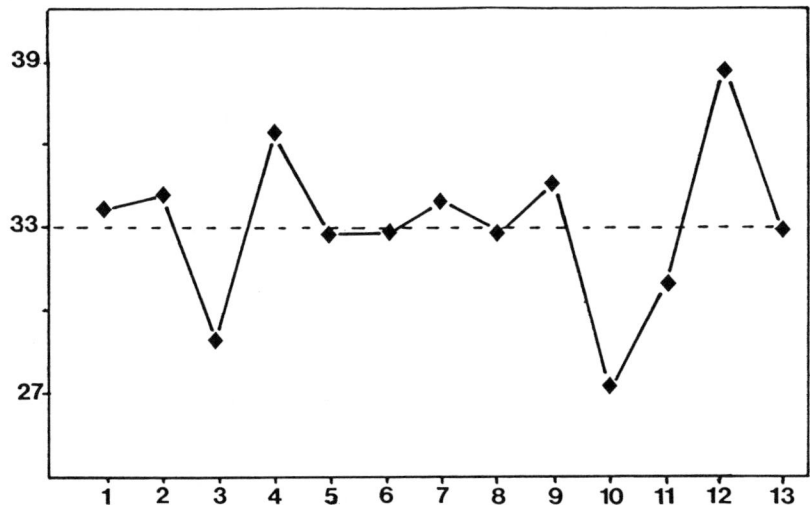

FIGURE 3. Variation of the twist angle along the duplex. Twist angle was calculated with a program written by J. Rosenberg and R.E. Dickerson[11].

The small value observed at step 10 between the pairs U10-A19 and A11-U18 corresponds to an unwinding of the helix. A strong rewinding with a twist of 38.9° is observed two steps further towards the end of the helix. Near U24, where the second kink is observed the situation is different. The twist variation shows first a winding of the helix followed by an unwinding in the following two steps.

These differences are probably linked to the amplitudes of the changes at the torsion angles α and γ (table 1) and to a difference in the stabilization of the two kinks.

A similar winding-unwinding effect is evident in the structure of the A-DNA oligonucleotide d(CCCCGGGG)[9], where an extended conformation has been observed at the CG step. The twist at this level is also smaller (25.1°) when compared to the average value of 33.5°. The resulting unwinding of the helix is very clear as this oligonucleotide crystallizes with only one strand in the asymmetric unit (the extended conformation is found on both strands at the central step). However, because of the symmetry between the two extended conformation steps, this DNA molecule is not kinked.

STABILIZATION OF THE KINKS

The kink at A11 is in the midst of a region of close intermolecular contacts. The packing involves contacts between minor grooves stabilized by H-bonds. Four H-bonds are formed : two between O2' hydroxyl groups, the remaining two between O2-uridine atoms and O2'-hydroxyls of adenines.Some solvent molecules are found in interesting locations, such as in the minor groove, between the N3-nitrogen atoms of A11 and A28, or between closely-packed backbones (figure 4).

The kink at U24 has a very different environment. It is located between two zones of packing contacts (figure 5). The interaction shown in the lower part of the figure ressembles that of the A11 region: a terminal base-pair (here A14-U15) of a symmetry-related molecule stacks on the minor groove with two H-bonds between O2-atoms of uridine bases and O2'-atoms of the adenine moieties riboses. There is no additional direct backbone-backbone interaction. A few solvent molecules are trapped in this minor groove contact. The second contact illustrated in the upper part of figure 5 involves major grooves and

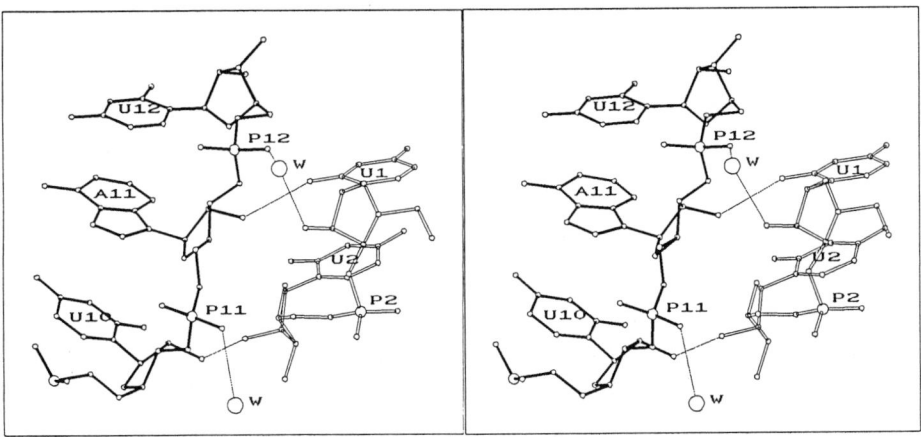

FIGURE 4. The packing contact near P11, where a modified conformation of the backbone is observed.

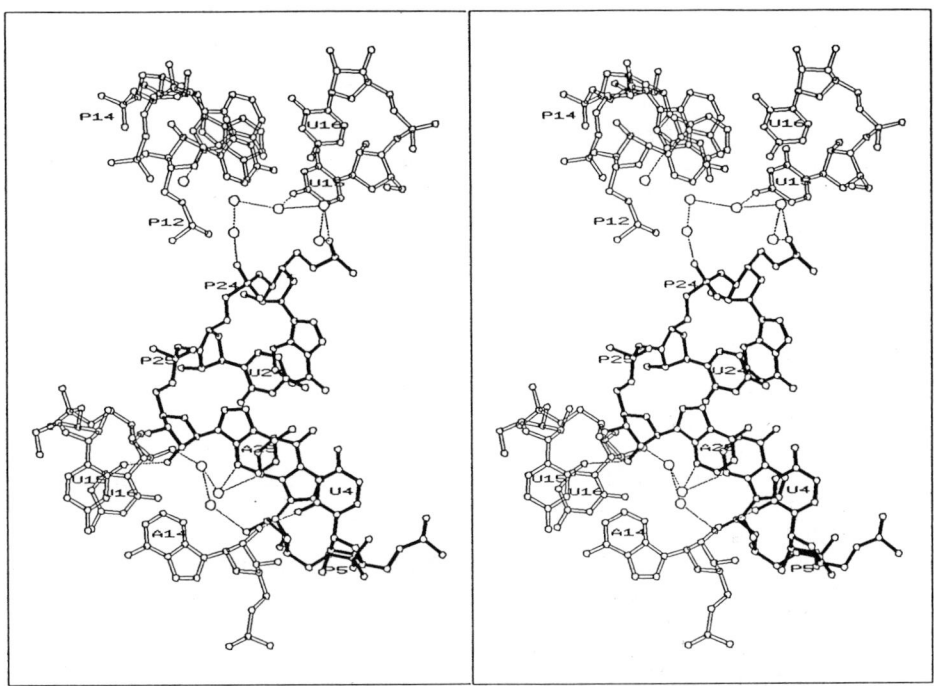

FIGURE 5 . The environment of P24 where a conformational change leads to a kink.

consists of a few Van der Waals interactions and some water bridges. There are no direct intermolecular H-bonds.

CONCLUSION

This molecular structure shows that the 2' hydroxyl groups participate extensively in interhelical contacts. In each case these contacts dramatically affect a nearby backbone P-O5' torsion angle by shifting it from *gauche* to *trans*.

The helix kinks at each of these locations in order to maintain base stacking. Thus if the 2' hydroxyl group restricts the RNA helix to the A form because of *intra*-helical contacts, it induces the remarkable kinks because of *inter*-helical interactions.

U(UA)$_6$A crystals have been obtained at 35°C, which is an unusually high temperature for crystallization of biological macromolecules. This may reflect the need of populating a less probable conformation, the kinked one, in order to build a stable lattice with strong intermolecular interactions. Therefore, this structure shows the potentiality of an RNA molecule to kink in order to adapt to its environment.

REFERENCES

1. Arnott S. and Hukins D.W.L. (1972). Optimised parameters for A-DNA and B-DNA. Biochem. Biophys. Res. Comm. 47:1504.
2. Shakked Z. and Rabinovitch D. (1986). The effect of base sequence on the fine structure of the DNA double helix. Prog. Biophys. molec. Biol. 47:159.
3. Arnott S., Hukins D.W.L. and Dover S.D. (1972). Optimised parameters for RNA double helices. Biochem. Biophys. Res. Comm. 48:1392.
4. Sussman J.L., Holbrook S.R., Warrant W.R., Church G.M. and Kim S.H.(1978). Crystal

structure of yeast phenylalanine transfer RNA. I. Crystallographic refinement. J. Mol. Biol. 123:607.
5. Hingerty B., Brown R.S. and Jack A.(1978). Further refinement of the structure of yeast tRNA^{Phe}. J. Mol. Biol. 124:523.
6. Westhof E., Dumas P. and Moras D.(1985). Crystallographic refinement of yeast aspartic acid transfer RNA. J. Mol. Biol. 184:119.
7. Watson J.D. (1987).Cold Spring Harbor Symposium Quant. Biology 52.
8. Nelson H.C.M., Finch J.T., Luisi B.F. and Klug A.(1987). The structure of an oligo(dA).oligo(dT) tract and its biological implications.Nature 330:221.
9. Haran T.E., Shakked Z., Wang A.H.J. and Rich A. (1987) . The crystal structure of d(CCCCGGGG):A new A-form variant with an extended backbone conformation. J. Biomol.Struc. Dyn. 5:199.
10. Heinemann U., Lauble H., Frank R. and Blöcker H. (1987). Crystal structure of an A-DNA fragment at 1.8 Å resolution: d(GCCCGGGC). Nucl. Acids Res. 15:9531.
11. Fratini A.V., Kopka M.L., Drew H. and Dickerson R.E. (1982). Reversible bending and helix geometry in a B-DNA dodecamer: CGCGAATT^{Br}CGCG. J. Biol. Chem. 257:14686.

PHYLOGENETIC COMPARISONS OF U2 SMALL NUCLEAR RNA SEQUENCES SUGGEST A PSEUDOKNOTTED STRUCTURE[1]

Manuel Ares, Jr. and A. Haller Igel

Department of Biology, University of California
Santa Cruz, California 95064

ABSTRACT U2 small nuclear RNA (snRNA) is a highly conserved component of the cell nucleus, where it recognizes and binds to the intron branchpoint during splicing of premessenger RNA (premRNA). To develop an understanding of the secondary and higher order structure of U2 snRNA, we compared the sequences of U2 RNAs from ten organisms, aligned them, and analyzed the pattern of nucleotide changes. The highly conserved 5' half of the molecule contains two helices that may participate in a pseudoknot, adjacent to the nucleotides involved in pairing to the intron branchpoint. Although most U2 RNAs contain two 3' stem-loop structures (stems III and IV), both trypanosome U2 and a deleted but functional yeast U2 appear to lack stem III.

INTRODUCTION

Splicing of messenger RNA precursors requires a subset of small nuclear ribonucleoprotein particles (snRNPs) that are assembled with premRNA into a complex called the spliceosome (for reviews see 1-3). Early in spliceosome assembly, the U2 snRNP recognizes the intron branchpoint (4). In the yeast

---

[1]This work was supported by institutional start-up funds from the University of California, Santa Cruz.

Saccharomyces cerevisiae, part of the specificity of U2 snRNP-intron branchpoint recognition is contributed by base pairing between U2 snRNA and the intron, where appropriate U2 mutations suppress intron branchpoint mutations in a fashion consistent with Watson-Crick type RNA-RNA base pairing (5).

To further understand the mechanism of U2 action during splicing, we are studying the structure and function of U2 RNA. To generate phylogenetically consistent models of U2 RNA secondary structure, we compared U2 RNA sequences from divergent organisms. We found that the highly conserved 5' half of U2 contains the potential to form a pseudoknot, a recently recognized RNA secondary structure with some intruiging features (6). In the 3' half of the RNA, a region known as stem III seems not to be essential for function of both trypanosome and yeast U2 (E. Shuster and C. Guthrie, pers. comm.; M.A. and H.I., submitted). Because of the extreme conservation of the 47 residues at the 5' end (including the branchpoint recognition region), structure models for this part of the molecule cannot yet be generated with confidence.

RESULTS AND DISCUSSION

Comparative Analysis Reveals Eight Conserved Pairing Interactions

Alignment. The ability to identify paired regions by phylogenetic comparisons depends on the alignment of homologous sequences (7). The 5' 100-120 nucleotides (nt) of U2 RNA are highly conserved, and were easily aligned. The 3' half of the RNA is less well conserved, and varies in length from more than 1000 nt in yeast (8) to as few as 40 in trypanosome (9). Excluding these extreme cases, most U2 RNAs contain 80 nt that are readily aligned. We used human U2 as a standard sequence, and introduced stars into it to reserve space where insertions relative to human U2 occur in the other sequences. The other sequences were aligned with the human sequence, and stars were placed where deletions relative to human U2 were apparent. A dash was placed where the sequence matched human U2. We did not attempt to include the entire yeast U2 sequence in our analysis; instead, we used the sequence of a 210 nt functional deletion derivative of yeast U2 constructed in our laboratory. We included trypanosome U2, even though it may not be functionally equivalent to the other U2 RNAs. The results of the alignment are shown in Figure 1.

Positions 1-6. U2 RNAs begin with an A residue blocked at the 5' end with a 2,2,7, trimethylguanosine cap. Positions 2-4 are conserved among the animals, but vary in bean, trypanosomes, and yeast, where there is also a single base insertion. Positions 5 and 6 are CU in all cases except yeast, where UC exists.

Stem I. Positions 7-26 have been proposed to form a stem-loop structure (10). There is phylogenetic support for the 12-14/19-21 interaction: all organisms have GCC/GGC here except worm and trypanosome, which have GCU/AGC. The lower part of the proposed stem would pair positions 7-9 with 24-26 (UCU/AGA). As yet, no variation in this sequence pairing has been found and it remains without phylogenetic support. In all organisms except worm and yeast, positions 10 and 11 are CG, while 22 and 23 are UA. The only difference in worm is an insertion of a U residue between positions 9 and 10, but this could affect the structure by pairing with A23. G11 could interact with U22, except that in yeast, all four positions are U. Perhaps canonical base pairs in this region are incompatable with function. The 15-18 loop consists of 4 U residues except for fly where position 17 is A, worm where position 16 is A, and trypanosome, where position 15 is A.

The branchpoint recognition region. The nucleotides from position 27 to 46 are the most highly conserved of U2, and include those that contact the intron branchpoint during splicing (5). Excluding trypanosomes, the only difference is the insertion of A between positions 30 and 31 in fly and worm. Guthrie and coworkers have shown that a G to U change at the yeast U2 equivalent of position 36 can suppress a UACUAAC to UAAUAAC mutation, and

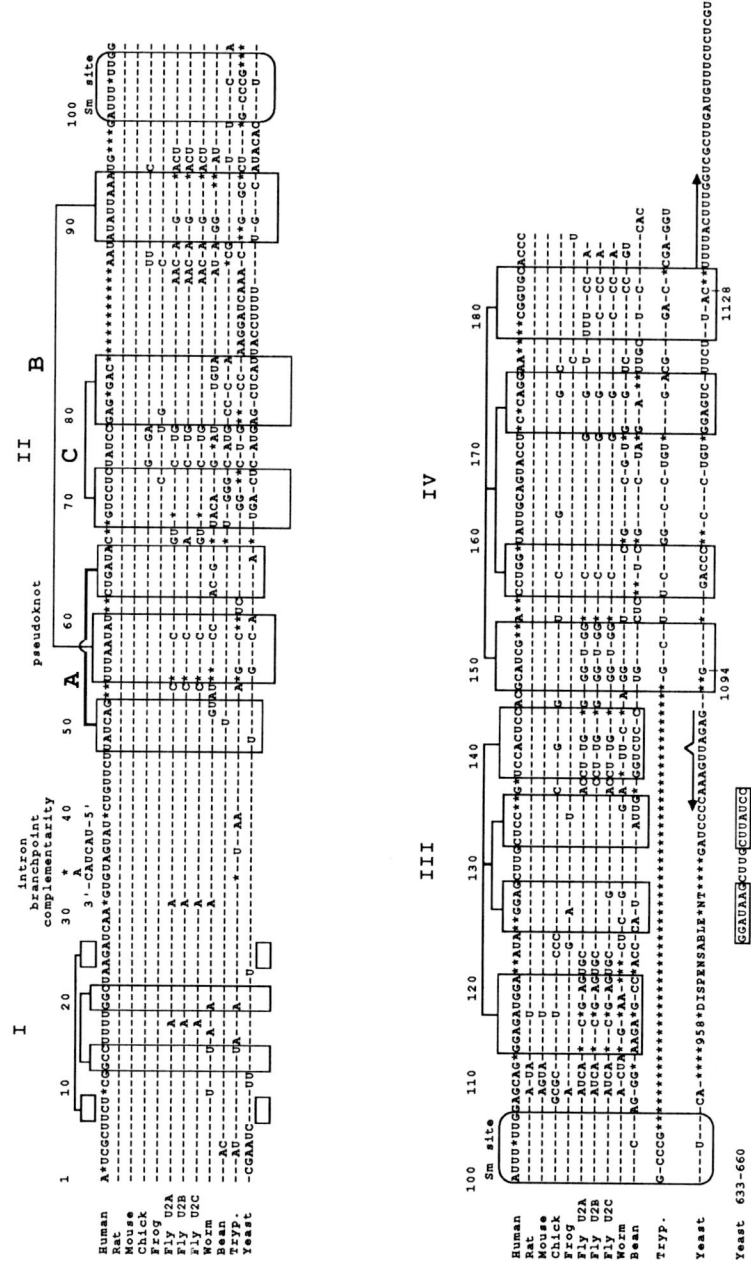

that a change equivalent to A35 to U suppresses a UACUAAC to UAC<u>A</u>AAC mutation in a yeast actin intron (5). These experiments show that at least in yeast, positions 35 and 36 interact with the pre-mRNA branchpoint region during splicing, and suggest the formation of an extended helix including positions 33-38. Trypanosomes have 4-6 differences (depending on alignment) between all other U2 RNAs in this region. We chose to introduce a deletion of position 33 and an insertion of an A between residues 39 and 40, so that only one nucleotide within the 34-38 segment is changed, G36 to U, the same change as in one of the yeast U2 suppressors. Because there are few nucleotide changes elsewhere in the region, the possibility that these nucleotides interract among themselves or with other highly conserved sequences cannot be asessed.

The Pseudoknot. The nucleotides 47-52 are complementary to those at 61-66. This was first noted by Doug Black, however, because of high conservation of the sequence, the pairing was not supported by compensatory base changes. In addition, some support existed for pairing between 53-60 and 88-95, and formation of the two helices simultaneously was considered unlikely. A theoretical treatment of the possibility of

---

FIGURE 1. (Facing page) An alignment and comparison of U2 snRNAs from different organisms. The name of the organism is to the left of the line. Top, the 5' half of the RNA. Bottom, the 3' half. Sequences are from the following references: human, rat, mouse, chick, frog (*Xenopus laevis*), fly (*Drosophila melanogaster*), (11); worm, *Caenorhabditis elegans*, (J. Thomas and T. Blumenthal, personal communication); Bean, *Vicia faba*, (12); Tryp., *Trypanosoma brucei*, (9); Yeast, *Saccharomyces cerevisiae*, (8). Boxed region connected by lines indicate pairings discussed in the text. Stars indicate nucleotide deletions, a dash indicates a match with human U2. The rounded box contains the Sm site, a sequence shared by anti-Sm antibody precipitable snRNPs (13), and which is essential for assembly of the RNA into RNP particles (14, 15). The position of this site in trypanosome U2 is not certain. Roman numerals refer to stem numbering used previously (10). Arrows above the yeast sequence indicate the possible extension of the stem IV pairing. Yeast 633-660 refers to a dispensable yeast U2 sequence with structure similar to stem III.

simultaneous formation and coaxial stacking of two such RNA helices was presented by Pleij and coworkers as part of an attempt to understand the recognition of 3' ends of plant RNA viruses by enzymes that recognize tRNA (6). The results of this treatment suggested that formation of such a structure, termed a pseudoknot, was possible.

With the addition of the yeast and worm sequences to the U2 database, phylogenetic support for both the 47-52/61-66 (labeled A in Figure 1) and the 53-60/88-95 helices (labeled B in Figure 1) appeared. Vertebrate, fly, and trypanosome U2 contain UAUCAG/CAGAUA in the A pairing, however in yeast the sequence reads UUUCAG/CAGAAA. Worm contains two sets of compensatory changes and a single change that converts an AU pair to GU: UAUCGU/ACGGUA. The pairing between 53-60 and 88-95 in vertebrates is UUUAAUAU/AUAUUAAA. This is supported by changes in yeast: UGUAACAA/UUGUUACA.

Because of the nature of pseudoknots, all the base pairs described above cannot be formed if the helices are to exist simultaneously. This is represented in Figure 2 and is predicted by computer graphics modeling experiments (16), that suggest that at least one or two unpaired nucleotides would be needed to span the deep groove of the discontinuous A-helix. Therefore, position 53 in vertebrates, bean, and yeast is likely not to be paired in the pseudoknotted structure. This is also suggested by the presence of unpaired nucleotides at the equivalent positions of fly, worm and trypanosome U2. The 53/88 pair may be important if the region undergoes a conformational switch during U2 function.

The pairing between 68-73 and 79-84 (C, Figure 1), and the 74-78 loop are quite variable and well supported phylogenetically. 0-2 nucleotides separate the stem-loop from the pseudoknot on the 5' side, and to the 3' side is a region of variable length (2-13 nt) that connects the stem-loop to the helix formed by the B pairing (see Figures 1 and 2). Though technically not part of the pseudoknot, this stem-loop is part of the RNA strand that spans the shallow groove to complete the pseudoknot, and is probably stacked on the 47-52/61-66 helix.

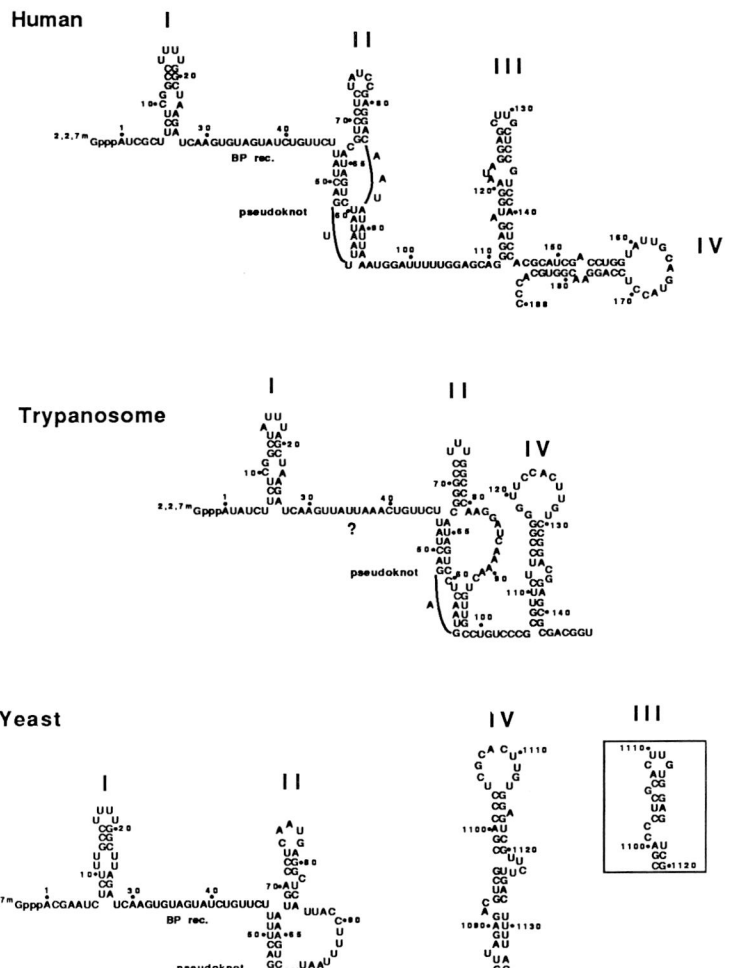

Figure 2. Secondary structure models for different U2 RNAs. Top, human U2; Middle, trypanosome U2; Bottom, a functional deletion derivative of yeast U2. Inset, an alternative folding of yeast U2 nucleotides 1098-1120 to form a stem III-like structure. Numbering is not standardized to human U2 RNA.

The Sm site. Positions 99-105 are occupied by the Sm site, a sequence (consensus $AU_{3-6}G$) shared by many of the snRNAs (13). Trypanosome lacks a clear match to the consensus, however there is a 5/6 run of pyrimidines bounded by a purine on each side. Two purine residues (105 and 106) usually follow the Sm site, and thereafter a stretch of 4-5 nt that is highly variable, even among vertebrates.

Stem III. Positions 112-144 compose stem III. The bottom portion of the stem is a 7-8 base pair helix formed by pairing 112-120 with 137-144, that may (as in vertebrates) or may not (as in fly, worm, and bean) contain a single bulged residue. The top helix is made up of 124-127 paired to 132-135. This helix varies from 4 to 6 base pairs in length. The loop region and the closing base pair are invariant and are related to the tetranucleotide loop described by Tuerk et al. (17). The two helices are connected by a 3-5 nucleotide stretch (121-123) to the 5' side, and by a 0-2 nucleotide stretch 3' of the top helix.

Stem III may not be present in all functional U2 molecules. Trypanosome U2, which may or may not be functionally equivalent to other U2 RNAs, does not contain stem III. In addition, a functional deletion derivative of yeast U2 we have constructed (submitted for publication) appears to lack stem III, although as diagramed in Figure 2, the sequence can be folded to resemble either stem III or stem IV, but not both. We prefer the stem IV folding for two reasons: First, trypanosome U2 lacks stem III, and second, the stem III-like folding presented in the inset of Figure 2 contains an AU pair to close the loop, inconsistent with both the conservation of GC at this position in U2 (Figure 1), and the use of CG at the equivalent position of related tetranucleotide loops found in other RNAs (17). Although dispensable, the sequence from positions 633 to 660 of yeast U2 could form a convincing stem III (Figure 1).

Stem IV. At the 3' end of U2 RNA is another two-helix stem topped by a 10-13 nucleotide loop. The bottom helix of 6-8 base pairs is formed by pairing 147-152 with 179-184. The top helix of 4-6 base pairs forms between 154-158 and 172-176. Yeast contains an unpaired A residue between 172 and 173, but the helix contains 8 pairs in this species. 0-3 unpaired residues can be found between the two helices to the 5' side of the top helix, while 2-4 unpaired residues separate them to the 3' side of the top helix.

The 159-171 loop is moderately well conserved. The most well conserved portion is the 163-167 region, which is GCAGU or GCACU. Trypanosomes have CCACU. To either side of these 5 nucleotides, there is greater variation, however it is not random. For example, yeast and trypanosomes are related, sharing 9 of 13 positions with each other, but only 5 of 13 with vertebrates.

The 3' end of U2 occurs 3-7 nucleotides past the base of stem IV, except in yeast. In yeast, stem IV can be extended an additional 9 base pairs (11 in our deletion derivative, see arrows, Figure 1), and the 3' end occurs 40 nucleotides past the end of this extension (22 in our deletion derivative).

ACKNOWLEDGEMENTS

We are grateful to Jeffrey Thomas and Tom Blumenthal for providing the worm U2 sequence prior to publication. We also thank Doug Black, Harry Noller, Kathy Parker, Ted Powers, Chris Tschudi and Elisabetta Ullu for stimulating discussions.

REFERENCES

1. Green, M (1986). Pre-mRNA splicing. Ann Rev Genet 20:671.
2. Sharp, P (1987). Splicing of messenger RNA precursors. Science 235:766.
3. Maniatis, T, Reed, R (1987). The role of small nuclear ribonucleoprotein particles in pre-mRNA splicing. Nature 325:673.
4. Black, D, Chabot, B, Steitz, J (1985). U2 as well as U1 small nuclear ribonucleoproteins are involved in pre-mRNA splicing. Cell 42:737.

5. Parker, R, Siliciano, P, Guthrie, C (1987). Recognition of the TACTAAC box during mRNA splicing in yeast involves base pairing to the U2-like snRNA. Cell 49:229.
6. Pleij, C, Reitveld, K, Bosch, L (1985). A new principle of RNA folding based on pseudoknotting. Nucl Acids Res 13:1717.
7. Woese, C, Gutell, R, Gupta, R, Noller, H (1983). Detailed analysis of the higher-order structure of 16S-like ribosomal ribonucleic acids. Microbiol Rev 47:621.
8. Ares, M (1986) U2 RNA from yeast is unexpectedly large and contains homology to vertebrate U4, U5, and U6 small nuclear RNAs. Cell 47:49.
9. Tschudi, C, Richards, F, Ullu, E (1986). The U2 analogue of *Trypanosoma brucei gambiense*: implications for a splicing mechanism in trypanosomes. Nucl Acids Res 14:8893.
10. Keller, E, Noon, W (1985) Intron splicing: a conserved internal signal in introns of *Drosophila* pre-mRNAs. Nucl Acids Res 13:4971.
11. Reddy, R (1986). Compilation of small RNA sequences. Nucl Acids Res 14:suppl. r61.
12. Kiss, T, Antel, M, Solymosy, F (1987). Plant small nuclear RNAs III. The complete primary and secondary structure of broad bean U2 RNA: Phylogenetic and functional implications. Nucl Acids Res 15:1332.
13. Branlant, C, Krol, A, Ebel, J-P, Lazar, E, Haendler, B, Jacob, M (1982). U2 RNA shares a domain with U1, U4, and U5 RNAs. EMBO J 1:1259.
14. Mattaj, I (1986). Cap trimethylation of U snRNA is cytoplasmic and dependent on U snRNP protein binding. Cell 46:905.
15. Hernandez, N, Weiner, A (1986) Formation of the 3' end of U1 snRNA requires compatible snRNA promoter elements. Cell 47:249.
16. Dumas, P, Moras, D, Florentz, C, Giege, R, Verlaan, P, van Belkum, A, Pleij, C (1987). 3-D graphics modeling of the 3' end of turnip yellow mosaic virus RNA: structure and functional implications. J Biomol Struct Dyn 4:707.

17. Tuerk, C, Gauss, P, Thermes, C, Groebe, D, Gayle, M, Guild, N, Stormo, G, D'Aubenton-Carafa, Y, Uhlenbeck, O, Tinoco, I, Brody, E, Gold, L (1988). CUUCGG hairpins: extraordinarily stable RNA secondary structures associated with various biochemical processes. Proc Natl Acad Sci USA 85:1364.

PSEUDOKNOTTED RNA OLIGONUCLEOTIDES[1]

Jacqueline R. Wyatt, Joseph D. Puglisi and Ignacio Tinoco, Jr.

Department of Chemistry
and Laboratory of Chemical Biodynamics,
University of California
Berkeley, California 94720

Pseudoknotting is a form of RNA tertiary structure in which nucleotides outside a hairpin fold back to pair with bases in the loop of the hairpin (Figure 1). The interaction leads to an extended helical structure involving stacking of the two stem regions. In order to simplify physical studies on this type of structure and identify stem and loop length requirements, we have synthesized a series of oligonucleotides designed to form pseudoknots. The structure of each has been probed using single- and double-strand specific enzymes. Coaxial stacking of the two stem regions of the pseudoknot has been deduced from proton NMR studies.

INTRODUCTION

The crystal structure of tRNA revealed some of the possible RNA tertiary interactions including triple base pairs, loop-loop interactions and intercalation. However, little is known about the primary or secondary structural requirements governing tertiary interactions in other RNAs. Oligonucleotide models of simple tertiary structures are needed in order that these interactions

---

[1]This work was supported in part by grants from the National Institute of Health (GM 10840) and the Department of Energy (DE-FG03-86ER60406).

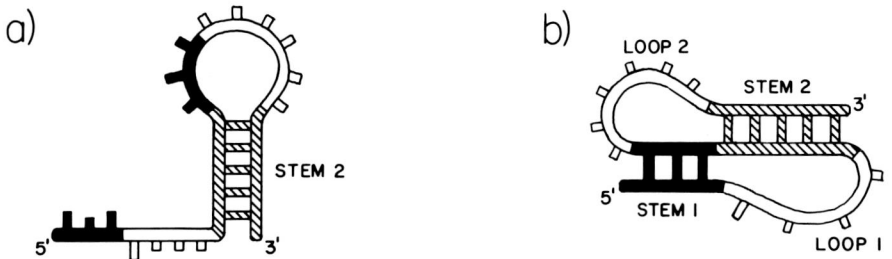

FIGURE 1. Schematic folding of oligonucleotide **GP1** into a pseudoknot. a) A five base pair stem (stem 2) forms at the 3' end of the molecule. b) A pseudoknot is formed when the single-stranded 5' end pairs with complementary bases within the loop to form a structure with two stem and two loop regions.

can be predicted in a manner analogous to predictions of secondary structure elements.

Pseudoknotting is a simple tertiary interaction originally proposed by Studnicka et al. (1). In general, a pseudoknot is formed when a single-stranded region outside a hairpin pairs in Watson-Crick fashion with bases in the loop of the hairpin to form a structure with two stem and two loop regions (Figure 1). Pleij and coworkers first described evidence for the occurrence of this type of structure at the 3' termini of certain plant viral RNAs (2). The pseudoknot found in the viral RNAs and considered in this paper is that in which stem nucleotides are adjacent. This type of folding allows formation of an elongated helix through coaxial stacking of the two stems of the pseudoknot.

We began investigation of this structural motif using a 19 nucleotide molecule similar in sequence to that of the pseudoknot found at the 3' end of turnip yellow mosaic virus. Enzymatic mapping and thermodynamic evidence supported a pseudoknot-hairpin equilibrium for this molecule; however, dimerization at concentrations above 10 µM prevented further characterization of the structure (3). The oligonucleotide **GP1**, $^{5'}$GCGAUUUCUGACCGCUUUUUUGUCAG$^{3'}$, discussed here, was designed to form a pseudoknot and to avoid dimerization. Preliminary NMR evidence indicates that oligonucleotide **GP1** forms a pseudoknot with coaxial stacking of the two stem regions. Variants of **GP1** which

TABLE 1
OLIGORIBONUCLEOTIDES

	stem 1	loop 1	stem 2	stem 1	loop 2	stem 2
GP1	5'GCG	AUUU	CUGAC	CGC	UUUUUU	GUCAG3'
GP2	5'GCG	AUUU	CUGAC	CGC	UUUUU	GUCAG3'
GP3	5'GCG	AUUU	CUGAC	CGC	UUUU	GUCAG3'
GP4	5'GCG	AUUU	CUGAC	CGC	UUU	GUCAG3'
GP5	5'GCG	AUU	CUGAC	CGC	UUUUUU	GUCAG3'

contain shorter loop regions (Table 1) have been investigated using enzymatic mapping.

METHODS

The RNA oligonucleotides were transcribed using T7 RNA polymerase and synthetic DNA templates as previously described (3, 4). Full length transcripts were isolated from denaturing (7 M urea) 20% polyacrylamide gels by electro-elution. Following ethanol precipitation, oligonucleotides to be used for spectroscopic studies were dialyzed 24 hours against 10 mM EDTA and 24 hours against water, then dried and resuspended in appropriate buffer. Transcripts were characterized by partial alkaline hydrolysis and partial RNase T_1 digestion of 5'-^{32}P-labeled molecules.

At 2 mM strand concentration, oligonucleotide **GP1** formed a monomeric structure in 5 mM $MgCl_2$, 50 mM NaCl, 10 mM sodium phosphate, pH 6.4 by comparison to retention times of molecules of known size on a Bio-Rad TSK-125 gel filtration column.

Enzymatic digestion experiments were performed as previously described (3), except that digestions with RNase V_1 were carried out at 4°C for 5 min in 5 mM $MgCl_2$, 60 mM NaCl, 5 mM Tris, pH 8.1 and digestions with nuclease S_1 were carried out at 4°C for 5 min in 5 mM $MgCl_2$, 60 mM NaCl, 5 mM MES, pH 6.3.

NMR spectra were recorded on a Nicolet GN-500 500 MHz spectrometer. Spectra of exchangeable protons of **GP1** were recorded in 90% H_2O/10% D_2O using a 1331 pulse sequence (5). Imino protons were assigned using one-dimensional NOEs with a 600 msec irradiation time at 4°C. Buffer conditions were 5 mM $MgCl_2$, 50 mM NaCl, 10 mM sodium phosphate, pH 6.4. Strand concentration was 1 mM.

In order to simplify the two-dimensional NMR spectrum, uridine triphosphate was deuteriated (6) and incorporated into oligonucleotide **GP1** during transcription. Approximately 100 mM rUTP was incubated in 2.5 M $(ND_4)_2SO_3$ (pH 8.0 at 50°C) for 24 to 30 hours at 50°C in a tightly sealed tube. The sulfonate intermediate was converted to d^5-UTP by incubation at pH 10.0 at room temperature for a minimum of 2 hours. After desalting, the deuteriated UTP was used in transcription reactions without purification from minor amounts of hydrolysis products.

RESULTS AND DISCUSSION

The oligonucleotide **GP1** was designed to form a pseudoknot with a three base-pair 5' stem (stem 1) and a five base-pair 3' stem (stem 2) as illustrated schematically in Figure 1. Before transcribing and purifying the large amounts of RNA necessary for NMR spectroscopy, **GP1** was probed with single- and double-strand specific enzymes (Figure 2). Under conditions specified in the Methods section, the proposed loop regions of **GP1** were accessible to single-strand specific nuclease S_1. Cleavage by S_1 occured after the third and fourth nucleotides of loop 1 and second, third and fourth nucleotides in loop 2. Significantly, no other bonds were accessible to S_1; if, for example, the molecules were in equilibrium between hairpin and pseudoknot, the dangling end of the hairpin should be cleaved to an extent dependent on the equilibrium population of hairpin. Double-strand specific RNase V_1 cut after each nucleotide in stem 1. Cleavage occured after each residue on the 5' side of stem 2. Only two cuts are observed on the 3' side of this stem. We have found that RNase V_1 leaves

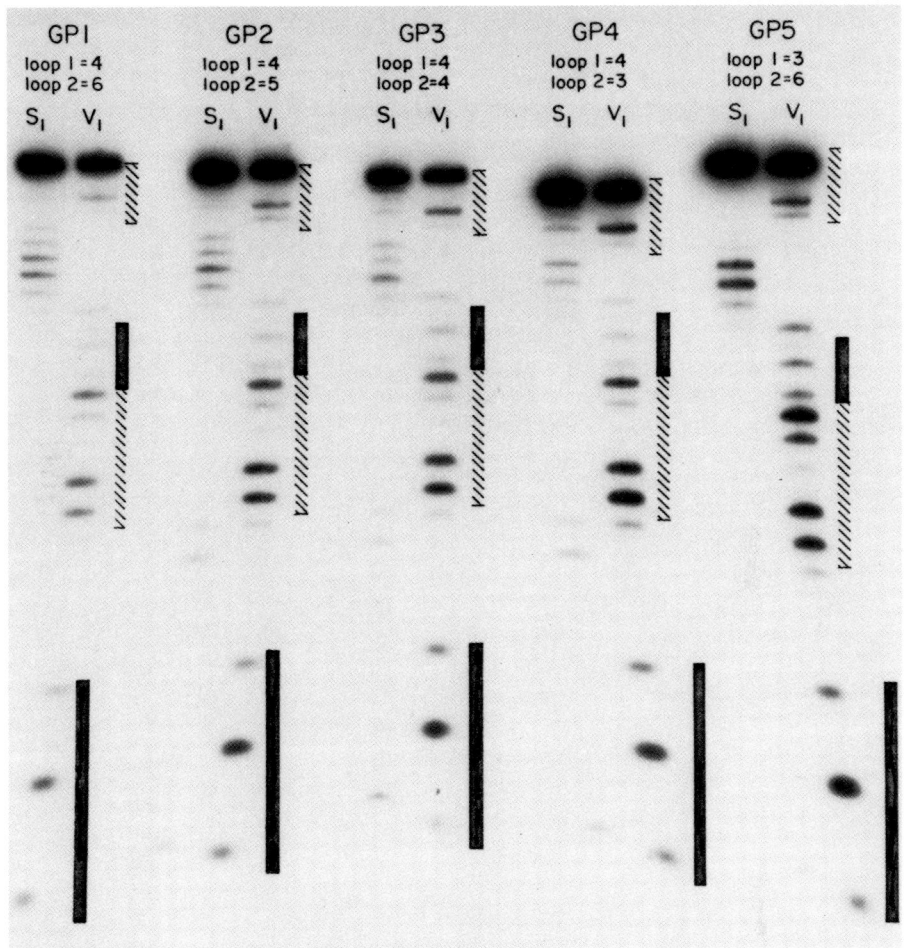

FIGURE 2. Autoradiographs of the gel-electrophoretic analysis of S_1 and V_1 digestions of oligonucleotides listed in Table 1. Sequences are listed in Table 1. Stem 1 regions are denoted to the right of the V_1 digestion lane by shading and those corresponding to stem 2 by hatching.

a trimer intact at the 3' end of duplexes (unpublished results) so lack of cleavage at the 3' end of **GP1** was expected.

Pleij et al. proposed that stacking of the two stem regions of a pseudoknot would allow formation of a continuous double helix (7). In the case of certain plant viral RNAs, the coaxially stacked stems of the pseudoknot allow the viral RNA to mimic the acceptor stem of tRNA (2). The imino region of the NMR spectrum of **GP1** provides evidence for stacking of the two stem regions in the pseudoknot (Figure 3). The seven peaks in the imino region were assigned to seven of the eight expected base-pairs of the pseudoknotted structure using one-dimensional nuclear Overhauser effect spectroscopy (NOE). From their chemical shifts, the two peaks furthest downfield correspond to the two A·U base-pairs. An NOE (which is only detectable for protons closer than 5Å) between imino 5 in stem 2 and imino 6 in stem 1 is evidence that the two stems form a

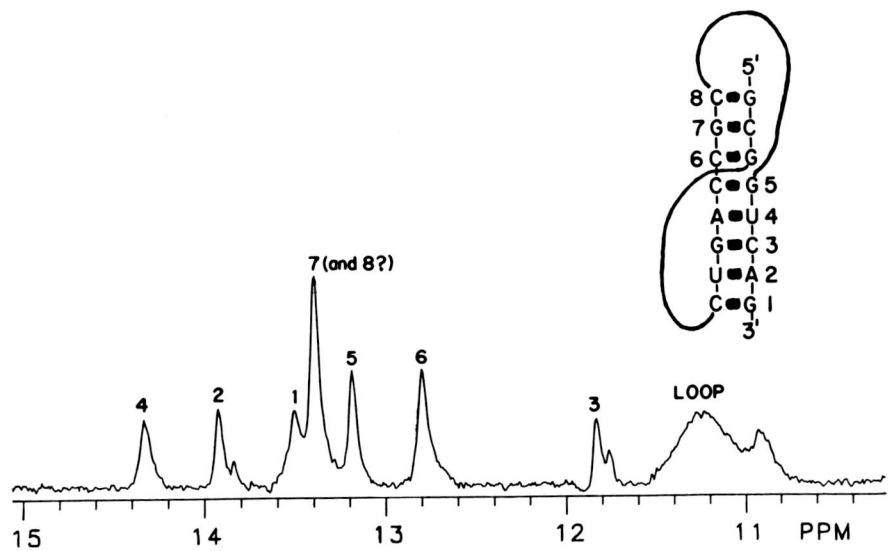

FIGURE 3. The imino spectrum of oligonucleotide **GP1**. The numbering of the resonances corresponds to the expected base pairs of the pseudoknotted structure as schematically illustrated. Assignments were made using one-dimensional NOEs.

continuous helix. Progress is being made towards assignment of the non-exchangable protons of the NMR spectrum. Incorporation of d^5-U within the oligonucleotide during transcription has considerably simplified the NMR spectrum of **GP1**.

If the two stems of the pseudoknot stack to form a continuous A-like helix, certain lower limits are placed on the loop size. Loop 1 crosses the major groove of the helix over the five base-pair stem: a distance of approximately 12Å if A-form geometry is assumed (7). This relatively short distance could conceivably be bridged by two nucleotides. In contrast, loop 2 must bridge the three base-pair 5' stem, a distance of at least 17Å across the minor groove. In order to determine the contribution of loop size to pseudoknot formation, variants of **GP1** which have shorter loops have been probed enzymatically.

Figure 2 shows autoradiographs of gel analysis of nuclease S_1 and RNase V_1 digestions of the variants listed in Table 1. At 4°C, similar patterns of digestion were obtained for all molecules. As with **GP1**, S_1 cleaved the variants in the proposed loop regions and V_1 cut after nucleotides in both stem regions. The 3' end of oligonucleotide **GP4**, which has only three residues in loop 2, was more susceptible to S_1 cleavage than other molecules in the series. Since RNase V_1 cleavage occured after G19 and U20 in stem 2, nucleotides which were also susceptible to S_1, we suggest that **GP4** may be in equilibrium between pseudoknot and 5' hairpin structures.

The results detailed above provide preliminary information about the three-dimensional structure of RNA pseudoknots. The imino region of the NMR spectrum of the oligonucleotide **GP1** shows that the two stems of the pseudoknot stack to form a continuous helix. Details of the geometry of the helix will require assignment of the non-exchangable protons. Minimum loop size will be dependent upon stem size and loop sequence. For the stem sequences used in this series, loop 2 can be shortened to at least three nucleotides although the structure may be in equilibrium between hairpin and pseudoknot. Loop 1 has been reduced to three nucleotides without disrupting pseudoknot formation. Variants containing even shorter loops will be characterized. Thermodynamic studies of the variants and of hairpins corresponding to the two stems of the pseudoknot are in progress. These studies will allow determination of the free energy contributions of stem and loop regions to the stability of the pseudoknotted structure.

REFERENCES

1. Studnicka GM, Rahn GM, Cummings IW, Salser WA (1978). Computer method for predicting the secondary structure of single-stranded RNA. Nucleic Acids Res 5:3365.
2. Rietveld K, van Poelgeest R, Pleij CWA, van Boom JH, Bosch L (1982). The tRNA-like structure at the 3' terminus of turnip yellow mosaic virus RNA. Differences and similarities with canonical tRNA. Nucleic Acids Res 10:1929.
3. Puglisi JD, Wyatt JR, Tinoco I, Jr.(1988). A pseudoknotted RNA oligonucleotide. Nature 331:283.
4. Milligan JF, Groebe DR, Witherell GW, Uhlenbeck OC (1987). Oligoribonucleotide synthesis using T7 RNA polymerase and synthetic DNA templates. Nucleic Acids Res 15:8783.
5. Hore, PJ (1983). A New Method for Water Suppression in the Proton NMR Spectra of Aqueous Solutions. J Magn Reson 54:539.
6. Wataya Y, Hayatsu H (1982). Effect of Amines on the Bisulfite-Catalyzed Hydrogen Exchange at the 5 Position of Uridine. Biochemistry 11:3583.
7. Pleij CWA, Rietveld K, Bosch L (1985). A new principle of RNA folding based on pseudoknotting. Nucleic Acids Res 13:1717.

DIMERS OF M1 RNA AS DETERMINED BY PSORALEN CROSSLINKING[1]

Samuel E. Lipson[2] and John E. Hearst

Department of Chemistry, University of California, Berkeley, CA 94720

ABSTRACT Crosslinked dimers of M1 RNA have been isolated and the base pairing interactions responsible for the dimerization determined. A comparison between mapped crosslinks and phylogenetic data is also presented.

INTRODUCTION

Maturation of biologically active tRNA in both eucaryotes and procaryotes involves the processing of a long RNA transcript. This processing includes both cleavage of the RNA transcript at specific sites and chemical modification of specific bases. All organisms investigated so far have RNase P activity which cleaves the tRNA transcripts specifically at the 5' end of the mature tRNA molecule. RNase P from *E. coli* is a ribonucleoprotein complex which is composed of one protein molecule and one RNA molecule, and this holoenzyme structure is similar in other organisms (1,2). The structure of the RNA of this ribonucleoprotein complex (M1 RNA) is of great interest because this RNA moiety is catalytic by itself under defined conditions (3). Previous structure predictions have been based on its known primary structure, on the thermodynamic parameters of double-strand helix formation (4) and on single strand and double strand specific nuclease data (5). Several structures have been proposed but unfortunately, with one exception, the conditions under which these structures were determined are not those under which the RNA alone is optimally catalytic. We have performed preliminary psoralen crosslinking experiments identifying four hairpin interactions within the M1 RNA (6). More recent phylogenetic studies (7,8) have shown compensatory base changes which suggest certain secondary structural features.

Other investigations into the structure and function of the M1 RNA

---

[1]This work was supported by NIH Grant # GM 11180
[2]Present Address, CODON, South San Francisco, CA 94080

include investigation of the metal ion requirements for activity (9), pH sensitive conformational changes (10), cross species reconstitution including RNA/protein from *E. coli*, *B. subtilis* and HeLa cells (3,11), and substrate effects on rates of reaction (5). Site specific mutagenesis and deletion mapping experiments (7) have also been performed. These studies have given us much insight into the mechanism of M1 RNA in solution, but the complete structure still eludes us.

RNase P from *B. subtilis*, which also has one RNA and one protein component, has been investigated. It has been found (12) that the metal ion requirements for catalysis by the RNA alone are similar but slightly more stringent than those for M1 RNA. The sequence of the RNA from *B. subtilis* is also known and phylogenetic arguments have been used to compare the structures of the two catalytic RNA molecules (13). The sequences are widely different, but similar hairpins can be drawn throughout the molecule.

A recent study by (9) has shown that the rate of catalysis by M1 RNA has a second order dependence on the concentration of the M1 RNA itself. These data imply that the M1 RNA might catalyze its reaction as a dimer. Our study addresses this issue from a structural standpoint.

To investigate the structure of the M1 RNA we have performed RNA crosslinking experiments with 4'-hydroxymethyl-4,5',8-trimethyl psoralen (HMT). Psoralens are tricyclic furocumarins which intercalate between base pairs of nucleic acid helices. They can then photoreact with pyrimidine bases to form covalent adducts. Most psoralens, including those discussed in this work, are bifunctional and can react with nucleic acids to form either mono- or diadducts. When a diadduct is formed, an interstrand crosslink is produced which covalently links the two strands of the nucleic acid. After intercalation, irradiation with long wavelength ultraviolet light (320-380 nm) produces monoadducts to pyrimidine bases. If another pyrimidine base is opposite and adjacent to the monoadducted pyrimidine, a crosslink can be formed with further irradiation at the same wavelength.

This photoreaction is photoreversible with irradiation of higher energy (260 nm) to produce the unmodified psoralen and nucleic acid. Reactions competing with photoreversal include breakdown of the psoralen and formation of pyrimidine dimers within the nucleic acid. Breakdown of the psoralen molecule itself also occurs with longer wavelength irradiation and must be considered when performing crosslinking experiments (14).

Psoralens react with nucleic acids under a wide variety of conditions (15,16). Calf thymus DNA has been reacted with psoralen to produce adducts in the presence of only 10 mM Tris buffer. At the other extreme, we have crosslinked RNA molecules in the presence of 100 mM $MgCl_2$ and 100mM NH_4Cl (7). Because psoralens only react with double stranded nucleic acids, are specific for crosslinking opposite and adjacent pyrimidines, and react only when activated photochemically, psoralens are powerful tools in elucidating secondary and tertiary structure in nucleic acids under whatever conditions are desired (For a review see 17). Psoralens are also thermally stable at neutral pH (Shi and Hearst, unpublished results), which allows the reaction to be performed over a wide range of temperatures.

Using a variation of the method originally described by Zweib and Brimacombe (18) for isolating UV induced crosslinks in RNA, we have used psoralen crosslinking of the M1 RNA to investigate the secondary and tertiary structure of the molecule. Our experiments were performed in buffers where the M1 RNA is optimally catalytic. Results obtained indicate that there are regions of the M1 RNA which are crosslinked to each other as hairpins and as long range secondary structure interactions. We have also identified 3 unique interactions within the M1 RNA which are intermolecular. We propose a new model for the secondary and tertiary structure folding of M1 RNA.

MATERIALS AND METHODS

All chemicals and reagents were of reagent grade unless otherwise noted.

Synthesis of M1 RNA.

Plasmid pDW27, which contains the template for T7 RNA polymerase transcription of an active M1 RNA, was the generous gift of Professor Norman R. Pace. The RNA produced by the plasmid contains 3 extra bases at the 5' end of the molecule and 7 extra bases at the 3' end of the molecule as well as two base changes at nucleotides 5 and 6 (AG to UC) within the sequence. The changes have no effect on the catalytic activity of the RNA (Pace, Norman R., personal communication). The plasmid was amplified in E. coli HB 101 and harvested as per (19), using ampicillin as the selection antibiotic.

The plasmid was restricted by dissolving 40 µg pDW27 in 200 µL SnaB1 buffer (50 mM NaCl, 7 mM Tris-HCl (pH 7.7), 10 mM $MgCl_2$, 5 mM 2-mercaptoethanol and 100 µg/mL BSA (Sigma)) and adding 40 Units of restriction endonuclease SnaB1 (New England Biolabs). The reaction was allowed to proceed for 2 hours at 37° C, at which time transcription was initiated. The solution was made 2.5 mM in each of the 4 ribonucleoside triphosphates, 2.5 mM 5'-HO-GpG-OH as initiator for transcription, 40 mM Tris-HCl (pH 8.0), 10 mM NaCl, 1 mM Spermidine (Cal Biochem), 10 mM DTT, 15 mM $MgCl_2$, 8% PEG (8000 MW, Sigma), and 4000 Units T7 RNA polymerase (generous gift of Chris Noren and Peter Schultz). This reaction was allowed to incubate at 37° for 2 hours. Samples of the plasmid DNA were analyzed on a 1.0% agarose gel to verify the restriction reaction.

The RNA produced was purified on a 15 x 40 x 0.05 cm, 5% polyacrylamide (BioRad Laboratories) gel containing 7 M urea (Swartz Mann) as denaturant. All gels described were 19:1 acrylamide:bisacrylamide with TBE (50mM Tris-borate, 1mM EDTA) as running buffer. The gel was run at 1000 V for 6 hours until the XC had run off the gel. The gel was removed from the glass plates and placed between plastic wrap (Reynolds). The RNA was visualized by UV shadowing. The band containing the RNA was cut from the gel with a razor blade. After removing the plastic wrap, the gel slice containing the RNA was cut into slices 2 x 10 mm and the RNA was eluted from the gel slices into 100 mM LiCl, 1 mM EDTA, 0.1% w/w SDS (LES), by shaking overnight at 37° C. The RNA was desalted by spinning through Centricon-30 microconcentrators (Amicon) with 4 dilutions with ddH$_2$O. The labelled product was purified on a 5% polyacrylamide gel as above.

Reactions with M1 RNA.

5' phosphorylation of M1 RNA. 100 µg M1 RNA was phosphorylated by dissolving the RNA in 100 µL kinase buffer (10mM MgCl$_2$, 5 mM DTT, 50 mM Tris-HCl, pH 8.5) containing 10 µM ATP (unlabelled, Boehringer Mannheim; 3000 Ci/mmol γ-^{32}P-ATP, Amersham), and 10 units polynucleotide kinase (New England Biolabs). The reaction was incubated for 1 hour at 37° C and quenched by extraction with one volume of phenol saturated with TE (50 mM Tris-HCl pH 7.5, 1 mM EDTA), followed by extraction with one volume of chloroform:isoamylalcohol (24:1), followed by Centricon-30 microconcentration with 4 dilutions with ddH$_2$O.

Crosslinking of M1 RNA. M1 RNA (5' end labelled with either cold ATP or 3000 Ci/mmol γ-^{32}P-ATP) was dissolved in activity buffer (100 mM MgCl$_2$, 100 mM NH$_4$Cl, 50 mM Tris-HCl, pH 7.5) to a concentration of 200 µg/mL or 0.5 µg/mL. HMT (either unlabelled or ^{3}H-HMT at 2.0 Ci/mmol; HRI Associates, Berkeley, CA) was added to a concentration of 30 µg/mL. The reaction was irradiated for 10 minutes at 37° C in a 1.5 mL Eppendorf centrifuge tube with a focused beam from a 2500 watt Hg-Xe lamp (Conrad-Hanovia, for details see Lipson et al. (6)). Unreacted HMT was removed by Centricon-30 microconcentrator dialysis as described above. Psoralen photoaddition was quantified by scintillation counting of the ^{3}H-HMT and UV spectroscopy of the RNA.

Partial photoreversal. The crosslinked RNA was dissolved in water to a concentration of 10 µg/mL in 1.5 mL Eppendorf microcentrifuge tubes. Partially photoreversed crosslinks were produced by exposure to a 40 W low pressure Hg vapor lamp (Sylvania) at a distance of 10 cm for periods ranging from 0 to 30 minutes. Products were analyzed by gel electrophoresis on a 34 x 40 x 0.05 cm, 5% polyacrylamide, 7 M urea gel system. Photoreversal products were visualized by autoradiography of the gel.

Probing with unmodified ^{32}P-M1 RNA. Unlabelled M1 RNA was photoreacted with HMT to a level of 0.5 HMT/M1 RNA, photoreversed for periods ranging from 0 to 30 minutes and mixed with an equal amount (w/w) of 5'-^{32}P-M1 RNA which had not been exposed to psoralen. The solution was brought to 1X in activity buffer at a final RNA concentration of 20 µg/mL. The solution was irradiated for 10 minutes as above. Re-crosslinked products were analyzed on 5% denaturing polyacrylamide gels.

<u>Preparative isolation of monomer and dimer crosslinks</u>. 300-500 µg 5'-^{32}P-M1 RNA was dissolved in activity buffer at 200 µg/mL with HMT added to a concentration of 30 µg/mL. The sample was irradiated for 10 minutes as described above except the sample was in a 1 cm x 1 cm quartz cuvette and the solution was constantly stirred. The sample was desalted and the HMT was removed by Centricon-30 concentrator dialysis as described above. An equal volume of weighting solution was added to the desalted sample and the crosslinks were resolved from the unmodified M1 RNA by 5% denaturing polyacrylamide gel electrophoresis in a lane 10 cm wide x 1 mm thick. The bands were visualized by autoradiography and were then cut out and eluted into 500 µL LES. The sample was again desalted as above. The samples were stored at -20° C.

Crosslink Mapping Procedures.

<u>Digestion of RNA</u>. 1-10 µg preparatively isolated M1 RNA crosslink was digested with RNase T_1 (Pharmacia, 10U RNase T_1/µg RNA) in a buffer containing 50 mM Tris-HCl pH 7.5, 200 mM NaCl for 15 minutes at 37° C. RNA concentration was 100 µg/mL. The reaction was quenched by adding 1 µg Proteinase K (Boehringer Mannheim) per unit RNase T_1 and incubating for another 5 minutes at 37° C. The mixture was ethanol precipitated and the RNA was recovered by spinning at 15,000 x g in a table top centrifuge for 15 minutes. The supernatant was removed and the pellet was dried. The digested RNA was 5' end labelled in kinase buffer by 5000 Ci/mmol γ-^{32}P-ATP (5 µM) using 3 units of polynucleotide kinase (total volume = 25 µL), for 2 hours at 37° C. The reaction was quenched by ethanol precipitation. The RNA was recovered as above. The supernatant was removed and the pellet dried in preparation for the two dimensional gel analysis. The samples were stored at -20° C.

<u>First dimension gel</u>. The following 2D gel system is that of Turner and Noller (21). The digested, labelled M1 RNA was dissolved in 10 µL weighting solution and heated to 90° C for 2 minutes to remove all secondary structure. The sample was then quick cooled on ice and immediately loaded onto a lane 1 cm wide on a 15 x 40 x 0.05 cm denaturing 20% polyacrylamide, 7 M urea gel and run at 2500 V. The gel was run until the BpB had migrated 33 cm.

<u>Photoreversal</u>. The top glass plate from the gel was removed and the gel was covered with plastic wrap. The RNA/HMT crosslinks were photo-reversed under a 40 W low pressure Hg germicidal lamp at a distance of 4 cm for 1 hour. The 30 cm of the lane containing the RNA, from immediately above the BpB to within 3 cm of the well, was excised using a straight edge and razor blade. The excised lane was placed horizontally on a glass plate 34 x 40 cm and covered with a second glass plate. The second dimension gel, which was also 20% acrylamide 7 M urea, was poured around the first

dimension lane and the gel was allowed to polymerize for a minimum of 2 hours.

Second dimension gel. The second dimension gel was run perpendicular to the first dimension at 2500 V until the BpB dye had run to the bottom of the gel. The top glass plate was removed from the gel and the gel was covered with plastic wrap. The labelled RNA was visualized by autoradiography for 1 hour at room temperature.

Isolation of below diagonal fragments. Fragments which migrated below the diagonal were cut from the gel and eluted for a minimum of 12 hours into 200 µL LES in 1.5 mL Eppendorf centrifuge tubes at 37° C with shaking. The LES was transferred into a new Eppendorf tube, the gel slice was washed two times with 100 µL LES and the washes were combined with the LES from the elution. The efficiency of elution was followed by Cerenkov counting.

Sequencing of isolated fragments. RNA fragments in LES were divided equally into three Eppendorf tubes. 3 µg unlabelled, carrier tRNA was added to each tube and the RNA was precipitated by adding 2.5 volumes 100% ethanol and cooling to -20° C overnight. The pellets were recovered by spinning at 15,000 x g for 15 minutes. The supernatants were removed and the pellets were vacuum dried at room temperature. The dried samples were enzymatically sequenced using the methods of Donis-Keller (20).

RESULTS

Unmodified ^{32}P M1 RNA Reacts with Photoreversed, Cold M1 RNA

We produced covalently bound HMT adducts on unlabelled M1 RNA by photoreacting unlabelled M1 RNA with HMT under conditions where the M1 RNA is known to be active. The excess psoralen was removed by flow dialysis using Centricon-30 concentrators until no measurable tritium (from the tritiated psoralen) passed through the membrane. The HMT photoreacted M1 RNA was then photoreversed for times ranging from 0 to 30 minutes under short wavelength UV light. This produced varying amounts of monoadduct from crosslinks as well as free psoralen. Excess photoreversed psoralen was again removed by flow dialysis. The addition of 5'-^{32}P labelled M1 RNA to the photoreversed sample followed by irradiation with 360 nm light to recrosslink any available psoralen monoadducts produced labelled bands which migrate much more slowly than the unmodified M1 RNA. As a control, irradiation of 5' end labelled M1 RNA under crosslinking conditions in the absence of HMT produces no slower moving bands detectable by autoradiography. Furthermore, without photoreversal very little recrosslinking to labelled RNA was observed, while photoreversal times up to about 20 minutes produce progressively more recrosslinking. Longer photoreversals (*i.e.* 30 minutes) produce less crosslinking, implying that photoreversal for 30 minutes under our conditions produces more free psoralen than recrosslinkable monoadduct. If the RNA was precipitated after crosslinking (to remove the HMT) the products produced were unable to enter the gels for analysis. We found that by keeping the RNA in solution during the removal of the HMT (by Centricon-30 flow dialysis) the different crosslinks could be separated by gel electrophoresis. This experiment indicates that two separate molecules of M1 RNA can be crosslinked by psoralen, because the only crosslinked species which can be seen on the autoradiogram must represent crosslinking of a labelled, unmodified M1 RNA and an unlabelled, monoadducted M1 RNA (see discussion for further details).

Identification of monomer bands vs. dimer bands. Irradiation of 5' end labelled M1 RNA with psoralen produces a series of 9 bands which migrate slower than the M1 RNA, as well as a smeared background. M1 RNA which is irradiated in the absence of psoralen produces no new bands. Our data indicates that each band represents a specific crosslink structure with a different electrophoretic mobility due to a different denatured shape caused by the crosslink. The dimer crosslinking products were identified as described above, and thus, the other products are due to monomer crosslinks. We have found 3 dimer interactions and 6 monomer interactions.

Photoreversal of probed adducts. Each band produced by the crosslinking reaction of M1 RNA and HMT was isolated from the purification gel, eluted into LES, desalted by Centricon-30 flow dialysis and photoreversed under 260 nm light for 30 minutes. On another purification gel all photoreversed structures produced bands corresponding to full length, unmodified M1 RNA.

Identification of interactions in monomer XL. Six bands did not appear upon recrosslinking of cold, HMT photoreacted/photoreversed M1 RNA to 5' end labelled M1 RNA, but they did appear on crosslinking 5' end labelled M1 RNA with HMT. This indicates that these bands are monomer M1 RNA crosslinked molecules. After purification by polyacrylamide gel electrophoresis, elution, and desalting, the samples containing monomer XLs were completely digested with RNase T_1 (cuts 3' to G) in 50 mM Tris-HCl pH 7.5, 100 mM NaCl. The RNase was quenched with Proteinase K and then the RNA was ethanol precipitated. The digested RNA was 5' end labelled with γ-^{32}P-ATP and the samples were immediately loaded onto a denaturing 20% polyacrylamide gel. After the BpB dye had migrated about 33 cm, one of the glass plates was removed from the gel. The gel was then covered with plastic wrap and the lane containing the RNA was placed under a germicidal lamp (254 nm) for 1 hour to photoreverse the crosslinked RNA fragments. The lane containing the RNA was then excised from the gel and the second dimension gel was poured around the first dimension lane. After polymerization, the gel was run until the BpB had migrated 40 cm. Unmodified RNA forms a diagonal on the two dimensional gel, as the mobility of each fragment is the same in each dimension. Fragments which had a psoralen crosslink photoreversed to either monoadduct or free psoralen between dimensions change their mobility in the second dimension. Hairpins which were crosslinked and then photoreversed between dimensions will be single bands below the diagonal, as their effective radius is larger when crosslinked than when uncrosslinked. Long range interactions (crosslinks which involve two different fragments of RNA) will photoreverse to two bands which will migrate directly above and below each other since they migrated as one molecule in the first dimension. Both bands are seen below the diagonal. Single bands observed below the diagonal may also be one fragment of two crosslinked together where the second fragment was less than 6 in length, and therefore was run off the bottom of the second dimension gel. Each of the 6 monomer crosslinked M1 RNA molecules produced photoreversal products which were identifed by enzymatic sequencing as indicated in table 1.

Identification of interactions in dimer XL. The identification of dimer interactions was essentially identical to the procedure for monomer crosslinks. Since the level of psoralen was determined to be ~0.4 HMT/M1 RNA, we expect that each dimer will have only 1 HMT. This aided our analysis in that only one intermolecular crosslink is expected per band. The crosslinks from bands 7-9 are also presented in table 1. Each band produced only 1 detectable interaction.

TABLE 1
MONOMER INTERACTIONS
Hairpins

Fragments	Proposed XL Site
136-170	U_{149} x U_{167}
199-218	U_{210} x C_{216}
301-314 x 317-329	U_{311} x U_{319}

Long Range Interactions

1-10 x 358-357	U_7 x U_{367}
65-82 x 231-243	C_{78} x C_{242}

DIMER INTERACTIONS

54-72 x 341-357	C_{71} x C_{355} (1)
112-125 x 351-377	C_{123} x U_{360} (2)
252-270 x 359-377	U_{267} x U_{360} (3)

3 has no corresponding region in *B. subtilis*.

Other lighter bands in the two dimensional gels. Other bands of lesser intensity were seen in the autoradiographs of the two dimensional gels of both monomer and dimer crosslinks of the M1 RNA. We sequenced many of these fragments and found that they contained the same sequence isolated as the main band but were longer in length. This is attributed to incomplete T_1 digestion of the M1 RNA before the two dimensional gel analysis. Twice, we noticed other sequences migrating below the diagonal. The number of counts in these bands was less than 1% of the major band and we attribute these bands to photodamage of the RNA caused by the photoreversal light.

Possibility of missed products of photoreversal. The 2D gel system employed has the potential of missing very short fragments produced during the photoreversal. Fragments shorter than 6 in length will be run off of the bottom of the gel. We believe this is the case in one of our analyses as the fragment produced upon photoreversal migrates with an apparent length of only three fewer nucleotides than it did before photoreversal.

DISCUSSION

RNase P from *E. coli* catalyzes the cleavage of the 5' terminus of precursor tRNA molecules as one step in the processing of tRNA to give a mature product. It is composed of an RNA 377 nucleotides in length (M1 RNA) and a protein (C5-protein). The RNA molecule can catalyze the same reaction as RNase P under specific conditions *e.g.* 60 mM Mg^{2+}, 100 mM NH_4^+ (3). Mapping of single stranded and double stranded regions in the molecule has yielded a model for the secondary structure of the M1 RNA (5). These analyses were performed under conditions where the nucleases and chemical reactants are most active. Unfortunately, these conditions do not correspond to those under which the M1 RNA shows optimal catalytic activity, with the exception of RNase A and RNase T1 mapping. We have investigated the structure of M1 RNA with psoralen crosslinking under conditions where the M1 RNA is active. Our data confirm two interactions proposed by the nuclease digestion and chemical mapping of the M1 RNA and contradict two other interactions. Recent phylogenetic studies by James *et al.* (8) indicate structural features which are different than those proposed by Guerrier-Takada *et al.* (15). Our crosslinking studies are confirmatory with one exception. Our data also show three unique intermolecular and 1 intramolecular interaction involving the 3' terminus of the M1 RNA. The kinetics of the catalytic reaction mediated by M1 RNA is second-order in M1 RNA concentration (9), implying the M1 RNA might act as a dimer. The dimer is one possible explanation for our multitude of interactions with the 3' terminus of the molecule. We have identified bands in a polyacrylamide gel which migrate at the positions where dimer M1 RNA should migrate only when M1 RNA is irradiated in the presence of HMT. We have isolated these crosslinked molecules and have elucidated the interactions involved. The following sections discuss the crosslinking of the M1 RNA.

Intramolecular Interactions.

Hairpin 136-170. This interaction was seen earlier (6). A complete discussion of this interaction is discussed there. This interaction also confirms the structure predicted by James *et al.* (8) with the proposed crosslink between U 149 and U 167.

Hairpin 199-218. This hairpin confirms the hairpin predicted by James *et al.* (8) with a possible crosslinking site between bases U 210 and C 216. It is contradictory to the structure predicted by Guerrier-Takada and Altman (5) which has this region involved as one of two strands in a long hairpin.

Hairpin 226-246. The mobility of the fragment isolated changed very little from its mobility before photoreversal (see figure 2). This is a concern as the interaction may not be a hairpin at all, but rather a long range interaction for which we found only one of the fragments. We cannot predict a stable structure for this fragment as a hairpin. We have also found an intramolecular crosslink between this region of the molecule and the fragment 65-82 (see discussion below), and we believe that this fragment is only one of two that were crosslinked together prior to photoreversal.

Hairpin 277-291. This is the same hairpin seen by Lipson *et al.* (6). Mutagenesis data (Lawrence and Altman, (7)) imply that the region between 282 and 292 is essential for activity of the M1 RNA. Bases 285-290 (5'-UGAAC-3') are complementary to the T-Ψ-C loop found in most tRNA molecules. As discussed in detail in Lipson *et al.* (6) deletion of this complementary region or destruction of the hairpin supporting this 5 base sequence would disallow the T-Ψ-C complement binding to tRNA. This hairpin is also discussed in the section dealing with the intermolecular interaction between bases 252-270 X 358-377.

Hairpin 301-329. This two strand intramolecular interaction describes the hairpin bounded by bases 304 and 327 in the James *et al.* (8) structure. We have identified this hairpin by isolating the two fragments of RNA making up the stem. The proposed crosslinking site is between U 311 and U 319.

Long range interaction 1-10 X 358-377. This long range interaction pairs the 5' end and the 3' end of the molecule. The one site for psoralen crosslinking is between U 6 and C 367. Many other stable RNA molecules (*e.g.* 5S rRNA, 16S rRNA, 23S rRNA, tRNA, etc.) have the ends of the RNA base paired together.

Long Range Interaction 65-82 X 231-243. The interaction between 65-82 X 231-243 is predicted by James *et al.* (8) as a feature of the M1 RNA which defines the two unique domains of the M1 RNA molecule. Our crosslinked fragments confirm this structure with the psoralen crosslink occuring between C 78 and C 242. This is a coaxial stack of two helices and is a hot spot for psoralen photocrosslinking (21).

Figure 1 summarizes our intramolecular interactions. The figure is based on James et al. (8). The two possible interactions in the 270-300 region are shown.

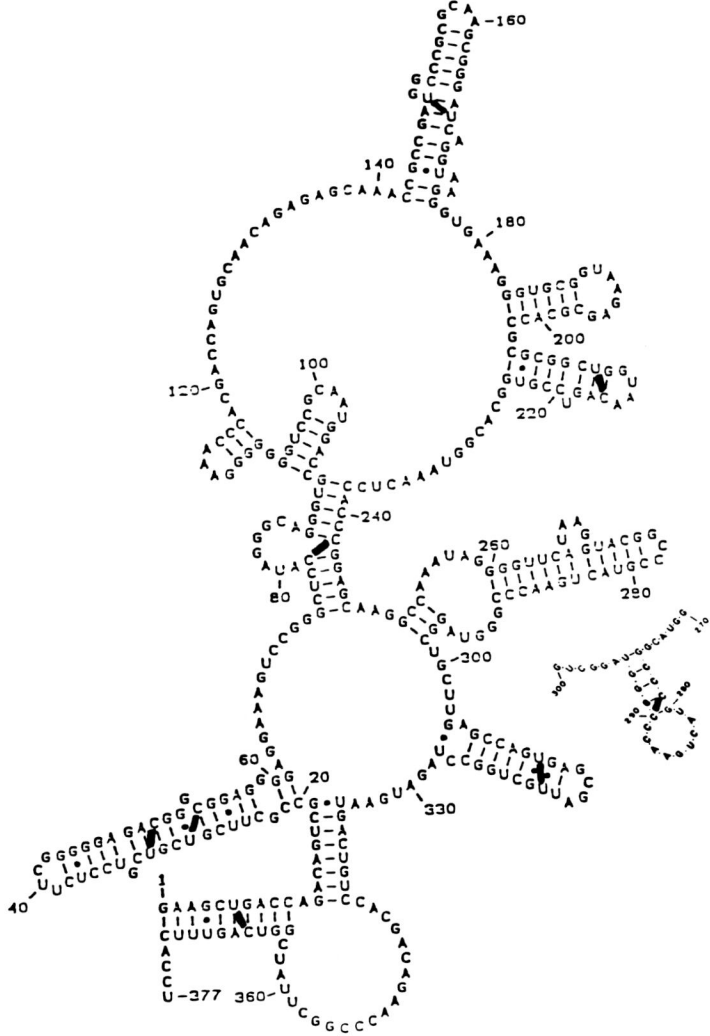

FIGURE 1. James et al. (8) structure with intramolecular crosslinks as blocks.

Dimer Interactions and their Implications.

The dimer M1 RNA was identified by probing for crosslinkable psoralen monoadducts on unlabelled M1 RNA with 5'-^{32}P end labelled M1 RNA. The only labelled material which will migrate slower than unreacted M1 RNA will be that which is crosslinked to unlabelled material through a psoralen. Control experiments where 5' end labelled M1 RNA was photoreacted with itself in the absence of HMT produced no material with a slower mobility, and thus no detectable photodamaged M1 RNA which might interfere with the isolation of dimer products.

Three bands were identified as dimer crosslinks. To further identify the crosslink bands which were due to monomer crosslinks, the 5' end labelled M1 RNA was crosslinked in the presence of HMT. This material was also run down the purification gel and 9 bands were seen. Bands 1-6 represent products which were not seen in the lanes produced by probing and thus are monomer crosslinks. Bands 7-9 correspond to the dimer crosslink bands. The smearing between the bands was isolated but we were not able to purify it to homogeneity on a gel and thus we did not attempt any two dimensional gel work on this material.

Dimer interaction 54-72 x 341-377. This interaction is postulated by James *et al.* (8) by phylogenetic methods as an interaction between these two regions of the M1 RNA. We have isolated this interaction in a complex formed by the interaction of two different M1 RNA molecules. The site of psoralen crosslinking is between C 71 and C 355. We therefore, agree with the proposed interaction of James *et al.* (8), but believe that this interaction is intermolecular.

Dimer interaction 112-125 x 351-377. This intermolecular interaction is shown below. We propose that the crosslink is between C 122 and U 360 or C 123 and U 359 both sites are in mismatched regions in the middle of a helix 6 long bounded by 120-125 base paired to 359-364. This is a potential hot spot for psoralen photoaddition. The regions involved in this intermolecular interaction are also conserved phylogenetically using the data of James *et al.* (8). The *B. subtilis* interaction actually has one more base pair than does the *E. coli* interaction.

Dimer interaction 252-270 x 358-377. The intermolecular interaction observed between 252-270 and the 3' end of the M1 RNA is explained by five base interaction between 266-270 and 358-362 with the psoralen crosslink occuring between U 267 and U 362 or U 267 and U 360. Our data also indicate that bases 277-291 also form a hairpin which we have isolated under two different sets of conditions. Our hairpin at 277-291 (present study; Lipson *et al.*, (6)) and this long range dimer interaction are compatible with each other, but together are mutually excluded with the model proposed by James *et al.* (8). This could be a dynamic interaction which is essential to catalysis by the M1 RNA.

How these intermolecular interactions might work. These three intermolecular interactions appear to occur in an asymmetric fashion between the two M1 RNA molecules interacting with each other. Specifically, the interaction between 120-126 and 356-363 precludes the interaction between

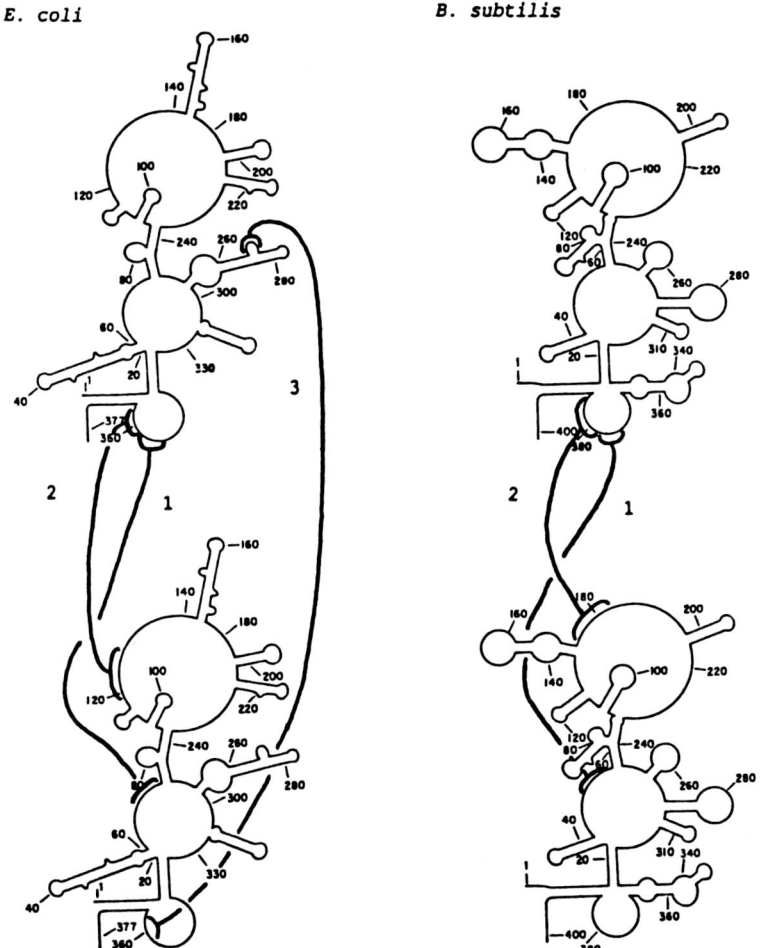

FIGURE 2. Intermolecular interactions as found in M1 RNA. For interactions 1 and 2, comparative interactions are shown on the *B. subtilis* structure.

261-269 and 359-367 as well as the interaction between 70-73 and 354-357. However, these last two interactions are not mutually exclusive and can occur simultaneously.

The dimeric interaction between 120-126 and 356-363 appears to be a consistent feature which could exist at all times. The feature is also conserved phylogenetically based on the data of James et al. (8). Defining the structure for labelling purposes, we assign the fragment 120-126 on molecule A and 356-363 on molecule B. When a tRNA precursor binds to the dimer M1 RNA, or by thermal equilibrium in the absence of tRNA precursor, the gross structure of the dimer changes to bring the region between 70-73 on molecule B in contact with 354-357 on molecule A. After these interactions establish themselves, the hairpin at 260-290 could dissociate, freeing the region at 261-269 on molecule B to intermolecularly pair with 359-367 on molecule A, thus exposing the T-Ψ-C complement loop (discussed above), to firmly set the substrate in position for reaction.

A total model of the M1 RNA dimer. Figure 2 is a schematic of the dimer interactions we have found within the M1 RNA. We have also identified the intramolecular interactions. The major difference between this model and the James et al. (8) model are the dimerization and the dynamic switch which we predict for M1 RNA catalysis. Otherwise, our data confirm the predictions of James et al. (8). It is possible that an active site which binds substrate is created by the intermolecular interaction occurring in a dimer of M1 RNA.

REFERENCES

1. Stark, BC, Kole, R, Bowman, EJ and Altman, S (1977). Ribonuclease P: An enzyme with an essential RNA component. Proc Natl Acad Sci USA 75:3719-3721.
2. Gardiner, KJ and Pace, NR (1980). RNase P of *Bacillus subtilis* has an RNA component. J Biol Chem 255:7507-7509.
3. Guerrier-Takada, C, Gardiner, K, Marsh, T, Pace, N and Altman, S (1983). The RNA moiety of ribonuclease P is the catalytic subunit of the enzyme. Cell 35:849-857.
4. Reed, RE, Baer, MF, Guerrier-Takada, C, Donis-Keller, H and Altman, S (1982). Nucleotide sequence of the gene encoding the RNA subunit (M1 RNA) of ribonuclease P from *Escherichia coli*. Cell 30:627-636.
5. Guerrier-Takada, C and Altman, S (1984). Structure in solution of M1 RNA, the catalytic subunit of ribonuclease P from *Escherichia coli*. Biochemistry 23:6327-6334.
6. Lipson, SE, Cimino, GD and Hearst, JE (1988). Structure of M1 RNA as determined by psoralen crosslinking. Biochemistry 27: 570-575.
7. Lawrence, NP and Altman, S (1986). Site-directed mutagenesis of M1

RNA, the RNA subunit of *Escherichia coli* ribonuclease P. The effects of an addition and small deletions on catalytic function. J Mol Biol 191:163-175.
8. James, BD, Olsen, GJ, Liu, J and Pace, NR (1988). The secondary structure of ribonuclease P RNA, the catalytic element of a ribonucleoprotein enzyme. Cell 52:19-26.
9. Guerrier-Takada, C, Haydock, K, Allen, L and Altman, S (1986). Metal ion requirements and other aspects of the reactions catalyzed by M1 RNA, the RNA subunit of ribonuclease P from *Escherichia coli*. Biochemistry 25:1509-1515.
10. Guerrier-Takada, C and Altman, S (1986). M1 RNA with large terminal deletions retains its catalytic activity. Cell 45:177-183.
11. Gold, HA and Altman, S (1986). Reconstitution of RNase P activity using inactive subunits from *E. coli* and HeLa cells. Cell 44:243-249.
12. Gardiner, KJ, Marsh, TL and Pace, NR (1985). Ion dependence of the *Bacillus subtilis* RNase P reaction. J Biol Chem 260:5415-5419.
13. Reich, C, Gardiner, KJ, Olsen, GJ, Pace, B, Marsh, TL and Pace, NR (1986). The RNA component of the *Bacillus subtilis* RNase P. Sequence, activity, & partial secondary structure. J Biol Chem 261: 7888-7893.
14. Kao, JPY (1984) PhD. Dissertation, University of California, Berkeley.
15. Hyde, JE and Hearst, JE (1978). Binding of psoralen derivatives to DNA and chromatin: influence of the ionic environment on dark binding and photoreactivity. Biochemistry 17:1251-1257.
16. Thompson, JF, Wegnez, MR and Hearst, JE (1981). Determination of the secondary structure of *Drosophila melanogaster* 5 S RNA by hydroxymethyltrimethylpsoralen crosslinking. J Mol Biol 147:417-436.
17. Cimino, GD, Gamper, HB, Isaacs, ST and Hearst, JE (1985). Psoralens as photoactive probes of nucleic acid structure and function: organic chemistry, photochemistry, and biochemistry. Ann Rev Biochem 54:1151-1193.
18. Zwieb, C and Brimacombe, R (1980). Localization of a series of intra-RNA cross-links in 16S RNA, induced by ultraviolet irradiation of *Escherichia coli* 30S ribosomal subunits. Nucleic Acids Research 8:2397-2411.
19. Maniatis, T, Fritsch, EF and Sambrook, J (1982). Molecular Cloning: A Laboratory Manual, Cold Spring Harbor Laboratory, New York.
20. Donis-Keller, H, Maxam, AM and Gilbert, W (1977). Mapping adenines, guanines, and pyrimidines in RNA. Nucleic Acids Research 4: 2527-2538.
21. Turner, S and Noller, HF (1983). Identification of sites of 4'-(hydroxymethyl)-4,5', 8-trimethylpsoralen cross-linking in *Escherichia coli* 23S ribosomal ribonucleic acid. Biochemistry 22:4159-4164.

TOWARDS DEFINING THE MINIMAL STRUCTURAL REQUIREMENTS FOR SELF-SPLICING OF THE PHAGE T4 td INTRON[1]

Marlene Belfort,[2] Jill L. Galloway Salvo, Karen Ehrenman[3] and Timothy Coetzee[3]

Wadsworth Center for Laboratories and Research
New York State Department of Health, P.O. Box 509
Albany, NY 12201-0509

ABSTRACT   Studies have been initiated with the td gene to derive a minimal functional group I intron. Experiments are based on information from comparative studies among the T4 introns and from genetic analyses of the td intervening sequence. After experimental verification of the P6 pairing a series of deletions extending from the L6a loop toward P6 was generated. As a result the 1016 nt td intron has been reduced to between 236 nt and 243 nt, while maintaining high levels of self-splicing activity. The deletion-tolerance of L6a suggested this region to be appropriate for dividing the td core ribozyme into two components. Indeed, the td intron, split at L6a, is able to function as a bimolecular complex, yielding ligated exons and a linear homologue of the intron cyclization product. The deletion experiments may eventually yield molecules small enough for physical analysis, while the trans-splicing of a group I intron stimulates questions about the evolution of RNA catalysts and split genes.

---

[1]This work was supported by NSF Grant DMB-8502961 and NIH grants GM-33314 and GM-39422 to M.B. J.L.G.S. is the recipient of an ACS post-doctoral fellowship.
[2]Joint affiliation: School of Public Health Sciences, The State University of New York, New York State Department of Health, Albany, NY 12201.
[3]Joint affiliation: Department of Microbiology and Immunology, Albany Medical College, Albany, NY 12208.

INTRODUCTION

The group I introns of phage T4 provide a unique system in which to delineate the critical functional components of this class of catalytic RNA. The system offers two main advantages: First, the existence of a family of three closely related introns (1) favors comparative studies (2). Highly conserved regions are considered to share important functions, whereas those regions of structural divergence may be either functionally disparate or dispensable (ref. 2 and Fig. 1). Second, the T4 introns provide an opportunity to apply the well developed methods of prokaryotic genetics to the study of these self-splicing RNAs. The intron-containing td gene, which encodes thymidylate synthase (TS), is particularly useful in this regard, as both its wild type and mutant forms confer selectable phenotypes to a TS⁻ (thyA) Escherichia coli host cell. These phenotypes have been exploited to isolate a wide range of non-directed td⁻ splicing defective mutations, which identify residues involved in the critical functioning of the intron (refs. 3-5; K.E., D. Hall and M.B., in preparation, and Fig. 1).

Figure 1 highlights identical sequences for the three T4 introns (td, nrdB and sunY) (2) and shows the 21 residues that have yielded splicing defective td mutations after random mutagenesis. Of the 14 pairings predicted by the td secondary structure model, the P2, P6a and P7.1 pairings show an absence of strict sequence similarity between the three T4 introns and also lack splicing defective mutations. These three elements are therefore attractive targets for deletion analysis.

As a first approach toward defining a minimal intron we have elected to investigate the P6-L6-P6a-L6a region (Fig. 1). On the one hand, mutational clustering in the P6 stem (K.E., D. Hall and M.B., in preparation) suggests the functional importance of this putative pairing. On the other hand, the dispensability of P6a and L6a is implied by the absence of mutations and the lack of sequence conservation (Fig. 1). Additionally, the 735 nt td intron open reading frame (ORF), that is looped out of the core structure at P6a, is not required for splicing (4). It was therefore of importance to verify the P6 pairing, to determine the extent of deletion of the ORF-containing L6a loop that is compatible with self-splicing, and to probe the requirement for P6a. In addition, it was of interest to split the td intron in L6a to test the ability of the td core structure to function as a bimolecular complex.

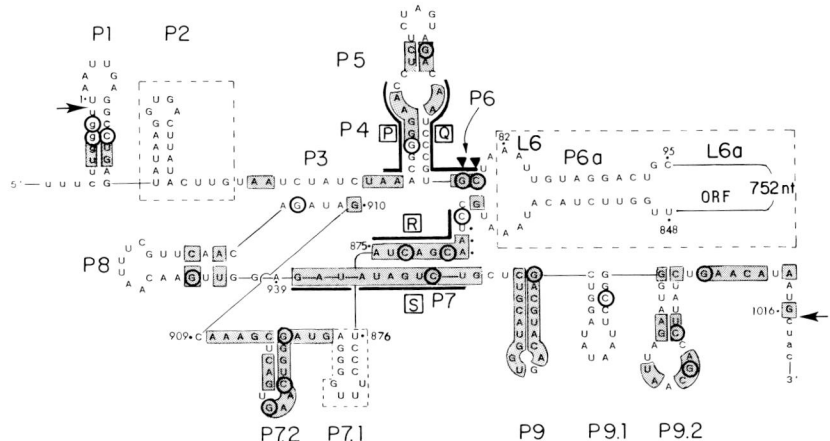

Fig. 1. The td intron secondary structure model. The model includes structure elements P1-P9.2, and the 21 sites of mutation (circled), representing a total of 33 mutations (3-5). Sites of 3-way identity with the nrdB and sunY introns are shaded. Conserved sequences P, Q, R and S are marked by solid lines. Occasional residue numbers are given (beside dots), to help trace the molecule. The mutational hotspots in P6[5'] are marked with black triangles. The splice sites are indicated by arrows. The mutation-free, non-conserved or loosely conserved elements are within dashed boxes.

RESULTS AND DISCUSSION

The P6 Pairing

The short P6 pairing is a conserved element and believed to be an integral component of the catalytic core structure (6-8), but its existence has not been verified experimentally. It was therefore important to test the P6 pairing before proceeding to delete the elements extending out from this short stem. Verification was of particular interest since the two 5' nucleotides of this dinucleotide bridge (G78 and C79) are mutational hotspots in the td intron. Of 33 independently isolated non-directed td mutations, 8 are in the G78-C79 dinucleotide (refs 3-5; K.E., D. Hall and M.B., in preparation). The mutagenesis strategy, outlined in Fig. 2A, was such that disruption of P6 was achieved either by the G78C-C79G inversion on

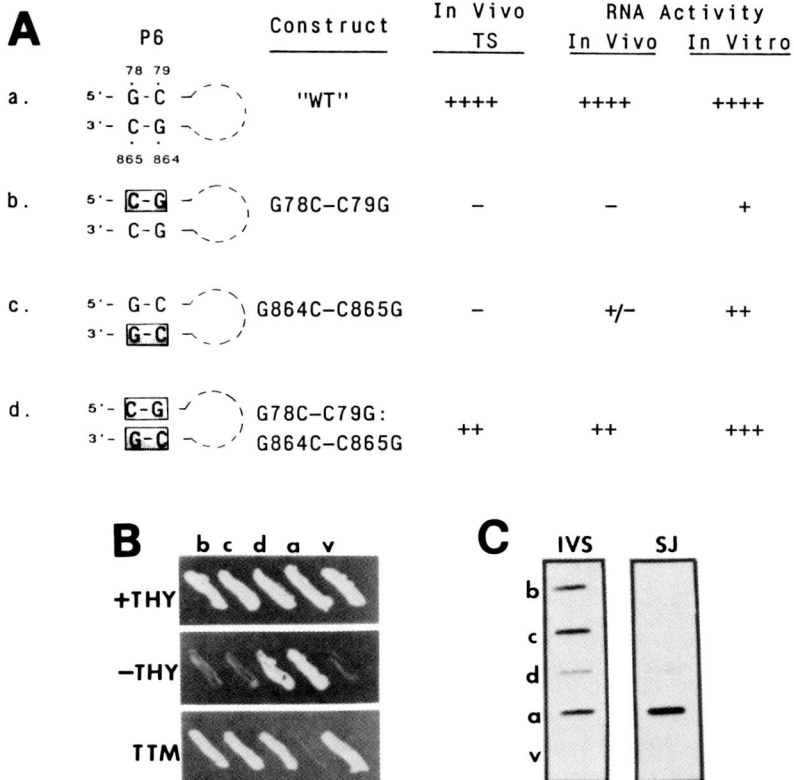

Fig. 2. The P6 stem. The splicing profiles of the wild type (a) and mutant constructs (b-d as indicated) are summarized in A. The wild type parent ("WT") is intron deletion Δ120-797, which has an intron size of 346 nt. "++++" = very active → "-" = no activity detectable. The in vivo TS phenotype was assessed at 37°C by the plating method shown in B (a-d as in A, v = vector; minimal media [-THY] was supplemented with thymine [+THY] or thymine and trimethoprim [TTM]). In vivo splicing activity was measured in slot blot assays shown in C, or by Northern hybridization analysis (not shown), with intron (IVS) or splice-junction (SJ) probes. Quantitation of blots by densitometry indicated that the compensatory mutant produces about 20% of wild type levels of mRNA in vivo. In the in vitro assays, conducted as in Fig. 3 at 42°C and 3 mM Mg^{++}, splicing of the compensatory mutant was >50% as efficient as the wild type.

the 5' side (P6[5']) or by the G864C-C865G inversion on the 3' side (P6[3']). Combining these two inversions (G78C-C79G:G864C-C865G) should restore a structure that approximates the original P6 pairing.

Inversion of either the 5' or the 3' elements of P6 resulted in greatly reduced splicing activity in vivo and in vitro, with the P6[5'] mutant displaying the more dramatic phenotype (summarized in Fig. 2A). The diminished activity in vivo is apparent from the inability of the P6[5'] and P6[3'] mutants to grow on minimal media lacking thymine and by their ability to grow on TS⁻-selective TTM media (Fig. 2B). Additionally, no ligated exons were detected (Fig. 2C). In contrast, the compensatory mutant G78C-C79G:G864C-C865G is able to grow in the absence of thymine and does produce mature mRNA, albeit at lower levels than wild type. Partial restoration of phenotype in vivo is also reflected in the ability of the compensatory mutant to grow on TTM media (growth in the absence of thymine as well as on TTM media indicates low TS levels). In vitro, however, the compensatory mutant splices at efficiencies approaching wild type (Fig. 2A and data not shown). Taken together, these results verify the P6 pairing and its importance in the catalytic activity of the intron. The difference in residual splicing activity of the P6[5'] and P6[3'] mutants, and the partial restoration of phenotype in the compensatory mutant are under investigation.

Deletion-tolerance of L6a-P6a

Previous experiments indicated that deletion of >80% of the ORF within L6a results in a fully functional intron (4). In considering extending the existing deletion (Δ167-797 in Fig. 3A, tdΔ1-3 in ref. 4) we noted from comparison with sunY and nrdB that P6a was increased in size when long sequences were looped out of this element (2). It seemed plausible that by deleting more of the apparently extraneous loop sequences from td we might also obviate the need for P6a.

Initially, L6a was reduced to a 5 nt loop by oligonucleotide-directed mutagenesis. This was achieved by deleting 129 nt from the parent construct Δ167-797 and inserting one G residue to create a SmaI site within L6a (Fig. 3A). This construct, Δ99-848, has 265 intron residues remaining (after deletion of a total of 752 nt)

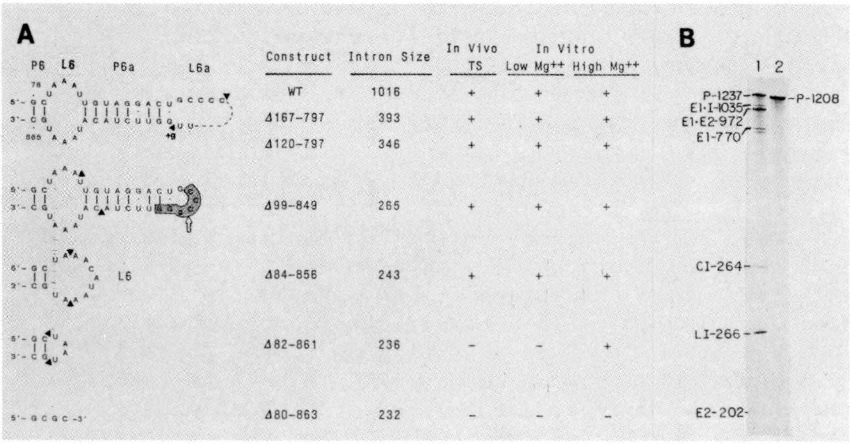

	Construct	Intron Size	In Vivo TS	In Vitro Low Mg++	In Vitro High Mg++	
	WT	1016	+	+	+	P-1237— —P-1208 E1-I-1035⁊— E1-E2-972 E1-770
	Δ167-797	393	+	+	+	
	Δ120-797	346	+	+	+	
	Δ99-849	265	+	+	+	
	Δ84-856	243	+	+	+	CI-264—
	Δ82-861	236	−	−	+	LI-266—
	Δ80-863	232	−	−	−	E2-202—

Fig. 3. P6a deletions. The constructs are described in A, with black triangles representing the sites of successive deletion endpoints. Constructs were assayed for in vivo TS phenotype by the plating assay (Fig. 2B). Assays in vitro were under low (3 mM) and high (8 mM) Mg++ conditions, in the presence of 40 mM Tris-HCl pH 7.5, 0.4 mM spermidine, and excess GTP (0.4-1.0 mM) for 30 min at 37°C. Examples of different responses under low Mg++ conditions are given in B: lane 1 (Δ99-849) shows a "+" response and lane 2 (Δ82-861) a "-" response as indicated in A.

and retains full splicing activity (Figs. 3A and B). By opening this construct at the SmaI site and making bidirectional deletions with exonuclease III and mung-bean nuclease, a set of deletions extending toward and beyond P6 was generated. A representative set of deletion constructs is shown in Fig. 3A, in which Δ84-856 contains the smallest intron (243 nt) capable of the full range of splicing activities in vivo and in vitro. The catalytic activity in Δ84-856 therefore occurs despite the absence of P6a. Furthermore, while deletion Δ82-861, with an intron of 236 nt, has a TS⁻ phenotype in vivo it is capable of splicing efficiently in vitro, but only under high Mg++ conditions. In theory the 4 nt L6 in Δ82-861 should favor the P6 pairing more than would the 11 nt L6 of Δ84-856, yet Δ84-856 splices under a broader range of conditions in vivo and in vitro. Thus, other requirements, such as specific residues within L6, must influence splicing ability. Predictably, the elimination of L6 in Δ80-863 leaves no possibility for the P6 pairing to form, and abolishes all catalytic activity.

These experiments corroborate the important role of P6 in splicing. Furthermore, they have generated a fully functional group I intron of 243 nt, and an intron of 236 nt that is catalytically active in vitro under high Mg^{++} conditions. What possibilities exist for further reducing intron size? Although those deletions of the P7.1-P7.2 region that have been tested abolish catalytic activity, deletion of P2, which is absent from the sunY intron (2), is consistent with an intermediate level of td splicing activity (J.L.G.S. and R. Schroeder, unpublished). Simultaneous deletion of different dispensable elements will hopefully result in a minimal functional intron that will yield to physical analysis.

Trans-splicing

We wished to determine whether the ribozyme core needed to be contiguous or whether two halves of the core could associate in trans to form a splicing proficient bimolecular complex. We chose to split the molecule within L6a for the following reasons: 1.) L6a is the site of the long td ORF (2,9) which naturally divides the core structure at this point; 2.) L6a is tolerant to both in vitro-generated deletions (above) and insertions (W.K. Chan and M.B., unpublished); 3.) Theoretically the two half-molecules could associate across the P3, P6 and P6a pairings, thus providing the most stable association possible between the two half-molecules (Fig. 4A); 4.) The P4 and P7 pairings each formed by the association of highly conserved sequences (P-Q and R-S, respectively) are not split, one pairing residing within each half-molecule (Fig. 4A).

We have shown that the two half-molecules (H1 and H2 in Figs. 4B and C) do indeed interact to form a catalytically active entity that results in exon ligation in trans. The reaction, which is accompanied by the release of the 5' and 3' intron components (I1 and I2, respectively, Fig. 4), is both Mg^{++}- and GTP-dependent (data not shown). Remarkably, I1 and I2 join to form a

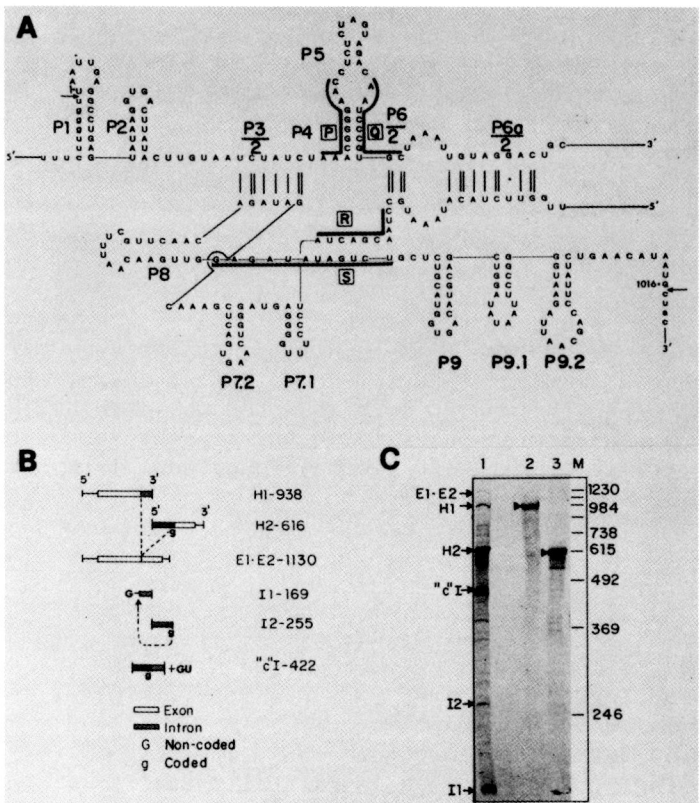

Fig. 4. <u>Trans</u>-splicing. The predicted core structure formed by the association of the half-molecules across P3, P6 and P6a is depicted in A. Conserved sequences P, Q, R and S are indicated by solid bars. The precursor half-molecules (H1 and H2) and splice products (a small amount of ligated exons, E1-E2; excised intron segments I1 and I2; and circle analog "c"I) are diagrammed in B, with predicted sizes in nucleotides. Reaction products are shown in C next to the sizes of DNA markers (M). Product identification was on the basis of hybridization analysis with probes specific for intron, exon, splice-junction and cyclization-junction sequences. The half-molecules H1 (lane 2) and H2 (lane 3) were mixed and incubated under self-splicing conditions with 30 mM Mg^{++} and 200 μM GTP for 30 min at 55°C (lane 1).

covalent linear homologue of the circular intron in a bimolecular reaction that is analogous to the normal cyclization reaction. From a practical standpoint, this system is convenient for generating active precursor molecules that do not splice during synthesis. Furthermore, preliminary experiments suggest that trans-splicing can also occur in vivo. It will not be surprising, therefore, to find naturally occurring trans-splicing group I introns.

The above experiments demonstrate the catalytic efficacy of a bimolecular group I ribozyme, while they stimulate questions on the evolutionary implications of trans-associative intron components. First, trans-splicing of group I and group II introns (10) supports the view that primitive RNA enzymes may have been assemblies of RNA molecules, with relatively simple component parts contributing to a catalytically active complex (11). Second, trans-splicing may be a mechanism for fueling the process of exon shuffling (12), in the absence of genetic recombination.

ACKNOWLEDGMENTS

We thank Peter DiMaria and Renee Schroeder for their input into the trans-splicing and deletion experiments. The helpfulness of our colleagues Debbie Bell-Pedersen, William Chan, Scott Chandry, Doris Dixon, Dwight Hall and Susan Quirk is much appreciated.

REFERENCES

1. Gott JM, Shub DA, Belfort M (1986). Multiple self-splicing introns in bacteriophage T4: Evidence from autocatalytic GTP labeling of RNA in vitro. Cell 47:81.
2. Shub DA, Gott JM, Xu M-Q, Lang BF, Michel F, Tomaschewski J, Pedersen-Lane J, Belfort M (1988). Structural conservation between three homologous introns of phage T4 and group I introns of eukaryotes. Proc Natl Acad Sci USA 85:1151.
3. Hall DH, Povinelli CM, Ehrenman K, Pedersen-Lane J, Chu F, Belfort M (1987). Two domains for splicing in the intron of the phage T4 thymidylate synthase (td) gene established by non-directed mutagenesis. Cell 48:63.

4. Belfort M, Chandry PS, Pedersen-Lane J (1987). Genetic delineation of functional components of the group I intron in the phage T4 td gene. Cold Spring Harbor Symp Quant Biol 52:Published.
5. Chandry PS, Belfort M (1987). Activation of a cryptic 5' splice site in the upstream exon of the phage T4 td transcript: Exon context, missplicing and mRNA deletion in a fidelity mutant. Genes and Development 1:1028.
6. Waring RB, Davies RW (1984). Assessment of a model for intron RNA secondary structure relevant to RNA self-splicing - a review. Gene 28:277.
7. Michel F, Jacquier A, Dujon B (1982). Comparison of fungal mitochondrial introns reveals extensive homologies in RNA secondary structure. Biochimie 64:867.
8. Kim S-H, Cech T (1987). Three-dimensional model of the active site of the self-splicing rRNA precursor of Tetrahymena. Proc Natl Acad Sci USA 84:8788.
9. Chu FK, Maley GF, West DK, Belfort M, Maley F (1986). Characterization of the intron in the phage T4 thymidylate synthase gene and evidence for its self-excision from the primary transcript. Cell 45:157.
10. Sharp PA (1987). Trans splicing: Variation on a familiar theme? Cell 50:147.
11. Uhlenbeck OC (1987). A small catalytic oligoribonucleotide. Nature 328:596.
12. Gilbert W (1978). Why genes in pieces? Nature 271:501.

A GROUP I INTRON IN *Bacillus subtilis* BACTERIOPHAGE SP01[1]

Heidi A. Goodrich, Jonatha M. Gott,[2] Ming-Qun Xu, Vincenzo Scarlato,[*,3] and David A. Shub

Department of Biological Sciences,
State University of New York, Albany
Albany, NY 12222
and
*Department of Biology
University of California, San Diego
La Jolla, CA 92093

ABSTRACT We have found a self-splicing group I intron in the *Bacillus subtilis* bacteriophage SP01. This is the first example of an intron in the Gram positive eubacteria. The intron has many structural features also found in the introns of phage T4. The implications of this finding, for both the origin of introns and their possible role in gene regulation, are discussed.

INTRODUCTION

The only examples of mRNA splicing observed in prokaryotes are the three self-splicing group I introns of the *E. coli* bacteriophage T4 (1) and its close relatives (2,3). Two of these introns are in genes that supply precursors for

---

[1] This work was supported by grants from the National Science Foundation (DMB8609066), National Institutes of Health (GM37746), and a Scholar Grant from the American Cancer Society to D.A.S. and a National Science Foundation grant (PCM8317847) to E. Peter Geiduschek.
[2] Present address: Department of Chemistry and Biochemistry, University of Colorado, Boulder, CO 80309.
[3] Present address: IIGB, C. P. 3061, Naples, Italy.

DNA biosynthesis (4-6), while the third is in a gene whose function is not known (6,7). Until the discovery of the intron in the *td* gene of T4 (4), group I introns had been found only in mitochondria, chloroplasts, and nuclear genes of simple eukaryotes. The T4 introns are structurally similar to the group I introns of eukaryotes: they contain the phylogenetically conserved sequences and elements of secondary structure that define this intron class (8,9). Each T4 intron, like many eukaryotic counterparts, contains an open reading frame (ORF). One of these has significant similarity to a set of related ORFs in mitochondria of filamentous fungi, prompting the suggestion that T4 might have recently undergone genetic recombination with DNA from one of these organisms (10). We were interested, therefore, to see whether the T4 introns were unique examples among the prokaryotes, or whether prokaryotic introns might be more generally distributed. Since they are obviously rare, we decided to start our search among bacteriophages whose structure, replication and regulatory strategies most resemble T4's. However, to minimize the possibility of recent genetic exchange with T4, we limited ourselves to phages of Gram positive bacteria.

Bacteriophage SP01 infects *B. subtilis*, but is remarkably similar to the T-even phages of *E. coli* (11). Among the similarities are: a large genome (140 kb), temporal control of transcription by a cascade of phage-encoded sigma factors, extreme virulence with respect to host macromolecular syntheses, and the presence of an unusual base completely replacing one of the normal DNA bases. In the case of SP01, hydroxymethyluracil replaces thymine; while for T-even phages, hydroxymethylcytosine replaces cytosine. Thus, it seemed plausible that if the introns play a role in gene regulation in T4, SP01 might have evolved a similar scheme.

We investigated whether SP01 has a self-splicing group I intron by the same method we used to discover additional introns in phage T4 (6). The splicing mechanism for these introns involves a series of transesterification reactions initiated by nucleophilic attack of the 3' OH of guanosine (or GTP) at the 5' splice site (12,13). A product of these reactions is a linear intron with an extra G at the 5' end. If the reaction is performed *in vitro* with $[\alpha-^{32}P]GTP$, the resultant radioactively end-labeled intron RNA can be used as a probe for intron sequences in phage DNA.

RESULTS

Total RNA was extracted from *B. subtilis* 168 before, and at 3, 10 and 20 minutes after infection by SPO1 at 37°. These times correspond to the early, middle and late transcription periods (14). After incubation with $[\alpha-^{32}P]GTP$, gel electrophoresis of the RNA revealed a discrete radioactive species, with an estimated size of 800-900 residues, only in the 10 minute RNA (unpublished data). Analogous molecules, generated by incubation of T4 RNA with $[\alpha-^{32}P]GTP$ *in vitro*, have been shown to be end-labeled group I introns (6).

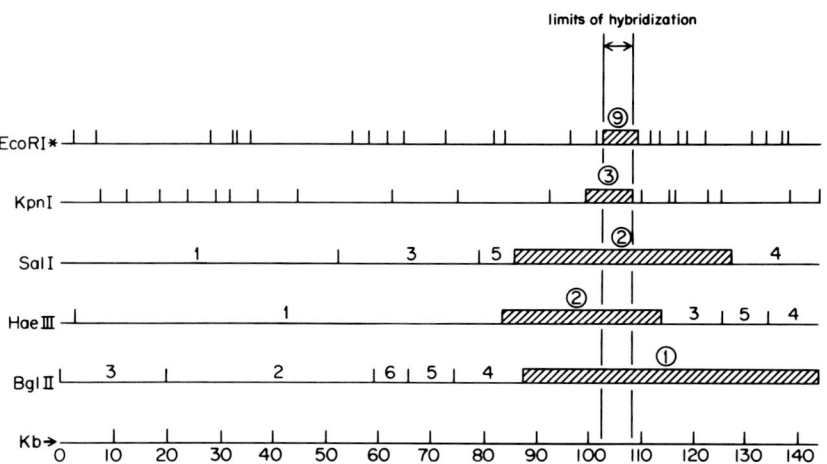

FIGURE 1. Hybridization of end-labeled RNA to SPO1 DNA. SPO1 DNA was digested to completion with the indicated enzymes ("star (*)", or reduced specificity conditions are required for digestion with EcoRI). After separation on an agarose gel, blot hybridization was performed with 10 min. SPO1 RNA, end-labeled with $[\alpha-^{32}P]GTP$ in vitro. Restriction fragments showing hybridization are indicated by crosshatched bars on the SPO1 restriction map (11,15).

In order to define the location of the putative intron in SPO1 DNA, Southern blot hybridization was performed on restriction digests of phage DNA, using this labeled RNA preparation as a probe. A summary of the results (Figure 1) shows a common zone of hybridization, roughly defined by the restriction fragment EcoRI-9. Plasmids containing EcoRI-9 and three subfragments were probed with the same RNA by dot

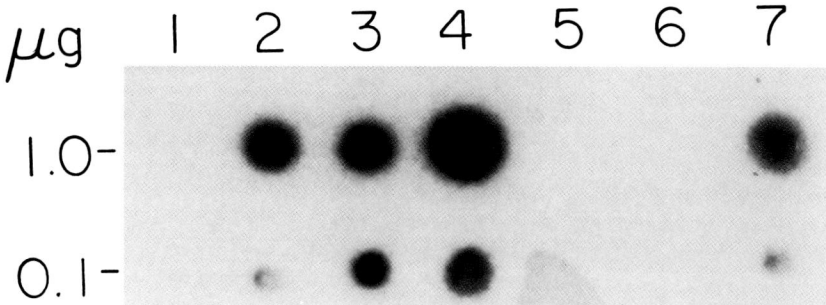

FIGURE 2. Dot hybridization of end-labeled SPO1 RNA to subclones of restriction fragment EcoRI-9. Portions of EcoRI-9 were cloned into pUC plasmids (see Figure 3) and 0.1 and 1.0 µg of each DNA was spotted onto nitrocellulose filters. SPO1 and pBR322 DNAs were included for comparison. Hybridization was carried out as in Figure 1. 1) pBR322, 2 & 3) pJK9, 4) α, 5) δ, 6) β, 7) SPO1.

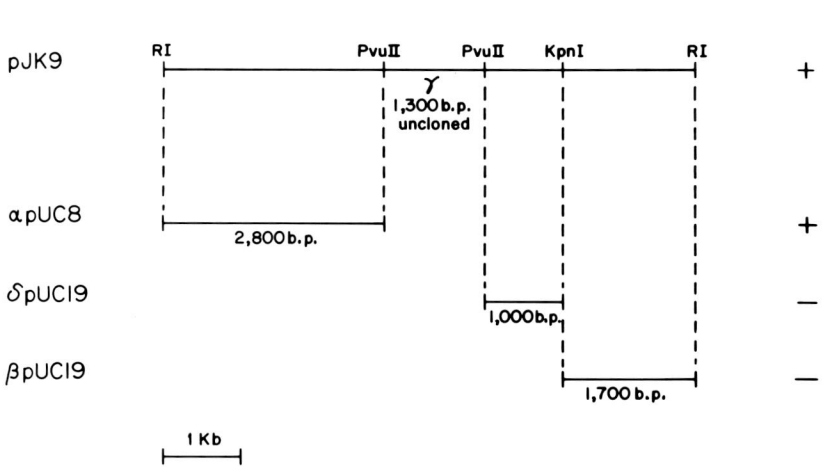

FIGURE 3. Construction of subclones of EcoRI-9. pJK9 was cut with the indicated restriction enzymes and ligated into the pUC plasmids shown. A subclone of the region between the two PvuII sites was not obtained. The results of the hybridization experiment in Figure 2 are summarized on the right.

hybridization (Figure 2). Only plasmids containing EcoRI-9 or its left end gave a positive signal (Figure 3). Probing of additional subclones allowed us to localize the 5' end of the intron quite close to the left end of EcoRI-9 (data not shown).

A 1.8 kb EcoRI-HindIII fragment from this region was cloned into a plasmid downstream of a phage T7 promoter and transcribed *in vitro* with T7 RNA polymerase. The product of this reaction was a mixture of RNA species typical of splicing (unpublished data). The nucleotide sequences of both the 1.8 kb DNA and of the ligated exon RNA species produced in the transcription reaction have been determined, and revealed the presence of a 883 nucleotide intervening sequence. These results will be published elsewhere.

The intron contains the sequences conserved in all group I introns in the appropriate positions of the folded secondary structure model (Figure 4). As in all other group I introns, splicing occurs after a U (which is paired with a G in the intron internal guide sequence) in the 5' exon, and after a G at the 3' end of the intron. A potential P10 pairing (9) of 3 bp exists (between the first 3 nucleotides of the 3' exon and the internal guide sequence) that could align the splice sites. Like the T4 introns, this SPO1 intron has the variations (in conserved sequences and extra nucleotides between P7 and P3) that typify the group IA subclass (8,9). An unusual feature is the presence of a large number of nucleotides between our assignments of the P3 and P4 helices. The usual number of nucleotides connecting these helices is usually three, and only two other group I introns, from fungal mitochondria, have been reported to deviate widely from this rule of intron structure (17). A number of alternative structures can be drawn in this interval, but since there is no phylogenetic proof for them we cannot predict which is most likely (N. Pace, personal communication).

As in the T4 introns (6), a long ORF is contained within the intron, starting within a loop, and ending within one of the highly conserved helical regions of the RNA structure (Figure 4). Thus, ribosomes traversing the intron ORF would disrupt the conformation necessary for splicing.

Finally, it was most intriguing that comparison of the predicted translation product of the ligated-exon sequence to a protein sequence data base, revealed a remarkable similarity to the COOH end of *E. coli* DNA polymerase I (to be published elsewhere). The region of similarity extends over most of the sequenced exons, and corresponds to the distal one-third of *E. coli* PolI. The remainder of SPO1 DNA polymerase presumably lies in the adjacent restriction fragment

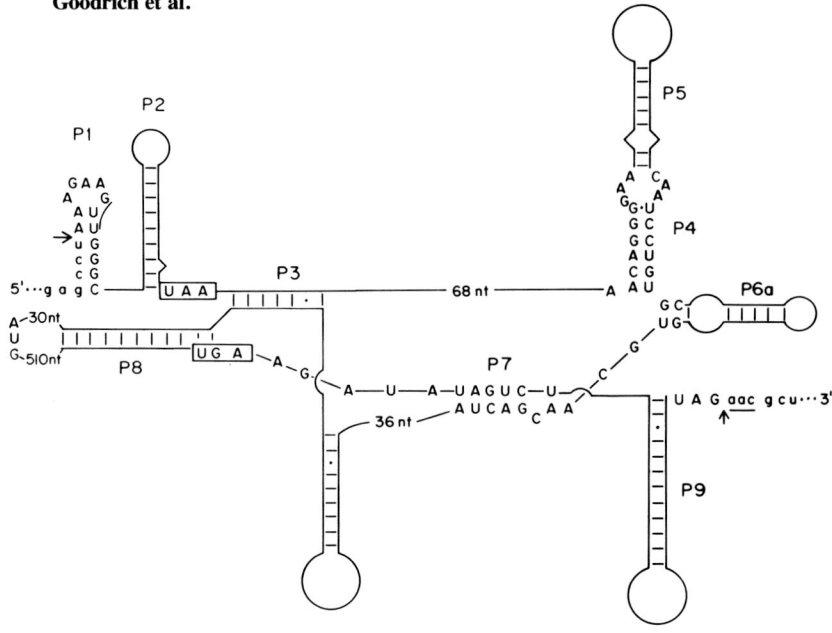

FIGURE 4. Secondary structure model of the SPO1 intron. The conventions of Burke et al. (16) are used. Exons are lower case and intron sequences are upper case. Pairing regions are indicated by "P." Base pairs are indicated by dashes or dots (G:U). Arrows show the splice sites. The conserved sequences surrounding P4 and P7 are shown and the nucleotides that could constitute P10 are underscored. The AUG that initiates an ORF is shown in the loop of P8, while the stop codons in phase with the ORF and with the 5' exon are boxed.

(EcoRI-23). Indeed, EcoRI-23 has been shown to contain part of the gene for SPO1 DNA polymerase by marker-rescue recombination with phage mutants (15).

DISCUSSION

We have proposed two ways that splicing might be used in regulatory networks by phage T4. One mechanism involves regulation of DNA synthesis by feedback regulation of splicing by small molecules (1,6), while the other (splice attenuation) proposes that translation of the intron ORF can alter the conformation of the intron to inhibit splicing

(6,7). It seems remarkable to us that SPO1, chosen for its similarity to T4, should actually have an intron. But even more remarkable is the location of the intron in a gene involved in DNA synthesis (DNA polymerase) and the placement of an ORF consistent with the splice-attenuation model. While these coincidences do not prove our speculations on the regulatory significance of phage introns, they are certainly encouraging.

This work has extended the distribution of group I introns to a gene that replicates (as far as we know) only in Gram positive eubacteria. It seems almost as unlikely that SPO1 and T4 could have exchanged genetic information with each other, as that they have both undergone genetic exchange with chloroplast or mitochondrial DNA. The alternative view: that group I introns are an ancient feature of gene structure, predating the divergence of Gram positive and negative bacteria or the invasion of eukaryotic cells by organelles, is simpler and is consistent with the data. Wherever the bacteriophage introns originated, their retention implies a selective advantage. Whatever the nature of this selection is, it seems likely that it would also be used in eukaryotic genomes that have retained these introns.

ACKNOWLEDGMENTS

We thank E. Peter Geiduschek for support (V.S.) and encouragement during the course of this work. D.A.S. gratefully acknowledges the help of Alice Sirimarco in typing the manuscript, and the hospitality of the laboratory of Tom Cech, where he was a visitor while it was written.

REFERENCES

1. Shub DA, Gott JM, Xu M-Q, Lang BF, Michel F, Tomaschewski J, Pedersen-Lane J, Belfort M (1988). Structural conservation among three homologous introns of bacteriophage T4 and the group I introns of eukaryotes. Proc Nat Acad Sci USA 85:1151.
2. Pedersen-Lane, Belfort M (1987). Variable occurrence of the nrdB intron in the T-even phages suggests intron mobility. Science 237:182.
3. Chu FK, Maley F, Martinez J, Maley GF (1987). Interrupted thymidylate synthase gene of bacteriophage T2 and T6 and other potential self-splicing introns in the T-even bacteriophages. J. Bacteriol. 169:4368.
4. Chu FK, Maley GR, Maley F, Belfort M (1984). Intervening sequence in the thymidylate synthase gene of bacteriophage T4. Proc Nat Acad Sci USA 81:3049.

5. Sjöberg B-M, Hahne S, Mathews CZ, Mathews CK, Rand KN, Gait MJ (1986). The bacteriophage T4 gene for the small subunit of ribonucleotide reductase contains an intron. EMBO J 5:2031.
6. Gott JM, Shub DA, Belfort M (1986). Multiple self-splicing introns in bacteriophage T4: evidence from autocatalytic GTP labeling of RNA in vitro. Cell 47:81.
7. Shub, DA, Xu M-Q, Gott JM, Zeeh A, Wilson LD (1987). A family of autocatalytic group I introns in bacteriophage T4. Cold Spring Harbor Symp Quant Biol 52:193.
8. Michel, F, Jacquier A, Dujon B (1982). Comparison of fungal mitochondrial introns reveals extensive homologies in RNA secondary structure. Biochimie 64:867.
9. Waring, RB, Davies RW (1984). Assessment of a model for intron RNA secondary structure relevant to RNA self-splicing - A review. Gene 28:277.
10. Michel F, Dujon B (1986). Genetic exchanges between bacteriophage T4 and filamentous fungi? Cell 46:323.
11. Stewart C (1988). Bacteriophage SP01. In Calender R (ed): "Bacteriophages, Vol. I," New York:Plenum Press, p 477.
12. Cech TR, Zaug AJ, Grabowski PJ (1981). In vitro splicing of the ribosomal precursor of Tetrahymena: Involvement of the intervening sequence. Cell 27:487.
13. Kruger K, Grabowski PJ, Zaug AJ, Sands J, Gottschling DE, Cech TR (1982). Autoexcision and autocyclization of the ribosomal RNA intervening sequence of *Tetrahymena*. Cell 31:147.
14. Gage LP, Geiduschek EP (1971). RNA synthesis during bacteriophage SP01 development. Six classes of SP01 RNA. J Mol Biol 57:279.
15. Curran JF, Stewart C (1985). Cloning and mapping of the SP01 genome. Virology 142:78.
16. Burke JM, Belfort M, Cech TR, Davies RW, Schweyen RJ, Shub DA, Szostak JW, Tabak HF (1987). Structural conventions for group I introns. Nucleic Acids Res 15:7217.
17. Trinkl H, Wolf K (1986). The mosaic *cox*1 gene in the mitochondrial genome of *Schizosaccharomyces pombe*: minimal structural requirements and evolution of group I introns. Gene 45:289.

A SITE-DIRECTED MUTATION IN THE *BACILLUS SUBTILIS* RIBONUCLEASE P RNA THAT SELECTIVELY INFLUENCES ITS PROTEIN-DEPENDENT REACTION[1]

David S. Waugh, Michael W. Sganga, Michael Raff and Norman R. Pace

Department of Biology and Institute for Molecular and Cellular Biology
Indiana University, Bloomington, IN. 47405

ABSTRACT We report the further characterization of a mutation in the *Bacillus subtilis* RNase P RNA that renders the molecule essentially inactive in the low salt, protein-dependent tRNA processing reaction, but has no influence on the activity of the RNA at high salt concentrations in the absence of protein. No difference in the abilities of the mutant and the wild-type RNase P RNAs to associate with the RNase P protein could be detected using a gel retardation assay for protein-RNA complexes. On the other hand, the mutant RNA differs subtly from the wild-type in its conformation, and it requires substantially higher ionic strength than does the normal RNA for activity in the absence of protein.

INTRODUCTION

The mature 5' ends of tRNA molecules are formed by the endoribonuclease RNase P. In all eubacteria so far inspected, RNase P consists of protein (ca. 14kD) and RNA (ca. 400 nucleotides)

[1]This work was supported by NIH grant GM34527 to N.R.P.

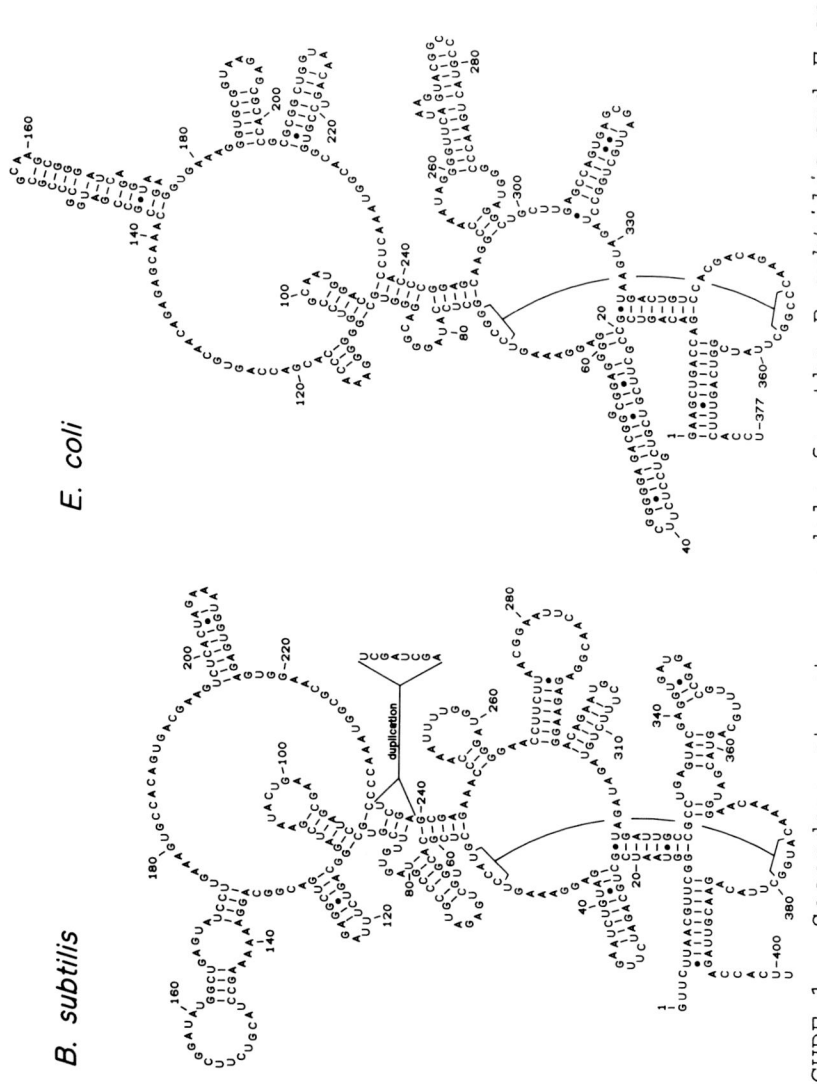

FIGURE 1. Secondary structure models for the *B.subtilis* and *E.coli* RNase P RNAs based on phylogenetic comparisons. (Adapted from 4.) The tandem duplication of *B.subtilis* residues 236-239 (see text) is indicated.

moieties. Although both components are essential *in vivo*, the *Bacillus subtilis* and *Escherichia coli* RNase P RNAs precisely cleave tRNA precursors in the absence of the protein at high salt concentrations *in vitro* (1). The RNase P RNA therefore contains features for binding of the substrate (pre-tRNA) and catalysis, as well as for the interaction with the RNase P protein. The extent to which specific elements of the RNase P RNA structure can be assigned to specific functions has yet to be determined.

Shiraishi and Shimura have found that a single nucleotide substitution (from G to A) at position 89 in the *E.coli* RNase P RNA interferes with the physical association of the RNase P protein and RNA moieties, but the mutation (ts 709) has no influence on the enzymatic activity of the RNA in the absence of protein (2). We have reported that a tandem duplication of nucleotides 236-239 in the *B.subtilis* RNase P RNA nearly abolishes its enzymatic activity in the protein- dependent reaction *in vitro*, but the mutation has no effect on the activity of the RNA alone at high salt concentrations (3). Although these mutations lie in disparate regions of the aligned *B.subtilis* and *E.coli* RNase P RNA sequences, they potentially disrupt homologous, long-range pairings in the proposed secondary structures for the RNAs (4, fig. 1). This raised the intriguing possibility that the homologous helices, *B.subtilis* 86-90/235-239 and *E.coli* 87-91/238-242, might offer a particularly crucial contact for the formation of the ribonucleoprotein holoenzyme. We undertook a more detailed characterization of the *B.subtilis* RNase P RNA containing the tandem duplication in order to explore further the possible relationship between helix 86-90/235-239 and the binding of the protein by the RNA. In addition, we have generated and characterized two point mutations in helix 86-90/235-239 of the *B.subtilis* RNase P RNA.

RESULTS

The *in vitro* cleavage of a tRNA precursor by the wild-type and mutant (containing a tandem duplication of nucleotides 236-239) *B.subtilis* RNase P RNAs is compared in figure 2. Although the two RNAs manifest comparable activity at high salt concentrations in the absence of protein, the mutant is essentially inactive at physiological salt concentrations in the presence of the RNase P protein.

A simple explanation for this phenotype, analogous to that of the *E.coli* ts 709 mutant, could be that the mutation prevents the association of the RNA and protein components of RNase P that is required for activity at physiological ionic strength. We therefore examined directly the binding of the mutant and wild-type RNAs to the RNase P protein using a gel retardation assay. The RNAs, labeled with ^{32}P, were incubated with *B.subtilis* RNase P protein under conditions optimal for the holoenzyme activity, then free RNA and RNA-protein complexes were separated by electrophoresis in non-denaturing, reaction buffer-containing, polyacrylamide gels and visualized by autoradiography. As shown in figure 3, incubation of the wild-type RNase P RNA with RNase P protein results in the formation of a distinct band, with retarded mobility compared to that of the RNA alone. An increase in the amount of protein added to a fixed amount of RNase P RNA leads to a corresponding increase in the amount of the complex, until all of the RNA occupies the more slowly migrating form. The unique character of the slowly migrating form suggests that it is a specific complex between the RNA and the protein, not a non-specific aggregate. The amount of RNase P protein required to titrate all of the RNase P RNA in this experiment is approximately the same amount needed for maximal enzymatic activity *in vitro* (data not shown); uncertainty in the absolute concentration of RNase P protein present precludes the extraction of quantitative binding information. The mutant RNA also is capable of forming a complex with the RNase P protein, with essentially

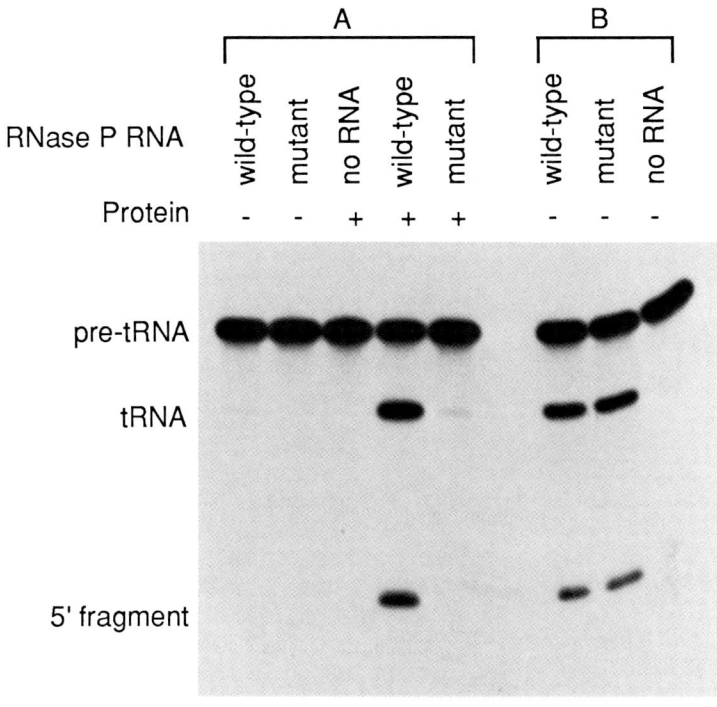

FIGURE 2. Enzymatic activity of mutant and wild-type forms of the *B.subtilis* RNase P RNA. Activity was assayed at 37°C by incubating uniformly ^{32}P-labeled pre-tRNA^{Asp} with RNase P RNA alone or in the presence of saturating amounts of purified *B.subtilis* RNase P protein as indicated. Substrate and enzyme RNAs were produced by *in vitro* transcription (3,5). RNase P RNA concentrations were 10^{-9} M and 10^{-8} M in the holoenzyme (A) and RNA alone (B) reactions, respectively. RNA alone reactions were carried out in 50 mM Tris-HCl (pH 8.0), 0.05% NP40, 0.8 M NH_4Cl, and 0.1 M $MgCl_2$. Holoenzyme reactions were carried out in 50 mM Tris-HCl (pH 8.0), 0.05% NP40, 0.1 M NH_4Cl, and 15 mM $MgCl_2$. The reactions were stopped after 20 minutes by adding three volumes of cold ethanol. The products were recovered as ethanol precipitates, resolved by electrophoresis in 8% polyacrylamide gels containing 8 M urea, and visualized by autoradiography.

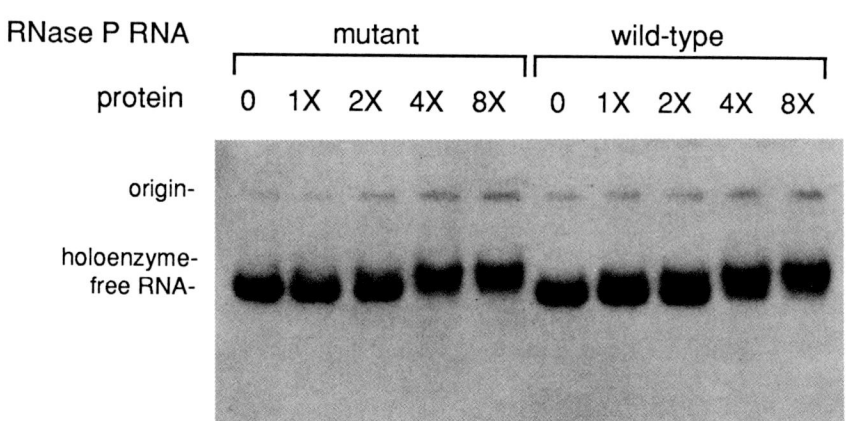

FIGURE 3. Formation of holoenzyme complexes with mutant and wild-type B.subtilis RNase P RNAs. Each 10 uL reaction contained approximately 2 ng of uniformly ^{32}P-labeled RNase P RNA and various amounts of RNase P protein (expressed in arbitrary units) as indicated above. Controls contained no protein. All reactions were incubated at 37^0 C for 15 minutes under conditions optimal for the holoenzyme-mediated processing of pre-tRNA in vitro, namely, 0.1 M NH_4OAc, 15 mM $Mg(OAc)_2$, 50 mM Tris-HOAc (pH 8.0), 0.05% NP40. Following incubation, 5 uL of reaction buffer containing 50% glycerol was added to each, and all the reactions were chilled on ice prior to electrophoresis in 4% polyacrylamide gels (0.8 mm thick) which were cast and run in reaction buffer containing 5% glycerol. Electrophoresis was conducted at 4^0 C for approximately 3 hrs at 4 V/cm. The products were visualized by autoradiography.

the same stoichiometery of association as the wild-type RNA. However, the mutant complex reproducibly migrates in the gel as a more disperse band than the normal complex, indicating that the mutant complex is more fluid in its overall conformation.

As a further test of the relative degree of specificity of the protein association with the normal and mutant RNAs, we examined the ability of an irrelevant RNA to compete with the RNase P RNAs for the protein in complex formation. Approximately the same amount of 16S ribosomal RNA (100-fold mass excess) was required to interfere with complex formation regardless of which RNase P RNA was present (data not shown). It therefore seems unlikely that an aberrant interaction between the mutant RNA and the RNase P protein is responsible for the substantially reduced catalytic activity.

The disperse character of the mutant RNA-protein complex in native gels (above) suggested that the four nucleotide insertion might disrupt the intramolecular folding of the RNA, reducing its ability to act on the substrate at physiological, but not elevated, ionic strength. For instance, the mutation might force structural elements of the RNA into closer proximity than in the native RNA. Consequent electrostatic repulsion of phosphates in the mutant RNA might cause more long-range disorder in its folding, resulting in diminution of activity. If this notion were correct, it was to be expected that the RNA alone activity of the mutant would require higher ionic strength than the wild-type RNA. The higher salt concentration might be required to titrate the disarranged repulsive elements. The results presented in figure 4 are consistent with this expectation: the activity of the mutant RNA diminishes much more rapidly than that of the wild-type RNA as the ionic strength of the *in vitro* reaction medium approaches levels that are optimal for the holoenzyme-mediated reaction.

Additional evidence for the conformational perturbation of the RNA by the four nucleotide insertion is seen in the behavior of the mutant RNA during electrophoresis in native gels in the absence of the RNase P protein. As shown in figure 5, the electrophoretic mobility of the mutant is retarded relative to that of the wild-type RNase P RNA or another mutant RNase P RNA that also contains a tandem duplication of four nucleotides (residues 28-31), but is indistinguishable in mobility from that of the wild-type RNA. Thus, the slower migration under native conditions of the RNase P RNA that contains the tandem duplication of nucleotides 236-239 is not simply a consequence of its increased length. Instead, its slower electrophoretic mobility must reflect a difference from the other RNAs in its higher order structure, resulting in an expanded hydrodynamic radius.

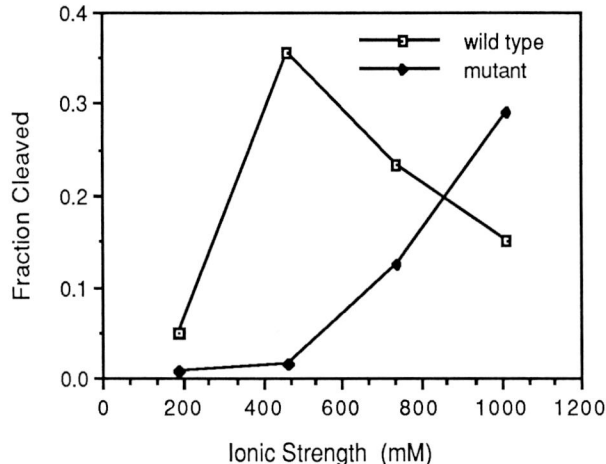

FIGURE 4. *In vitro* enzymatic activity of mutant and wild-type *B.subtilis* RNase P RNAs as a function of ionic strength. The highest ionic strength tested in this experiment corresponds to the conditions used for the RNA-alone assays in figure 2. Lower ionic strengths were obtained by reducing the overall concentration of NH_4OAc and $Mg(OAc)_2$ in the reactions without changing their relative proportions. The assays were conducted as detailed in the legend to figure 2, except that the gel was fixed and dried prior to autoradiography. The bands subsequently were excised and monitored by scintillation counting to determine the extent of the reactions, which is reported as the fraction of pre-tRNA^{Asp} cleaved.

We attempted to mimic the *E.coli* ts 709 mutation by using oligonucleotide mutagenesis to change nucleotide 90 in the *B.subtilis* RNase P RNA from G to either A or U. Because the 89/236 base pair, U/U, is noncanonical in *B.subtilis* (every other known RNase P RNA utilizes a standard, Watson-Crick base pair at this position), both mutants would have two consecutive mismatches within the 86-90/235-239 helix, potentially disrupting its formation. Although in the absence of the RNase P protein the *in vitro* processing activity of these two mutant RNase P RNAs was indistinguishable from that of the wild-type at 37°, at higher temperatures (>45°) the activity of both mutants was significantly less (ca. 4-fold) than that of the wild-type B.subtilis RNase P RNA. However, neither mutation had any measurable effect on the enzymatic activity of the RNA in the low salt, protein-dependent reaction (data not shown).

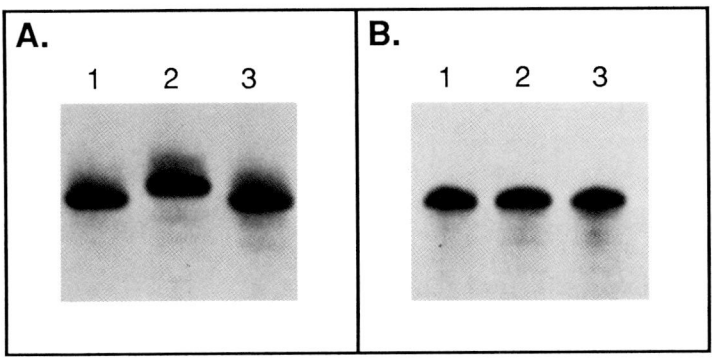

FIGURE 5. Electrophoretic mobilities of mutant and wild-type B.subtilis RNase P RNAs in non-denaturing (panel A) and denaturing (panel B) 4% polyacrylamide gels. Conditions for non-denaturing electrophoresis were as described in the legend to figure 3. Denaturing gels contained 8 M urea, 89 mM Tris-borate, 89 mM boric acid, and 2 mM EDTA (pH 8.0). Lane 1: Wild-type B.subtilis RNase P RNA. Lane 2: RNase P RNA containing a tandem duplication of residues 236-239. Lane 3: RNase P RNA containing a tandem duplication of nucleotides 28-31.

DISCUSSION

In summary, although the duplication of residues 236-239 in the B.subtilis RNase P RNA selectively influences the protein-dependent reaction, we have been unable to establish a specific relationship between helix 86-90/235-239 and the ability of the RNase P RNA to associate with the RNase P protein. Despite the difference in the enzymatic activities of the wild-type and mutant holoenzyme complexes *in vitro*, the insertion mutant appears to bind to the RNase P protein with the same stoichiometery and specificity as does the wild-type RNase P RNA. Although the normal and mutant holoenzyme complexes differ slightly in their behavior in native gels, this does not necessarily indicate that they differ with respect to their association with the RNase P protein. Rather, the difference is likely the result of a conformational defect in the mutant RNase P RNA. This is suggested by the observation that the mobilities of the mutant and wild-type RNase P RNAs in non-denaturing gels also differ in the absence of protein. The fact that higher salt concentrations are required for the RNA alone activity of the mutant RNA than are required by the normal RNA also is consistent with such a view. The elevated ionic strength could compact the mutant RNA, forcing it into an active conformation. A requirement for such a salt-induced conformational change could also account for the comparatively low level of activity manifested by the mutant holoenzyme at physiological salt concentrations.

The two point mutations introduced at position 90 in the *B.subtilis* RNase P RNA also reduce the enzymatic activity of the RNA in the absence of protein at high salt concentrations, but we can detect no effect on their protein-dependent activity at physiological salt concentrations. Hence, of the three mutations we have analyzed which have the potential to interfere with helix 86-90/235-239, none has any measurable, specific influence on the ability of the RNase P RNA to respond to the RNase P protein. On the other hand, all three mutations reduce the enzymatic activity of the RNA under some conditions in the absence of protein.

ACKNOWLEDGEMENTS

We thank G. J. Olsen for figure 1, and B. Pace for providing *B.subtilis* RNase P protein.

REFERENCES

1. Guerrier-Takada C, Gardiner K, Marsh T, Pace N, Altman S (1983). The RNA moiety of ribonuclease P is the catalytic subunit of the enzyme. Cell 35:849.
2. Shiraishi H, Shimura Y (1986). Mutations affecting two distinc functions of the RNA component of RNase P. EMBO J 5:3673.
3. Pace NR, Reich C, James BD, Olsen GJ, Pace B, Waugh DS (1987). Structure and catalytic function in ribonuclease P. Cold Spring Harbor Symp. Quant. Biol. 52:239.
4. James BD, Olsen GJ, Liu J, Pace NR (1988). The secondary structure of ribonuclease P RNA, the catalytic element of a ribonucleoprotein enzyme. Cell 52:19.
5. Reich C, Olsen GJ, Pace B, Pace N (1988). Role of the protein moiety of ribonuclease P, a ribonucleoprotein enzyme. Science 239:178.

PROCESSING OF A SYNTHETIC tRNA PRECURSOR MODEL
BY E. coli RNase P AND M1 RNA[1]

Christopher K. Surratt,[2] Barbara J. Carter, and Sidney M. Hecht

Departments of Chemistry and Biology,
University of Virginia, Charlottesville, VA 22901

ABSTRACT A synthetic "dimeric" tRNA molecule was constructed as a possible substrate for RNase P and M1 RNA. This dimeric tRNA was prepared by ligating mature E. coli tRNA^{Phe} (acceptor molecule) to nucleotides 1-36 of yeast tRNA^{Phe} (donor molecule) with T4 RNA ligase, and then annealing this oligomeric tRNA molecule to the 3'-half of yeast tRNA^{Phe} (nucleotides 38-76). E. coli RNase P and M1 RNA both effected processing of the "dimeric" tRNA substrate constructed from E. coli tRNA^{Phe} (5'-tRNA) and yeast tRNA^{Phe} (3'-tRNA); when analyzed on a denaturing polyacrylamide gel, processing reactions yielded products that comigrated with authentic E. coli tRNA^{Phe} and the denatured 5'-half of yeast tRNA^{Phe}. By utilizing dimeric tRNA substrates radiolabeled at the phosphodiester groups immediately preceding or following the putative cleavage site, processing by both RNase P and M1 RNA was shown to occur exclusively at the phosphate ester linkage joining the mature tRNA molecules.

[1]This work was supported by National Institutes of Health Research Grant GM-27815
[2]Present address: Department of Biochemistry, University of California, Berkeley, California

INTRODUCTION

Ribonuclease P (RNase P) is an enzyme composed of an RNA subunit 377 nucleotides in length (M1 RNA) and a protein subunit (C5 protein) of M_r 13,700 (1, 2, 3). RNase P catalyzes the processing of the 5'-end of precursor transfer RNA molecules (tRNA's) affording mature tRNA's as products. Several organisms have been shown to elaborate transfer RNA precursor transcripts containing the sequences of two (4) or more (5, 6) tRNA's. The "spacer" regions separating the tRNA's are of variable length and composition (7, 8), yet RNase P mediates cleavage to produce mature tRNA's in all cases (4). Surprisingly, M1 RNA was found to be the catalytically active constituent of the holoenzyme (9).

In vivo, RNase P processed the dimeric tRNA^{Pro}-tRNA^{Ser} molecule encoded by T4 bacteriophage, but only after the 3'-terminal CCA sequence was added by 3'-end processing (10). In vitro, RNase P exhibited only a three-fold preference for the dimeric precursor having a 3'-terminal CCA (8), whereas M1 RNA exhibited a strict requirement for the 3'-terminal CCA (11) as noted for certain in vivo processing reactions. To facilitate a more detailed analysis of the structural features that permit RNase P and M1 RNA processing of dimeric tRNA precursors, we have prepared synthetic tRNA precursor models by the ligation of structural constituents of mature tRNA's (12). As shown here for the E. coli tRNA^{Phe}-yeast tRNA^{Phe} dimer, processing occurred in a time dependent manner, at the expected phosphate ester linkage joining the two mature tRNA's, and, in the case of M1 RNA, was dependent on the presence of the ultimate 3'-terminal CA dinucleotide.

METHODS

The 5'-half (nucleotides 1-36) and 3'-half (nucleotides 38-76) of yeast tRNA^{Phe} were prepared by modification of a previously published method (13, 14). The 5'-half of yeast tRNA^{Phe} was then ligated to the 3'-end of intact E. coli tRNA^{Phe} using T4 RNA ligase. After purification on a 10% polyacrylamide gel, the oligomeric tRNA product was excised from the gel, electroeluted, and purified on a Du Pont NENSORB 20 nucleic acid purification cartridge (15). This oligomeric tRNA molecule was then annealed to the 3'-half of yeast tRNA^{Phe} to afford a "dimeric" tRNA as described (12), and recovered by precipitation with ethanol. Processing reactions were carried out as described by Guerrier-Takada et al (9) for RNase P and M1 RNA.

RESULTS

The dimeric tRNA substrate was prepared as diagrammed in Scheme I. Since the 5' nucleotide of a tRNA molecule is recessed, it is inaccessible to T4 RNA ligase; thus the two tRNA molecules could not simply be ligated together. Synthesis of the tRNA dimer involved two steps: first, ligation of the 5'-half of yeast tRNA^{Phe} (2) (nucleotides 1-36) to the 3'-end of intact E. coli tRNA^{Phe} (1) using T4 RNA ligase produced oligomeric tRNA 3.

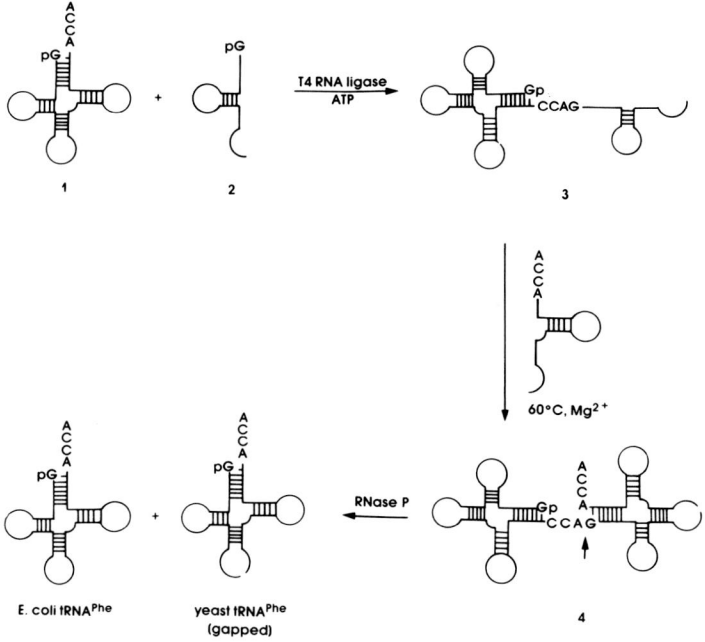

SCHEME I. Preparation of a dimeric tRNA precursor model

Second, the 3'-half (nucleotides 38-76) of yeast tRNA^{Phe} was annealed to oligomeric tRNA 3 to afford "dimeric" tRNA 4. This dimer was employed as a substrate for RNase P and M1 RNA.

Figure 1 is an autoradiogram that illustrates the time-dependent processing of dimeric tRNA 4 by RNase P. The dimeric substrate contained a ^{32}P-label in the yeast 5'-half tRNA; appearance of the 5'-half molecule indicated that successful processing of the dimer had occurred. The 5'-half molecule was readily apparent after 10 min. reaction time; the majority of the dimeric tRNA substrate was processed after 30 min. under these experimental conditions.

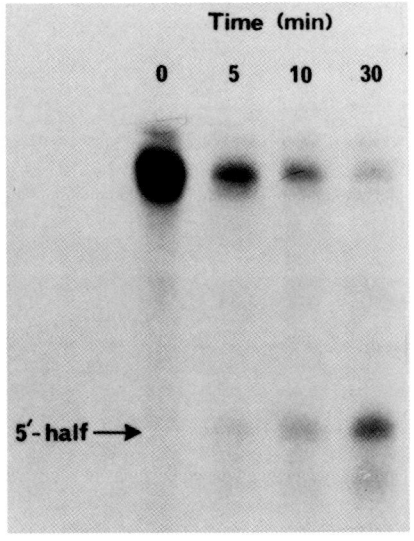

FIGURE 1. Time dependence of tRNA precursor processing by E. coli RNase P. Reactions were carried out at 37° C in incubation mixtures (5 μL total volume) containing 10 mM Tris-HCl (pH 8.0), 10 mM NH$_4$Cl, 5 mM MgCl$_2$, 0.1 mM EDTA, and 0.1 mM β-mercaptoethanol.

Synthetic tRNA Precursor Model   83

In order to determine the exact site at which processing of dimer 4 occurred, we prepared dimeric tRNA 4 having a ^{32}P label at the phosphate group either immediately preceding or following the expected phosphate ester cleavage site. Scheme II illustrates the positions at which the radiolabel was incorporated, and the cleavage products that could be expected from these two dimeric substrates were processing to proceed as anticipated. One radiolabeled substrate would give a radiolabeled mature tRNA when processed; the other would give radiolabeled 5'-half tRNA^{Phe}.

The results of processing of these precursors by RNase P is illustrated in Figure 2. As shown, the dimeric precursor containing the radiolabel between C_{75} and A_{76} of E. coli tRNA^{Phe} afforded a single product that comigrated with authentic tRNA^{Phe}; when the ^{32}P radiolabel was present at the 5'-terminus of

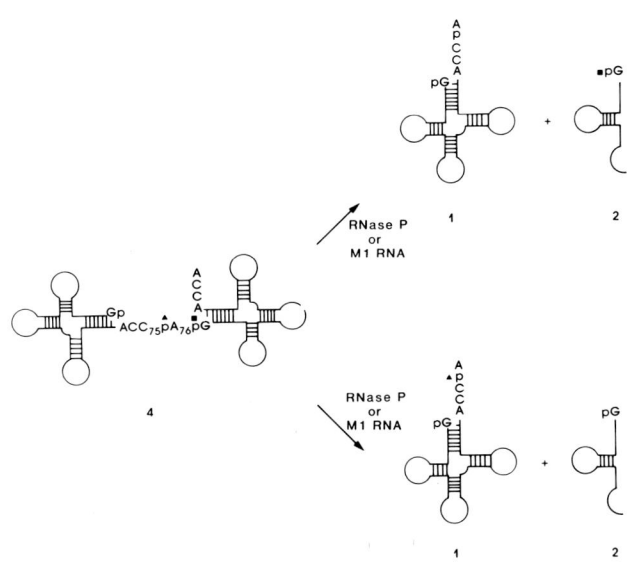

SCHEME II. Internally radiolabeled dimeric tRNA's designed to identify the site of RNase P processing.

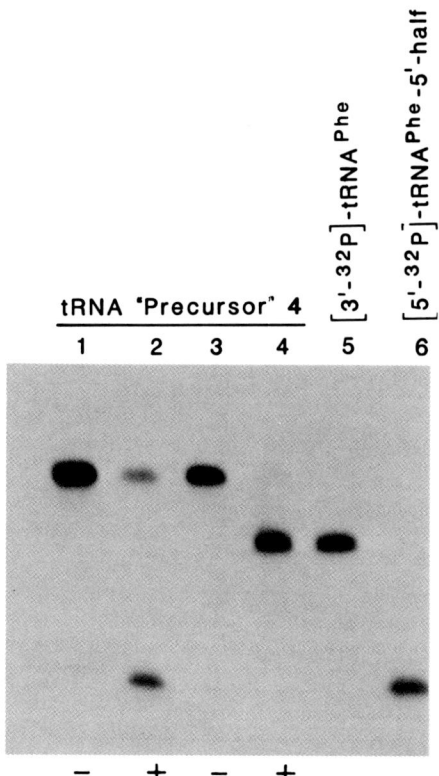

FIGURE 2. RNase P processing of internally radiolabeled tRNA precursor models. Reactions were carried out at 37° C for 60 min in incubation mixtures (5 μL total volume) containing 10 mM Tris-HCl (pH 8.0), 10 mM NH$_4$Cl, 5 mM MgCl$_2$, 0.1 mM EDTA, and 0.1 mM β-mercaptoethanol. Identical results were obtained when M1 RNA was used in place of RNase P.

the 5'-half molecule (i.e., derived from yeast tRNA^{Phe}) processing afforded a single product that comigrated with authentic yeast tRNA^{Phe} 5'-half molecule. Identical results were obtained when E. coli M1 RNA was used in place of RNase P as the catalyst.

The effect on processing of the ultimate 3'-terminal CCA sequence was investigated for the synthetic tRNA dimer by comparison of the processing of intact dimeric tRNA 4 and dimer lacking the ultimate 3'-CA dinucleotide sequence. Figure 3 shows the time course of processing of these two tRNA precursor models by M1 RNA. As illustrated, in the absence of the 3'-CA sequence processing was incomplete even after 90 min; under the same experimental conditions, the intact dimeric precursor was processed to the extent of ~80% within 15 min.

DISCUSSION

There is considerable evidence that E. coli RNase P and M1 RNA recognize tRNA-like structures present within tRNA precursor transcripts. To facilitate a study of the chemical mechanism of tRNA precursor processing, it would be useful to be able to vary the

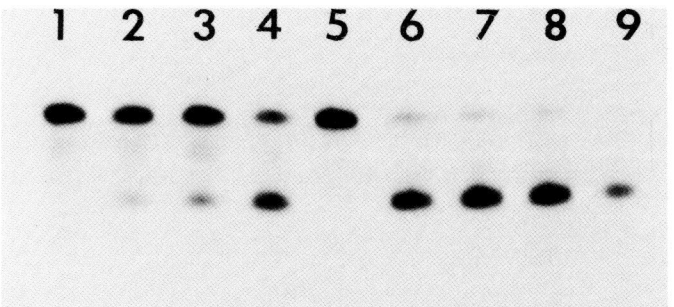

Figure 3. M1 RNA processing of dimeric tRNA's lacking the 3'-terminal CA sequence. Reactions were carried out at 37° C for either 15, 45, or 90 min in a total reaction volume of 5 μL containing 20 pmol M1 RNA, 50 mM Tris-HCl, pH 7.5, 0.1 M NH_4Cl, 60 mM $MgCl_2$, and 5% glycerol. Lane 1 contained tRNA precursor lacking the CA sequence; lanes 2, 3 and 4 illustrate processing by M1 RNA after 15, 45 and 90 min, respectively. Lane 5 contained tRNA precursor with the CA sequence; lanes 6, 7 and 8 illustrate its processing after 15, 45 and 90 min, respectively. Lane 9 contained authentic $tRNA^{Phe}$.

structures of the substrate RNA's, including the introduction of modified nucleotides at key positions. We have approached this problem by constructing tRNA precursor models from mature tRNA's; in the present case we describe the preparation of a dimeric precursor containing structural elements of E. coli tRNA^{Phe} and yeast tRNA^{Phe}. The feasibility of this approach was suggested by the findings that at least one natural dimeric tRNA precursor can be processed after the introduction of most of the post-transcriptional modifications have been made to the individual nucleotides (16), and that the dimeric precursors need not be separated by "spacer" nucleotides (10).

In fact, the synthetic tRNA precursor model described here was processed readily both by RNase P and M1 RNA under reaction conditions similar to those employed for natural pre-tRNA substrates. Further, processing by M1 RNA proceeded readily only when the ultimate 3'-CCA sequence was present, as has been found for natural precursor tRNA substrates. The actual site of processing was also found to be the same as for natural dimeric tRNA's, and the processing of the synthetic substrates proceeded essentially to completion. That dimeric tRNA precursor 4 was processed readily indicates that the presence of an intact anticodon within the 3'-tRNA is not required for effective processing.

The ability to manipulate the structures of tRNA precursor substrates readily should facilitate the analysis of a number of parameters pertinent to RNase P and M1 RNA-mediated processing of tRNA precursors. These might be thought to include the participation of spacer regions of unusual structure in the processing reaction, as well as the actual chemical mechanism of tRNA processing.

ACKNOWLEDGEMENTS

We thank Dr. Francis Schmidt, University of Missouri, for his generous gifts of E. coli RNase P and plasmid pTZ19-M1, which encodes pre-M1 RNA.

REFERENCES

1. Kole R, Altman S (1979). Reconstitution of RNase P activity from inactive RNA and protein. Proc Natl Acad Sci USA 76:3795.
2. Motamedi H, Lee Y, Schmidt FJ (1984). Tandem promoters preceding the gene for the M1 RNA component of *Escherichia coli* ribonuclease P. Proc Natl Acad Sci USA 81:3959.
3. Vioque A, Altman S (1986). Affinity chromatography with an immobilized RNA enzyme. Proc Natl Acad Sci USA 83:5904.
4. Altman S (1975). Biosynthesis of transfer RNA in *Escherichia coli*. Cell 4:21 and references therein.
5. Ilgen C, Kirk LL, Carbon J (1976). Isolation and characterization of large transfer ribonucleic acid precursors from *Escherichia coli*. J Biol Chem 251:922.
6. Hollingsworth MJ, Hallick RB (1982). *Euglena gracilis* chloroplast transfer RNA transcription units. J Biol Chem 257:12795.
7. Guthrie C (1975). The nucleotide sequence of the dimeric precursor to glutamine and leucine tRNAs coded by bacteriophage T4. J Mol Biol 95:529.
8. Schmidt FJ, Seidman JG, Bock RM (1976). Transfer ribonucleic acid biosynthesis. J Biol Chem 251:2440.
9. Guerrier-Takada C, Gardiner K, Marsh T, Pace N, Altman S (1983). The RNA moiety of ribonuclease P is the catalytic subunit of the enzyme. Cell 35:849.
10. Seidman JG, Schmidt FJ, Foss K, McClain WH (1975). A mutant of *Escherichia coli* defective in removing 3' terminal nucleotides from some transfer RNA precursor molecules. Cell 5:389.
11. Guerrier-Takada C, McClain WH, Altman S (1984). Cleavage of tRNA precursors by the RNA subunit of E. coli ribonuclease P (M1 RNA) is influenced by 3'-proximal CCA in the substrates. Cell 38:219.
12. Surratt CK, Lesnikowski Z, Schifman AL, Schmidt FJ, Hecht SM (1988). Construction and processing of transfer RNA precursor models. Biochemistry submitted.
13. Nishikawa K, Adams BL, Hecht SM (1982). Chemical excision of apurinic acids from RNA. A structurally modified yeast tRNA^{Phe}. J Am Chem Soc 104:326.

14. Nishikawa K, Hecht SM (1982). A structurally modified yeast tRNA^{Phe} with six nucleotides in the anticodon loop lacks significant phenylalanine acceptance. J Biol Chem 257:10536.
15. Johnson MT, Read BA, Monko AM, Pappas G, Johnson BA (1986). A convenient new method for desalting, deproteinizing and concentrating DNA or RNA. BioTechniques 4:64.
16. Barrell BG, Seidman JG, Guthrie C, McClain WH (1974). Transfer RNA biosynthesis: the nucleotide sequence of a precursor to serine and proline transfer RNAs. Proc Natl Acad Sci USA 71:413.

CLEAVAGE OF THE tRNA-LIKE RNA OF THE TURNIP YELLOW MOSAIC VIRUS BY THE CATALYTIC RNA COMPONENT OF RNASE P[1]

Christopher J. Green and Barbara S. Vold

Department of Molecular Biology
SRI International, Menlo Park CA 94025

ABSTRACT The tRNA-like pseudoknotted structure at the 3'-end of the Turnip Yellow Mosaic Virus (TYMV) genomic RNA can be cleaved by the catalytic RNA component of *Bacillus subtilis* RNase P. The optimal ionic conditions for this reaction are much lower than those reported for the cleavage of monomeric tRNA precursors. The TYMV conditions are comparable to those for the cleavage of tRNA precursors by the RNase P holoenzyme. These data indicate that the putative role of the protein component of RNase P is more than that of an ionic shield and that the structure of the substrate has an important influence on the reaction.

INTRODUCTION

RNase P from eubacteria is a ribonucleoprotein with an RNA component of approximately 400 nucleotides in length (1,2). This RNA is the actual catalyst for the cleavage of the 5'-leader sequence from tRNA precursors (3). *In vitro* reactions in the presence of high salt buffers allow the RNase P RNA to accurately perform the cleavage of tRNA precursor substrates. The addition of the RNase P protein to form the holoenzyme allows the

---

[1]This work was supported by Public Health Service Grant GM29231

cleavage to occur at much lower ionic strengths. Typical tRNA precursors can be cleaved by the *Bacillus subtilis* RNase P holoenzyme in 100 mM NH$_4$Cl and 15 mM MgCl$_2$. With the catalytic RNA alone, efficient cleavage requires at least 0.8 M NH$_4$Cl and 100 mM MgCl$_2$ (4,5). The requirement for high ionic strength buffers in tRNA precursor processing by the RNA alone has been attributed to ionic shielding needed to bring together the charged substrate and the catalytic RNA (5). The protein component of RNase P also appears to facilitate the release of the processed tRNA from the catalyst, which results in a k_{cat} for the holoenzyme that is approximately 20 times higher than with the catalytic RNA alone (5).

We are interested in the structural features of the substrates of RNase P RNA. We have recently found that different substrates can have radically different optimal ionic requirements for cleavage. In this paper, we report the conditions necessary for the cleavage of the tRNA-like structure of TYMV by the catalytic RNA component of *B. subtilis* RNase P. The 3'-end of the TYMV RNA genome can be folded into a tRNA-like structure by the formation of a pseudoknot as is shown in Figure 1 (6). This structure acts as a substrate for several tRNA associated enzymes including nucleotidyl transferase, elongation factors, and valyl-tRNA synthetase (7-11). It was also reported that a partially purified enzyme preparation of RNase P could cleave the tRNA-like structure of TYMV RNA (12).

METHODS

RNA Transcription Templates.

The TYMV tRNA-like substrate was made by *in vitro* SP6 RNA transcription from the pGEM20 plasmid constructed by Morch *et al.* (13). When the plasmid cut with *Eco*RI was used as template, an RNA 377 nucleotides in length was produced with the pseudoknotted tRNA-like structure at the 3'-end. The plasmid could also be linearized at a *Pvu*II site immediately downstream from the *Eco*RI site, generating an RNA with 42 extra nucleotides at the 3'-end. The

latter was used for reverse transcriptase sequencing. The
B. subtilis histidine tRNA precursor was also generated
by SP6 RNA polymerase transcription (14). The catalytic
RNA from *B. subtilis* was transcribed from a vector with a
T7 RNA polymerase promoter (15). Both unlabeled and ^{32}P-labeled transcriptions and purifications were carried out
as described (14).

Processing of the TYMV tRNA-Like RNA.

All reactions contained 50 fmol of substrate and 44
ng of the catalytic RNA from *B. subtilis* RNase P in 20 µl
of 50 mM Tris-HCl, pH 8.0, and were incubated at 37°C for
1 h. The concentrations of $MgCl_2$ and NH_4Cl in the buffers
were varied and are given in Figure 2.

Reverse Transcriptase-Dideoxy Sequencing of the Processed
and Unprocessed TYMV RNAs.

The *Pvu*II cut template described above was used to
generate an RNA with a 42 nucleotide 3'-extension. This
RNA was cleaved as described above using 100 mM NH_4Cl and
100 mM $MgCl_2$. A 21 nucleotide DNA oligomer with the
sequence GGCTTATCGAAATTAATACGA was synthesized and purified as described (14). The DNA was complementary to the
3'-end of the RNA transcribed from the *Pvu*II cut pGEM20
plasmid. This DNA oligomer was used as the primer for the
reverse transcriptase sequencing of the TYMV RNA before
and after cleavage by the catalytic RNA. Sequencing was
performed as previously described (14).

3' and 5'-End Labeling of the Processed TYMV RNA.

The 3'-end labeling of the cleavage site and identification of the terminal nucleotide was essentially as
described (14). The 5'-end labeling was performed by
treating the cleaved TYMV RNA with bacterial alkaline
phosphatase to remove the 5'-phosphate followed by

incubation with polynucleotide kinase and [γ-^{32}P] ATP. The purified labeled RNA was then digested with RNase T2 (Bethesda Research Labs). The resulting nucleotide 3',5'-(bisphosphate) was identified by thin layer chromatography (TLC) on polyethyleneimine plates (Sybron-Brinkman) with 0.55 M $(NH_4)_2SO_4$ in the presence in unlabeled nucleotide 3',5'-(bisphosphate) standards. The standard compounds were detected under UV light. The radioactive compounds were detected by autoradiography.

RESULTS

Ionic Conditions of Processing.

The TYMV tRNA-like RNA generated from the *Eco*RI cut pGEM20 plasmid is shown in Figure 1. Not shown in this figure are the first 136 nucleotides at the 5'-end. The pseudoknotted structure proposed to adapt this RNA into a tRNA-like conformation (7) creates a structure analogous to the aminoacyl stem, however this is formed by a folding of the 3'-end upon itself and no base pairing with what would be the 5'-terminus of a tRNA occurs. Cleavage of this structure at a place analogous to the 5' cleavage site of a tRNA precursor releases a short 3'-fragment from the longer 5'-fragment under denaturing conditions. The *B. subtilis* histidine tRNA precursor sequence has been reported (14).

Conditions for processing of *B. subtilis* precursor histidine tRNA have been established previously by us (14). These were similar to the conditions used by others for monomeric precursor tRNAs using the *B. subtilis* catalytic RNA without the protein (5,16).

Electophoresis on a 12% polyacylamide-7 M urea gel was used to separate the products of this cleavage. The unprocessed and 5'-processed RNA ran near the top of the gel. The small 3'-fragment ran to the bottom. All of these reactions contained varying amounts of $MgCl_2$ and NH_4Cl in the buffer as indicated in Figure 2. In 1.2 M NH_4Cl, the TYMV substrate has a very rapid response to increasing $MgCl_2$ concentration, from no detectable

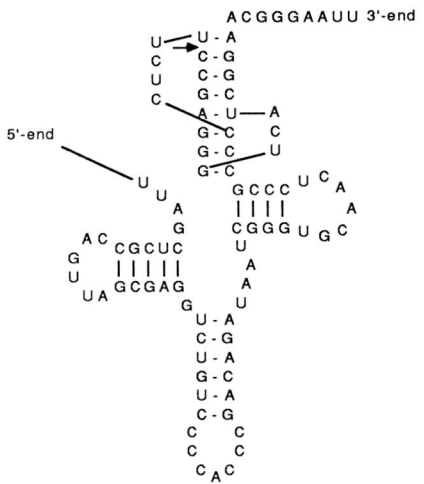

Figure 1. Sequence of the precursor tRNA-like structure from TYMV.
The figure shows the last 91 nucleotides of the transcript from the EcoRI cut pGEM20 template. The structure is folded into a pseudoknot according to the model of Rietveld et al. (6). The arrow denotes the site of cleavage by the catalytic RNA.

cleavage without any $MgCl_2$ to almost optimal cleavage at 10 mM. Processing of the histidine tRNA precursor required a much higher $MgCl_2$ concentration, with an optimum level of over 100 mM.
  The TYMV tRNA-like structure could also be processed by the catalytic RNA component of B. subtilis RNase P at much lower NH_4Cl concentrations than are needed for optimal cleavage of the histidine tRNA precursor. The TYMV substrate could be efficiently cleaved in 100 mM NH_4Cl and 20-60 mM $MgCl_2$. While the tRNA^{His} precursor showed very little if any cleavage by the B. subtilis RNase P RNA. This is also true of other tRNA precursors (4,16).

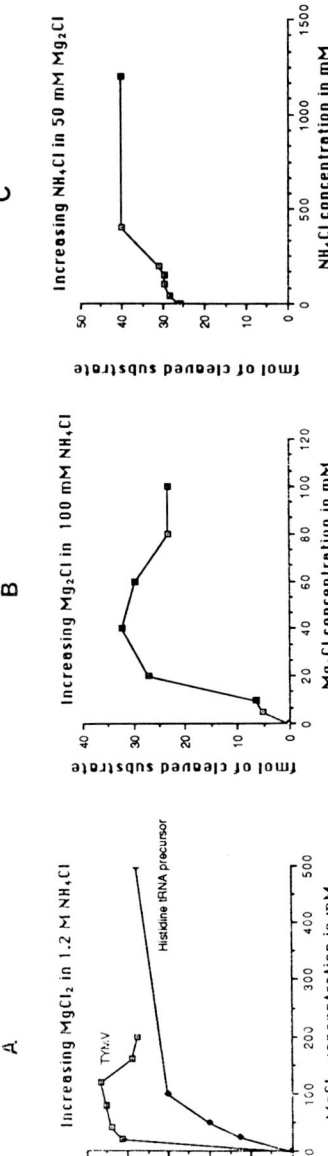

Figure 2. Ionic requirements for the cleavage of the precursor tRNA-like structure of TYMV and the precursor of *B. subtilis* histidine tRNA. Femtomoles of the released 3'-fragment are plotted against various concentrations of NH_4Cl or $MgCl_2$ while holding the other salt concentration constant. The 5'-fragments were not plotted because they migrate very close to the uncleaved substrate while the 3'-fragment electrophoresed to near the bottom of the gel and was easy to quantitate cleanly. Products shown in A were cleaved in the presence of 1.2 M NH_4Cl and various concentrations of $MgCl_2$. Products in B were cleaved in the presence of 100 mM NH_4Cl and various concentrations of $MgCl_2$. Products in C were cleaved in the presence of 50 mM $MgCl_2$ and various concentrations of NH_4Cl.

Sequencing.

Sequencing of the pseudoknotted TYMV structure was difficult, probably due the unusual secondary structure. The processed TYMV RNA generated a sequence with two dark bands across all lanes (data not shown), indicating a cleavage site that was either between the two U residues or between the C and U residues at the begining of the structure corresponding to an aminoacyl stem of a tRNA precursor. To determine the exact cleavage site, an unlabeled TYMV RNA generated from the EcoRI cut template was cleaved with the catalytic RNA. Both the 3'- and 5'-ends of the cleavage site were labeled with ^{32}P and the terminal nucleotides identified by TLC. This analysis conclusively demonstrated that the catalytic RNA cleaved the TYMV tRNA-like RNA between a U and a C residue as shown in Figure 1, leaving a 3'-hyroxyl and a 5'-phosphate

CONCLUSIONS

The demonstration that the TYMV tRNA-like structure can be processed by the catalytic RNA component of B. subtilis RNase P provides further evidence for the pseudoknot structure in the TYMV conformation. The processing site is in an area comparable to that found in tRNA precursors. This site is just inside of the structure analogous to the "aminoacyl stem" (7). The exact cleavage site of RNase P is known to be dependent on the sequence of the region near the aminoacyl helix (14,17).
The TYMV tRNA-like structure can be recognized as a substrate by nucleotidyl transferase and a 3'-terminal CCA sequence can be added (8,9). However, the TYMV substrate used in these experiments did not have this feature. Therefore, even in very low ionic strength buffers, the CCA is not necessary for the TYMV cleavage reaction by the catalytic RNA component of RNase P.
Although it was originally proposed that processing by the catalytic RNA alone required a high ionic strength buffer (4,5), our data indicate that this is not true for all substrates. It has been recognized that precursor

tRNAs differ in their optimal ionic conditions for processing by the catalytic RNA (18; Vold and Green, unpublished data). However the TYMV RNA can be processed efficiently at the lowest of any of the reported conditions, equivalent to those of the holoenzyme. It has been proposed that the RNase P protein acts as an ionic shield to reduce the anionic repulsion between the substrate and the catalyst (5). However, the protein component of RNase P could also function by inducing a more favorable conformation in some substrates at low ionic strength that otherwise can only be caused by the presence of a high ionic strength buffer. Other substrates, such as the tRNA-like structure of TYMV, may already have the correct conformation for processing even in low ionic strength conditions.

ACKNOWLEDGMENTS

We wish to thank M.D. Morch, R. L. Joshi and A-L. Haenni for the gift of the pGEM20 plasmid and C. Reich and N. Pace for the *B. subtilis* RNase P RNA plasmid.

REFERENCES

1. Stark, BC, Kole, R, Bowman, EJ, Altman, S (1978). Ribonuclease P: An enzyme with an essential RNA component Proc Natl Acad Sci, USA 75:3717.
2. Gardiner, KJ, Pace, NR (1980).RNase P of B. subtilis has an RNA component J Biol Chem 255:7507.
3. Guerrier-Takada C, Gardiner K, Marsh, T, Pace, N, Altman, S (1983). The RNA moiety of ribonuclease P is the catalytic subunit of the enzyme. Cell 35:849.
4. Marsh, TL, Pace, B, Reich, C, Gardiner, K, Pace, NR (1985). Processing of tRNA precursors by an RNA catalyst from *B. subtilis*. In Calendar R ,Gold L, (eds.): "Sequence Specificity in Transcription and Translation," New York: Alan R. Liss, p. 441.

5. Reich C, Olsen, GJ, Pace B, Pace, NR (1988). Role of the Protein Moiety of Ribonuclease P, a Ribonucleoprotein Enzyme. Science 239:178.
6. Rietveld, K, Van Poelgeest, R, Pleij, CWA, Van Boon, JH, Bosch, L (1982).The tRNA-like structure at the 3' terminus of TYMV RNA. Nucleic Acids Res. 10:1929.
7. Joshi S, Chapeville F, Haenni AL (1982). Length requirements for tRNA-specific enzymes and cleavage specificity at the 3'-end of TYMV RNA. Nucleic Acids Res 10:1947.
8. Yot, P, Pinck, M, Haenni, AL, Duranton, HM, Chapeville, F (1970). Valine-specific tRNA-like structure in TYMV RNA. Proc. Natl. Acad. Sci. USA 67:1345.
9. Litvak, S, Carr, DS, Chapeville, F (1970). TYMV RNA as a substrate of the tRNA nucleotidyl transferase. FEBS Lett, 11:316.
10. Joshi, S, Chapeville, F, Haenni, AL (1982). TYMV RNA is aminoacylated *in vivo* in Chinese cabbage leaves. EMBO J. 1:935.
11. Joshi S, Ravel JM, Haenni AL (1986). Interaction of TYMV Val-tRNA with eukaryotic elongation factor EF-1. EMBO J. 5:1143.
12. Silberklang M, Prochiantz A, Haenni AL, Rajbhandary VL (1977). Studies on the sequence of the 3'-terminal region of TYMV RNA. Eur J Biochem 72:465.
13. Morch, MD, Joshi, RL, Denial, TM and Haenni, AL (1987). A new 'sense' RNA approach to block viral RNA replication *in vitro*. Nuc Acids Res 15:4123.
14. Green CJ and Vold BS (1988). Structural Requirements for the Processing of Synthetic tRNA^{His} Precursors by the Catalytic RNA Component of RNase P. J Biol Chem 263:652.
15. Reich, C, Gardiner, KJ, Olsen, GJ, Pace, B, Marsh, TL, Pace, NR (1986). The RNA component of *B. subtilis* RNase P. J Biol Chem 261:7888.
16. Gardiner, KJ, Marsh, TL, Pace, NR (1985). Ion dependence of *B. subtilis* RNase P reaction. J Biol Chem 260:5415.

17. Burkard V, Willis I, and Söll D (1988). Processing of Histidine Transfer RNA Precursors. J Biol Chem 263:2447.
18. Nichols L Schmidt FJ (1988). Dependence of M1 RNA substrate specificity on magnesium ion concentration. Nucleic Acids Res. 16:2931.

THE SELF-CLEAVING RNAs OF HUMAN HEPATITIS DELTA VIRUS[1]

John Taylor, Lamia Sharmeen, Mark Kuo, and Gail Dinter-Gottlieb*

Fox Chase Cancer Center, Philadelphia, Pennsylvania 19111 and
*Drexel University, Philadelphia, Pennsylvania 19104

ABSTRACT Hepatitis delta virus is unique among known animal viruses, but is in many ways similar to certain subviral plant agents, in particular, the viroids, virusoids and satellite RNAs. Because of this analogy we asked if the HDV RNA, like that of the plant agents can undergo *in vitro* self-cleavage. We report that both the genomic RNA of HDV and its complement, the antigenomic RNA, can undergo site-specific self-cleavage.

INTRODUCTION

Rizzetto *et al*. (1) first showed that *in vivo* replication of hepatitis delta virus (HDV) is dependent upon the presence of an hepadna virus, such as hepatitis B virus (HBV). It would appear that HBV provides at least the coat protein needed for the packaging of progeny HDV particles. HDV, itself, appears to express only one protein, the delta antigen, a species of about 24 kDa that is found inside the HDV particle and in the nucleus of the infected liver cell. Its function is unknown.

In the last three years work from several laboratories has made it clear that both the structure

---

[1] This work was supported by grants CC-22651, CA-06927 and RR-05539 from the National Institutes of Health, grant MV-7 from the American Cancer Society and by an appropriation from the Commonwealth of Pennsylvania.

and the replication mechanism of the HDV genome are unique among the known animal viruses (2-4). However such studies have at the same time revealed that there are a number of strong similarities to certain pathogenic RNAs of plants (5). The analogy is summarized in Table 1.

Studies by Prody et al. (6) and by Forster and Symons (7) have shown that the genomes of certain plant viroids, virusoids and satellites RNAs can undergo self-cleavage in vitro. In one case, Buzayan et al. have even demonstrated a reverse reaction, self-ligation (8). We therefore undertook to search for similar processing events using the RNAs of HDV. Our expectation was that there should be at least one site on the genomic RNA and another on the antigenomic RNA, which would be sufficient to explain the generation of monomer RNAs from a rolling-circle mechanism of replication.

TABLE 1
COMPARISON OF CERTAIN SUBVIRAL PLANT AGENTS AND HDV

	Plant agent[a]	HDV
Genomic RNA:		
single-stranded	+	+
length less than 2 kb	+	+
circular rather than linear conformation	+/-	+
ability to fold into unbranched rod	+/-	+
Replication of genomic RNA:		
via antigenomic RNA	+	+
circular forms of unit length RNA	+/-	+
linear forms of multimeric RNA	+	+
replicative structures	+	+
possible rolling-circle model	+	+
occurs in the nucleus	+/-	+
Proteins:		
either none or one	+	+

[a]The plant agents considered include viroids, virusoids and satellite RNAs and viruses (5).

RESULTS AND DISCUSSION

Cleavage of RNA Synthesized in vitro.

In order to look for self-cleavage of HDV RNA sequences we made use of recombinant clones of the complete HDV genome. As recently described by Kuo et al. (9) we initially created a cDNA library using RNA from the liver of an infected woodchuck. HDV-related subgenomic clones were obtained by screening and then assembled to make a clone of the full 1679 b genome. Such cloned DNA was transferred to an RNA transcription vector, pGEM4Blue, and used to generate HDV genomic and antigenomic RNA species.

To search for possible self-cleavage of the RNA transcription reaction products we used the following approach. The RNA transcripts of expected size were gel purified and then submitted to the self-cleavage conditions of Uhlenbeck (10). Briefly, the RNA in 0.1 mM EDTA was heated to $68^{o}C$, snap-cooled on ice, adjusted to 10 mM $MgCl_2$ and 50 mM Tris-HCl (pH 7.5), and incubated at $50^{o}C$ for 1 h. The products were then examined by agarose gel electrophoresis after denaturation with glyoxal (11).

Our first indication of self-cleavage was obtained with a transcript of antigenomic RNA that was 180 b longer than unit length. Actually it was 1679 b together with a direct terminal repeat of 180 b. As recently reported (11), we found two cleavage sites in this molecule; they were both present in the terminal repeated sequence we had generated on the RNA. Primer extension studies were used to show that the cleavage site was at position 900/901 (using the genomic numbering system of Wang et al. (2)). In subsequent studies we detected a similar cleavage site for the genomic HDV RNA; this was precisely located at position 685/686.

The site on the antigenomic RNA showed the following intriguing relationship to that on the genomic RNA. We know that the antigenomic RNA, like the genomic RNA, has the ability to fold on itself to form an unbranched rod structure (9). More specifically the region of antigenomic RNA spanning 900/901 has the potential to base pair with the region 685/686. Thus, on the complementary strand, the genomic RNA, the region spanning 685/686 resembles the cleavage site seen on the antigenomic RNA. In this way we were able to predict, to the nucleotide, the site on the genomic RNA independent of the actual experimentation.

Cleavage of RNA Synthesized *in vivo*.

After the self-cleavage reaction was found for the *in vitro* synthesized genomic and antigenomic HDV RNAs it seemed reasonable to determine whether a similar process occurred for *in vivo* RNA. We began by using the same primer extension method (11) to map "cleavage" sites on RNA from the livers of two experimentally infected chimpanzees. For both sources of RNA, cleavage of genomic RNA was detected at precisely the same site as previously found for *in vitro* RNA. This finding could be considered as evidence that the self-cleavage has relevance in the life cycle of the virus. Further studies are needed however, to determine when the self-cleavage of *in vivo* RNA occurs; because of the nature of the primer extension reaction we have yet to determine whether the specific cleavage occurs before and/or during the assay process.

Studies are also underway with antigenomic liver RNA, which is 5-20 times less abundant than genomic RNA (4), and with the genomic RNA isolated from virions.

Nature of the Cleavage Junction.

With the self-cleaving *in vitro* RNA we have begun studies on the nature of the cleavage junction using the linear, *in vitro*, transcript. We found that the 5' terminus generated by *in vitro* cleavage could be directly labeled with [gamma-P32]ATP and T4 kinase. The 3'-terminus, on the other hand, could not be labeled with [alpha-P32]pCp and T4 RNA ligase, whereas the 3'-terminus of the uncleaved RNA was readily labeled. Even after treatment with calf intestinal alkaline phosphatase the 3'-terminus generated by cleavage could not be labeled. Our interpretation of this observation is that cleavage releases a 5'-OH and a 2'-, 3'-cyclic monophosphate. This interpretation is based on the work of others on the cleavage junctions of several plant agents (6,7).

Sequences Needed for Cleavage.

Using the RNA species transcribed in vitro it was possible to determine the sequences surrounding the cleavage site that are necessary for cleavage. Preliminary studies with the antigenomic RNA, which cleaves at 900/901 are summarized in Figure 1. We found that no more than 20 b were needed on the 5'-side of the cleavage site. In contrast on the 3'-side 26 b was not enough (panel 1) but 115 b was sufficient (panel 2). A puzzling finding was that when the sequence on the 3'-side was increased further, to about 400 b, the RNA failed to cleave (panel 3). As shown in Figure 1 this allowed formation of the unbranched rod structure in the region spanning the cleavage site. An hypothesis we wish to test is that formation of a rod structure in this region inhibits self-cleavage, presumably by blocking the formation of another folding conformation that is essential for self-cleavage. Interestingly, the terminally redundant antigenomic RNA that is longer than unit length but with the self-cleavage site in the redundancy, can cleave (panel 4). Further tests of our hypothesis are now being carried out with bimolecular reactions (panels 5 and 6).

Preliminary studies using another approach indicate that the requirements for self-cleavage of genomic and

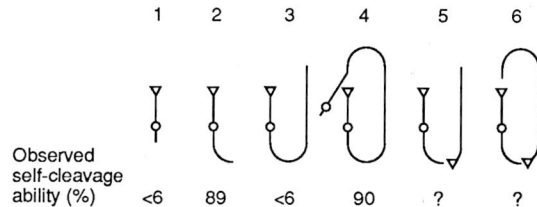

FIGURE 1. Ability of antigenomic HDV RNA species to undergo self-cleavage. The RNAs were transcribed in vitro from cloned DNA. The lengths and locations of the RNAs are indicated relative to a circular monomer; the vertical elongation of the circle represents the unbranched rod structure. The sites of initiation and potential self-cleavage are indicated by small triangles and circles, respectively. In 1-4 are represented single RNA species whereas in 5-6 are shown reactions containing two species.

antigenomic RNAs may be met by a sequence including about 20 b 5′ and about 75 b 3′ of the cleavage site. More specific definition is needed however for both genomic and antigenomic RNAs. It should be possible to obtain data such as been reported for others with RNAs of plant agents (7,10). We have so far made a preliminary comparison of HDV to the sequences surrounding the cleavage sites of the five known virusoids (5) and seen limited sequence homology. However, we have not detected the consensus "hammerhead" structure first noted by Symons and coworkers(5,7) and refined by Uhlenbeck (10).

There are two features of the HDV cleavage sites that are especially worthy of further study. Firstly, the cleavage site in the antigenomic RNA is just downstream of the termination codon of the open reading frame of the only known HDV protein, the delta antigen (9) suggesting a role of self-cleavage in mRNA formation. Secondly, immediately upstream of the sites is a pyrimidine-rich sequence. On the antigenomic RNA it begins with an unbroken sequence of 16 pyrimidines. This is much larger than the pyrimidine sequence usually found upstream of splice acceptor sites (15).

Sequence Needed for Ligation.

Based upon the analogy of HDV to the plant agents, we expect HDV replication will require a ligation process to produce circular monomers. At this time only one of the plant agents has been shown to carry out an in vitro self-ligation reaction (8). As noted in the previous section, we are testing the hypothesis that the predicted unbranched rod structure is a conformation which blocks self-cleavage. A related hypothesis, currently being tested, is that the rod structure favors self-ligation. Our thinking is that even though the cleavage and ligation reactions may use the same primary sequence, they nevertheless depend upon different secondary structures.

Rolling-circle Model.

As discussed at the outset there are major similarities between the structure and replication of the HDV genome and that of certain plant agents (Table 1). Rolling-circle models have been proposed by others for these plant agents (12,13) and at this time there is need for a similar working model for HDV. As shown in Figure 2 we propose such a model which includes some of our information on self-cleavage and our hypothesis on the role of the rod structure in limiting self-cleavage and promoting self-ligation. In terms of the latter it should be noted that self-cleavage of a linear multimer releases unit length linear monomers. Then, the released monomer folds on itself into the rod structure to provide the substrate for self-ligation. The model also shows that upon infection the first RNA transcribed is antigenomic (panels 2-7) and that the <u>first</u> self-cleavage event could release a specific antigenomic RNA. This short RNA (ca. 0.8 kb) could not fold into the rod structure. Also, because of the

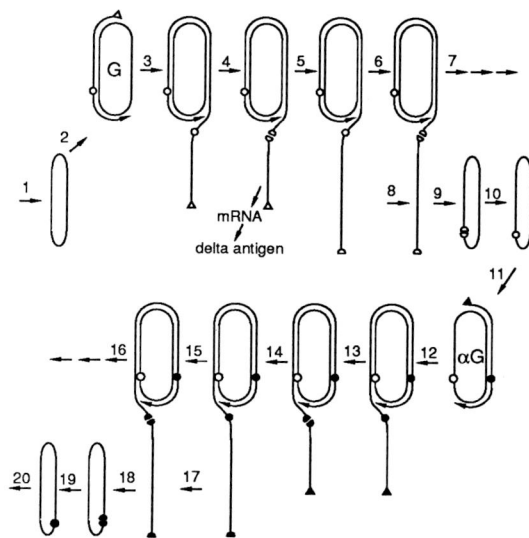

FIGURE 2. Rolling-circle model of HDV genome replication. The sites for the initiation and self-cleavage are indicated by small triangles and circles, respectively. The closed and open symbols, refer to genomic (G) and antigenomic (αG) RNA species, respectively.

presumed nature of its 5'-end, it could not ligate. The location of this RNA on the antigenomic sequence is such that it could go on to become the mRNA for the delta antigen (14). In such a model the delta antigen product may have a relatively early role in genome replication.

OUTLOOK

In this article we have concentrated on one aspect of the replication of HDV, namely, the possible involvement of self-processing reactions. The RNAs of HDV may be the first mammalian RNAs to be found to undergo self-cleavage. However, a clear rationale behind our experiments was to test an extrapolation from those plant agents that have many similarities to HDV. The list of those similarities, as presented in Table 1, should now be increased to include self-cleavage. In the immediate future we envisage testing whether we can also add self-ligation. And, as our understanding of the processing of HDV RNA becomes clearer, we would not be surprised to find another analogy, namely to the normal processing of events of eucaryotic RNAs, especially splicing (15) and the processing of 3'-termini associated with polyadenylation (16).

ACKNOWLEDGMENTS

We thank William Mason for valuable criticism of the manuscript. Laura Coates provided technical assistance in this work, Karen Trush prepared the art work and the word-processing was carried out by Arlene Capriotti. Infected liver samples were provided by John Gerin and Colin Howard.

REFERENCES

1. Rizzetto M, Canese MG, Aricó J, Crivelli O, Bonino F, Trepo CG, Verme G (1977). Immunofluorescence detection of a new antigen-antibody system (δ/anti-δ) associated to the hepatitis B virus in the liver and in the serum of HBsAg carriers. Gut 18:997.
2. Wang K-S, Choo Q-L, Weiner AJ, Ou H-J, Najarian RC,

Thayer RM, Mullenbach GT, Denniston KJ, Gerin JL, Houghton M (1986). Structure, sequence and expression of the hepatitis delta virus genome. Nature 323:508.
3. Kos A, Dijkema R, Arnberg AC, van der Meide PH, Schellekens H (1986). The hepatitis delta virus possesses a circular RNA. Nature (London) 332:558.
4. Chen P-J, Kalpana G, Goldberg J, Mason W, Werner B, Gerin J, Taylor J (1986). The structure and replication of the hepatitis delta virus genome. Proc Natl Acad Sci USA 83:8774.
5. Keese P, Symons R (1987). The structure of viroids and virusoids. In Semancik JS (eds): "Viroids and Viroid-like Pathogens," Florida: CRC Press, p 1.
6. Prody GA, Bakos JT, Buzayan JM, Schneider IR, Bruening G (1986). Autocatalytic processing of dimeric plant virus satellite RNA. Science 231:1577.
7. Forster AJ, Symons RH (1987). Self-cleavage of viroid RNA is performed by the proposed 55-nucleotide active site. Cell 50:9.
8. Buzayan JM, Hampel A, Bruening G (1986). Nucleotide sequence and newly formed phosphodiester bond of spontaneously ligated satellite tobacco ringspot virus RNA. Nucl Acids Res 14:9729.
9. Kuo, MY-P, Goldberg J, Coates L, Mason W, Gerin J, Taylor J (1988). Molecular cloning of hepatitis delta virus RNA from an infected woodchuck liver: Sequence, structure and application. J Virol, in press.
10. Uhlenbeck OC (1987). A small catalytic oligoribonucleotide. Nature 328:596.
11. Sharmeen L, Kuo MY-P, Dinter-Gottlieb G, Taylor J (1988). The antigenomic RNA of human hepatitis delta virus can undergo self-cleavage. J Virol, in press.
12. Branch AD, Robertson HD (1984). A replication cycle for viroids and other small infectious RNAs. Science 223:450.
13. Hutchins CJ, Keese P, Visvader JE, Rathjen PD, Melnnes JL, Symons RH (1985). Comparison of multimeric plus and minus forms of viroids and virusoids. Plant Mol Biol 4:293.
14. Taylor J, Kuo M, Chen P-J, Kalpana G, Goldberg J, Aldrich C, Coates L, Mason W, Summers J, Gerin J, Baroudy B, Gowans E (1987). Replication of

hepatitis delta virus. In Robinson W, Koike K, Will H (eds): "Hepadna Viruses," New York: Alan R. Liss, p 541.
15. Fu X-Y, Ge H, Manley JL (1988). The role of the polypyrimidine stretch at the SV40 early pre-mRNA 3' splice site in alternative splicing. EMBO J 7:809.
16. Takagaki Y, Ryner LC, Manley JL (1988). Separation and characterization of a poly(A) polymerase and a cleavage/specificity factor required for pre-mRNA polyadenylation. Cell 52:731.

DISCUSSION SUMMARY: RNA - PROTEIN INTERACTIONS

David E. Draper

Department of Chemistry
Johns Hopkins University
Baltimore, MD 21218

How proteins recognize and bind specific RNA sites is still an open and intriguing question. A protein may make a variety of contacts with an RNA (i.e. hydrogen bonds, hydrophobic interactions, ionic contacts). It is plausible to think that specificity is achieved by matching a three dimensional array of contacts in the RNA with appropriate protein structures. We do not yet know whether some kinds of contacts are more frequently used to derive specificity, or whether some kinds of RNA structural motifs are preferred protein recognition sites. A number of protein - RNA complexes from a variety of sources are now under study, and we may be able to discern some "rules" for RNA binding in the near future.

In this discussion session, two kinds of work were represented. A number of groups are attempting to define the precise RNA features recognized by a protein by measuring protein affinity for a series of closely related RNA substrates. The ultimate goal is a detailed map of the nucleotides important for contacting the protein or for maintaining the correct RNA folding. Other speakers emphasized the protein features required for binding an RNA. Much less work has been done from the protein point of view, and these talks illustrated the kinds of approaches that need to be taken to eventually understand the details of protein - RNA recognition.

Paul Romaniuk (Univ. Victoria) described an extensive series of measurements designed to locate RNA sequences essential for recognition between a Xenopus transcription factor (TFIIIA) and 5S rRNA. In a "block mutagenesis" approach, almost all helical or loop segments of the RNA have been systematically altered in sequence (maintaining Watson-

Crick pairing in helices). The largest effect on TFIIIA specificity is a set of changes in a single loop which reduces the binding affinity by a mere factor of three. This particular sequence variant may alter the coaxial stacking of two helices. The origin of specificity in this system is therefore puzzling: "footprints" and deletion mutants have localized TFIIIA recognition to a specific region of the RNA, but the particular sequences appear to make only a very weak contribution to the specificity.

David Draper (Johns Hopkins Univ.) described work with the ribosomal protein S4, which recognizes the *E. coli* 16S rRNA and also binds its own messenger RNA as part of a translational control mechanism. The hope is that similar features in each RNA will become apparent and point the way to specific recognition features. The mRNA recognition site is ≈130 nucleotides in length, and has been shown to have a complex, "pseudoknot" folding. The minimal rRNA site is ≈460 nucleotides, though some internal deletions can be made without affecting protein binding. Binding studies with smaller fragments or with RNAs containing internal deletions suggest that two separate regions within the 460 bases are required for full binding. With both the mRNA and rRNA, S4 appears to recognize complex RNA structures containing long range secondary or tertiary interactions. Only minor similarities can be detected between the two RNAs, suggesting that very different overall foldings of the RNAs form structures which are similar in three dimensions.

John Milligan (Univ. Colorado) reported work designed to detect individual RNA phosphate residues which might be important for R17 coat protein recognition of its small hairpin binding site. Hairpins containing thiophosphates replacing some of the normal phosphates were prepared by transcription with T7 RNA polymerase and one α-thio-NTP. Thiophosphates were located at unique sites by ligating RNA fragments. Substitution at one position in the loop enhances protein binding ten fold, while thiophosphates at several stem sites weaken binding three to 10 fold. The salt dependence of the binding constants are unchanged from the unmodified hairpin, showing that the number of ionic contacts has not been altered.

Ignacio Faus (Indiana Univ.) discussed efforts to determine features of the cro mRNA required for rho protein activity. Two deletions were examined. An upstream deletion decreased the number of ionic interactions with rho by two, but had no effect on the ATPase activity of rho. A second deletion further downstream decreased the protein binding affinity five fold and drastically reduced the ATPase activity, without affecting the number of ionic interactions. In this way it is possible to distinguish different kinds of contacts that different parts of the RNA make with the protein.

Barbara Merrill (Yale Univ.) described work on the A1 protein associated with hnRNA. The protein has two domains: the N-terminus contains two repeats of a 92 amino acid sequence, and the C-terminus is glycine rich. Trypsin digestion in vitro separates the two domains. The N-terminal domain is able to bind to single strand DNA-cellulose, and is crosslinked to p(dT)8 by UV irradiation. Four phenylalanine residues were identified as the crosslinking sites, and derive from two homologous regions of the repeated sequence. These results are very similar to those obtained previously with single strand specific DNA binding proteins, which are known to intercalate aromatic amino acid side chains between nucleotide bases. This work shows that a similar mechanism may take place with RNA binding proteins.

Iwona Wower (Univ. Massachusetts) described a major effort to genetically map the regions of a ribosomal protein which are important for RNA binding. S8 protein from E. coli regulates the translation of ribosomal proteins in the spc operon. The S8 gene has been cloned in a plasmid under lac promoter control; its overexpression slows cell growth. After mutagenesis of the S8 gene, many mutations in S8 were picked up which no longer affect the cell growth rate; these presumably have an altered affinity for the spc mRNA. The mutations tend to fall into regions of the protein amino acid sequence which have been conserved in evolution.

RIBOSOMAL PROTEIN S4 RECOGNIZES COMPLEX MESSENGER AND RIBOSOMAL RNA STRUCTURES[1]

David E. Draper, Careen Tang, and Jailaxmi Vartikar

Department of Chemistry, Johns Hopkins University
Baltimore, MD 21218 USA

ABSTRACT The ribosomal protein S4 binds directly to the 16S rRNA to initiate ribosome assembly, and also to the leader region of its own mRNA as part of a translational regulation system. The structures of the protein binding sites in the two RNAs have been determined by measuring S4 affinity for RNA fragments with differing endpoints, with deletions, and with compensatory mutations. Three helices are required in a 139 base mRNA fragment; two of these compose a pseudoknot. Two separate regions within the 5' 500 bases of the rRNA contribute to recognition. Similarities between the two RNAs are noticeable. S4 appears to recognize a complex, three dimensional RNA structure.

INTRODUCTION: AUTOREGULATION OF THE α OPERON

Coordination between the synthesis of ribosomal RNA and ribosomal proteins in *E. coli* is achieved (in part) by translational regulation (1). In the α operon, four ribosomal proteins are synthesized. One of these, S4, can repress translation of all the ribosomal proteins in the operon (2). Therefore synthesis of S4 is limited to a small excess over that required to assemble ribosomes. (A fifth protein in the operon, the α sub-

[1]This work was supported by NIH grant GM29048 and a Research Career Development Award to DED.

unit of RNA polymerase, is regulated independently of the ribosomal proteins.)

All ribosomal proteins which act as translational repressors are known to bind directly to the large or small subunit rRNA, implying that mRNA binding takes advantage of the same RNA recognition site used by the protein in ribosome assembly (1). Nuclease protection studies carried out with S4 and the 16S rRNA suggest that S4 recognizes a complex structure comprising the 5' third of the RNA, ≈550 nucleotides (3). To determine the RNA features recognized by S4, we have been exploring its binding to both the messenger and ribosomal RNA recognition sites. Our hope has been that similarities between the two sites will come into focus and give us substantial insight into the RNA features recognized by the protein.

S4 RECOGNIZES A COMPLEX PSEUDOKNOT STRUCTURE IN THE mRNA

Specific binding of S4 with the leader sequence of the alpha messenger RNA can be detected by a filter binding assay (4,5). A series of mRNA fragments with different 5' and 3' termini have been prepared by in vitro transcription with T7 RNA polymerase. The smallest fragment giving the maximum binding constant is 139 nucleotides, although a fragment extending from 19 to 69 still binds S4 with detectable specificity. A secondary structure for this RNA fragment has been proposed on the basis of "structure mapping" experiments (6), and is shown in Figure 1.

To confirm this RNA folding scheme, a series of compensatory mutations have been made, and their affinities for S4 measured by the filter binding assay (7). An example is shown in Figure 2. Changing G49-G50 to CC lowers the S4 binding affinity by about ten fold, as does an alteration of C100-C101 to GG. A combination of the two sequence changes in the same RNA restores the S4 affinity to wild type levels, providing strong evidence that nucleotides 49-50 base pair with 100-101. Similar sets of compensatory sequence changes have confirmed the existence of the three helical regions

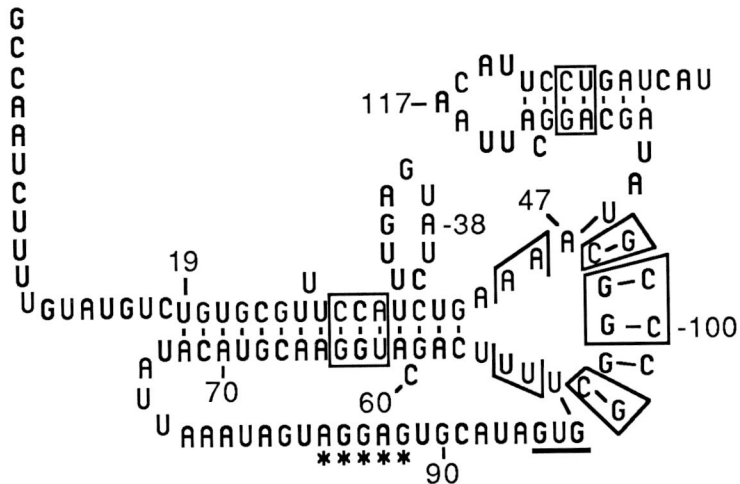

FIGURE 1. Secondary structure of the α mRNA leader sequence. Boxed base pairs have been confirmed by compensatory changes. Open boxes indicate bases which can be changed with no effect on S4 binding. Asterisks and underlining mark the Shine-Dalgarno sequence and initiation codon.

shown in Figure 1.

Two of the helices form a so-called "pseudoknot" structure, i.e. the 3' tail of a hairpin folds back to base pair with the hairpin loop. The Shine-Dalgarno sequence and initiator GUG codon are contained within the single stranded region stretching between the two helices defining the pseudoknot. The structure is highly suggestive in terms of the translational repressor function of S4: stabiliza-tion of the pseudoknot base pairing must prevent the binding of tRNA to the second codon of the message. It is also possible that the S4 - mRNA complex could occlude the ribosome binding site and slow the initial ribosome - mRNA binding step.

Within the pseudoknot structure, bases C41-A47 may potentially pair with U53-G59. If all of these pairs form it is sterically impossible for more than four C-G pairs to form within the pseudoknot helix A47-U53 - G97-U103. The compensatory changes we have made show that at least five pseudoknot base pairs do form, and we also find that mutations at bases A45-A46 or U54-U55 have virtually no effect on S4 binding. Therefore we have drawn the potential C41-A47 – U53-G59 helix as partially disrupted; the extent of pairing within this helix is not yet known. A third helix, entirely within the coding region of the mRNA, is also required for S4 recognition.

The sets of compensating base changes which we have made in the mRNA establish the secondary structure of the S4 recognition site. We do not yet know what specific contacts the protein makes with the RNA within this region. It does appear that a complex, three dimensional RNA structure is required to maintain the recognition site.

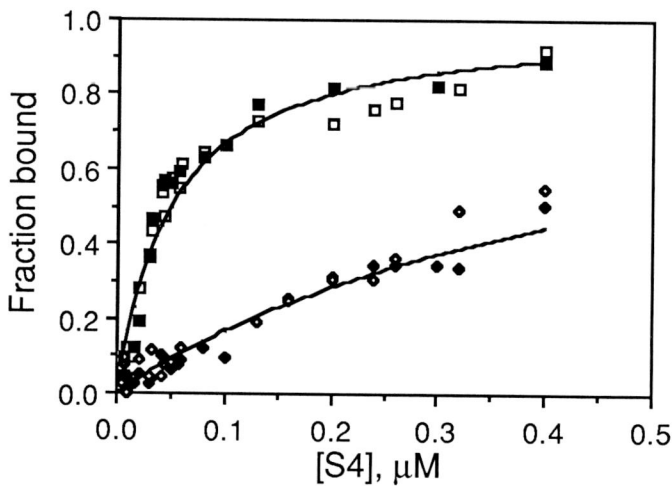

Figure 2. Binding of S4 to wild type and mutant α mRNA leader sequences. S4 – RNA complexes were detected by a nitrocellulose filter binding assay as described (4). , 139 nucleotide wild type sequence. , G49-G50 → CC. ♦, C100-C101 → GG. ■, G49-G50 → CC and C100-C101 → GG. The lines are binding isotherms calculated with equilibrium constants of 20 μM^{-1} (upper curve) or 2.0 μM^{-1} (lower curve).

S4 REQUIRES TWO NON-CONTIGUOUS DOMAINS IN THE RIBOSOMAL RNA

We have measured S4 binding affinities for a number of 16S rRNA fragments. As expected from earlier nuclease protection studies (3), fragments containing the 5' domain bind as well as the intact 16S rRNA ($K \approx 2 \times 10^7$ M^{-1}). The region between nucleotides 38 and 500 appears to contain all the recognition features (Figure 3). Binding of fragments with internal deletions of specific helices suggests that two regions of the rRNA are important for recognition: the helix centered on base 465, and the structure extending from 116 to 288. These regions must either stabilize a structure required for recognition, or themselves be sites of direct protein-RNA contact. In either case, S4 must be stabilizing the three dimensional structure of a large RNA domain.

Expert-Bezançon and Wollenzien have proposed a folding for the 16S rRNA within the 30S subunit, based on extensive crosslinking data (8). The "upper" and "lower" subdomains are in parallel planes, which are hinged by the helix containing base pair 52-358 and sandwich S4 in between. This model brings the two regions defined by our studies as critical for S4 binding into close proximity.

SIMILARITIES CAN BE MISLEADING

After proposing the mRNA folding shown in Figure 1, we carefully searched for structures it might have in common with the 5' domain of the 16S rRNA. The striking similarity shown in Figure 4 was discovered. Six base pairs are nearly identical between the two helices, and both helices have an unusual feature at the same position (a single base bulge or a G-A juxtaposition). However, we have now deleted the indicated helix from the 16S rRNA and find no change in the S4 affinity, and three base pairs of the messenger helix have been changed, again with no effect on S4 binding. We conclude that the similarity is entirely

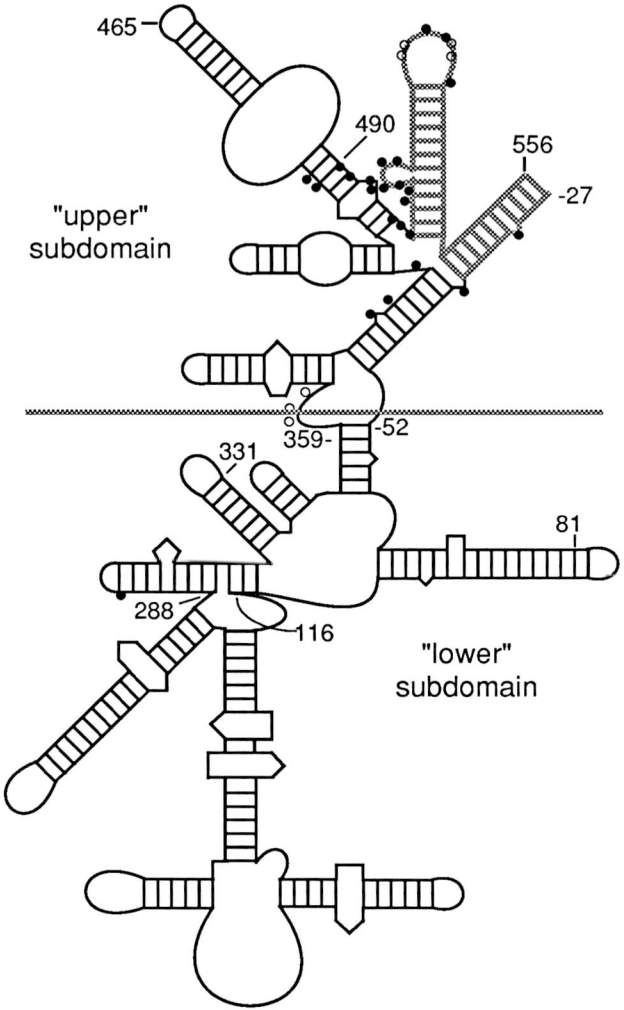

Figure 3. The 5' domain of the 16S rRNA. The phylogenetically conserved secondary structure is shown (9). Dark lines indicate the region retaining full S4 binding affinity. Dots are placed by nucleotides protected (filled) or enhanced (open) in their reactivity towards cleavage reagents in the presence of S4, as determined by Stern, Wilson, & Noller (10).

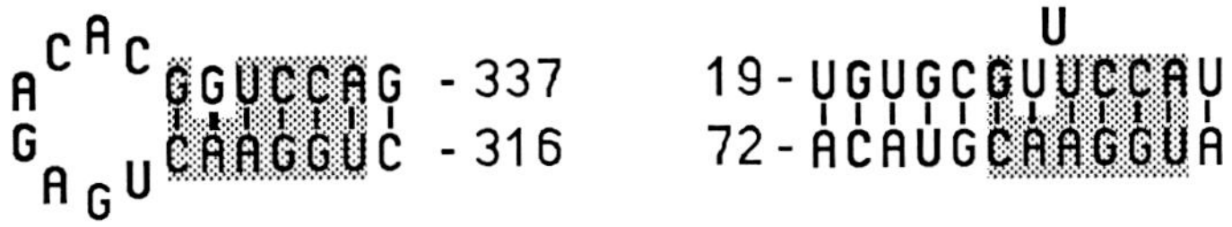

```
     C A                                     U
   A   C  GGUCCAG  -337      19- UGUGCGUUCCAU
  G   A U CAAGGUC  -316      72- ACAUGCAAGGUA
      G
```

Figure 4. Similar helices from the mRNA and rRNA. The structure on the right, from the α mRNA (see Figure 1), is compared with a helix from the 16S rRNA (left). Similar bases are shaded.

coincidental. We are able to find other impressive similarities between mRNA and rRNA regions required for S4 recognition, and are currently testing their significance.

FOOTPRINTING DOES NOT ACCURATELY MAP S4 - RNA CONTACTS

The S4 interaction with intact 16S rRNA has been extensively studied by "footprinting" techniques (10). The V1 nuclease, specific for helical RNA structures, and a variety of chemical reagents reacting with single stranded nucleotides were used. A cluster of protections were observed in a limited region of the 5' domain, and only one protection elsewhere in the 5' domain (Figure 3). These data suggested that S4 interacts principally, if not exclusively, with the cluster of strong protections. Our binding studies with rRNA fragments also indicate that S4 binds within the 5' domain, but nucleotides 500 to 544 can be deleted with no effect S4 binding affinity, even though this sequence is strongly protected. The footprinting and fragment binding experiments are obviously sensing different aspects of the S4 - rRNA complex. Possible reasons for the discrepancy between the two sets of experiments are:
Size exclusion- A protein may protect a larger area from cleavage than it actually contacts. The reagent may be too large to cleave right up to the edge of the protein - RNA contact area (very likely in the case of a nuclease) or the protein may restrict the diffusion of the reagent near the RNA.

Conformational effects- A binding protein may induce conformational changes in an RNA over a substantial distance, and these changes may alter the reactivity of nucleotides towards footprinting reagents. Substantial changes in RNA cleavage several base pairs distant from an ethidium intercalation site have been documented (11). If S4 is stabilizing a large RNA structure, extensive protections may be induced for this reason.
Silent interactions- While a number of nuclease and chemical reagents are available for footprinting experiments, they certainly do not probe all possible RNA - protein contact sites. The single protection in the "lower" subdomain could therefore reflect substantial interactions.

ARE THERE SIMILARITIES BETWEEN THE mRNA AND rRNA BINDING SITES?

Although the S4 recognition sites in the mRNA and rRNA have very different secondary structures, sequence and structural similarities can be found. We propose that the two RNAs use different overall secondary and tertiary foldings to place a similar array of recognition features in the same three dimensional configuration in each. It should also be kept in mind that specific S4 binding is only a factor of ≈ 50 tighter than non-specific. This level of specificity may be achieved with only a few interactions, distributed sparsely in a large RNA structure. Similarities may therefore be difficult to detect. We expect that fundamental questions about RNA - protein recognition mechanisms will be answered as we resolve the precise features recognized by S4 in each RNA.

REFERENCES

1. Nomura, M, Gourse, R, Baughman, G (1984). Regulation of the synthesis of ribosomes and ribosomal components. Ann Rev Biochem 53:75.
2. Thomas, MS, Bedwell, DM, Nomura, M (1987). Regulation of α operon gene expression in E. coli: a novel form of translational coupling. J

Mol Biol 196:333.
3. Ehresmann, C, Stiegler, P, Carbon, P, Ungewickell, E, Garrett, RA (1980). The topography of the 5' end of 16-S RNA in the presence of absence of ribosomal proteins S4 and S20. Eur J Biochem 103:439.
4. Deckman IC, Draper DE (1985). Specific interaction between ribosomal protein S4 and the α operon messenger RNA. Biochemistry 24:7860.
5. Deckman IC, Draper DE (1987). S4-α mRNA translation repression complex: thermodynamics of formation. J Mol. Biol. 196:313.
6. Deckman IC, Draper DE, Thomas, M (1987). S4-α mRNA translation repression complex: secondary structures of the RNA regulatory site in the presence and absence of S4. J Mol Biol 196:323.
7. Tang CK, Draper DE, ms. in preparation.
8. Expert-Bezançon, A, Wollenzien, PL (1985). Three dimensional arrangement of the Escerichia coli 16 S ribosomal RNA. J Mol Biol 184:53.
9. Noller, HF (1984). Structure of ribosomal RNA. Ann Rev Biochem 53:119.
10. Stern, S, Wilson, RC, Noller, HF (1986). Localization of the binding site for protein S4 on 16S ribosomal RNA. J Mol Biol 192:101.
11. White, SA, Draper, DE (1987). Single base bulges in small RNA hairpins enhance ethidium binding and promote an allosteric transition. Nucleic Acids Res 15:4049.

THE SPECIFICITY OF THE RNA BINDING ACTIVITY OF XENOPUS TRANSCRIPTION FACTOR IIIA[1]

Paul J. Romaniuk, Isabel Leal de Stevenson and Qimin You

Department of Biochemistry and Microbiology,
University of Victoria,
Victoria, BC Canada V8W 2Y2

ABSTRACT Xenopus transcription factor IIIA (TFIIIA) acts not only as a positive enhancer of 5S RNA gene transcription, but also binds to the resulting 5S RNA to provide a stable storage particle in immature oocytes. We have been studying the interaction of TFIIIA with 5S RNA, specifically to determine what features of the 5S RNA sequence and structure are necessary for TFIIIA binding. This paper summarizes the results that we have obtained in the past several years.

INTRODUCTION

Transcription factor IIIA binds to an internal control region (ICR) in the Xenopus 5S RNA gene in the first step of a process that leads to the expression of these genes (1). In addition, TFIIIA binds to the 5S RNA transcript to form a 7S ribonucleoprotein storage particle in the cytoplasm of immature oocytes, which stabilizes the 5S RNA until it is required for ribosome assembly later in development (2,3). Interest in how the protein is able to bind to two conformationally distinct nucleic acid species has been enhanced by recent discoveries about the structure of TFIIIA. Analysis of the cDNA sequence for the protein indicates the presence of nine imperfect sequence repeats that have two invariant cysteine and two invariant histidine residues (4,5). The fact that the protein also contains nine molar equivalents of Zn^{2+} ions has lead to the proposal that co-ordination of the cysteines and histidines to a Zn^{2+} ion results in the formation of a "finger" structure (5). Each finger is thought to interact in an independent fashion with DNA, a feature that is essential to the role of TFIIIA as a transcription factor. The unusual conformation of the protein

[1]This work was supported by a grant from NSERC

fingers may also aid in the binding of the protein to RNA (6). Recently, it has become obvious that finger structures are a common motif in the putative nucleic acid binding domains of a wide range of eukaryotic regulatory proteins (6,7).

Our lab has been interested in studying the interaction of TFIIIA with 5S RNA. We have determined the thermodynamic and kinetic parameters for the equilibrium leading to the formation of 7S RNP using a nitrocellulose filter binding assay (8). The protein binds to *Xenopus* oocyte 5S RNA with a dissociation constant (K_d) of 1 nM; complex formation is favoured by both enthalpy and entropy. The ionic strength dependence of the K_d indicates that approximately 5 ionic bonds are formed between TFIIIA and 5S RNA in the complex, although almost 70% of the free energy of binding is derived from non-electrostatic interactions (8). The binding of TFIIIA to RNA is highly specific: the affinity of the protein for tRNA is over 100 times lower than it is for *Xenopus* 5S RNA. However, TFIIIA does bind to a variety of eukaryotic 5S RNA molecules with roughly equal affinity (8-10), suggesting that the protein is interacting with highly conserved features of 5S RNA structure and/or sequence.

Footprinting experiments using a variety of structure probes (8,11-14), revealed that TFIIIA primarily interacts

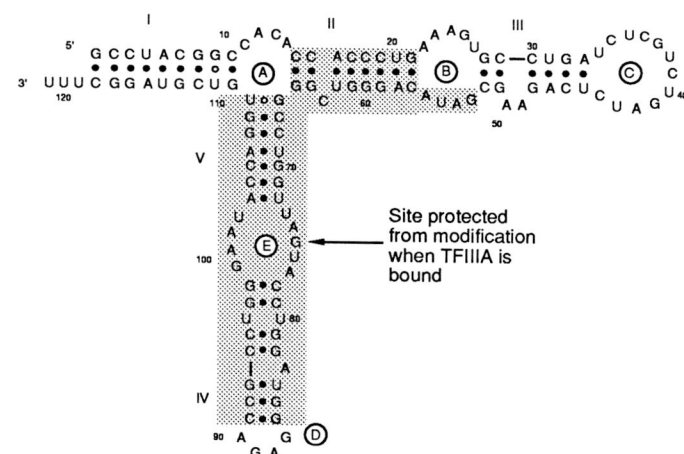

Figure 1. Secondary structure of *Xenopus* 5S RNA showing the TFIIIA footprint area.

with the helix II/loop B and helix IV/loop E/helix V structural domains of the 5S RNA (Figure 1). However, relatively little is known about functionally important regions within this binding site in comparison to knowledge about similar regions of the TFIIIA binding site on the 5S gene. Using site-directed mutants of 5S RNA, we have been investigating the RNA molecule in order to determine those structural features which are essential for the binding of TFIIIA.

RESULTS

The Borders of the TFIIIA Binding Site on 5S RNA

In order to determine the relative importance of the two "halves" of the TFIIIA binding site, we measured the association constants for TFIIIA binding for a series of 5' and 3' deletions of the 5S RNA (15). As the data in Figure 2 show, deletion from either terminus past helix I results in a significant decrease in TFIIIA binding affinity. How-

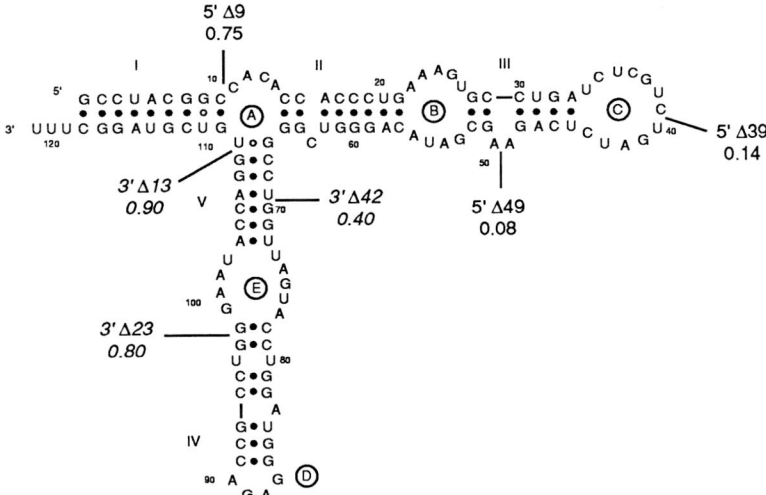

Figure 2. Relative TFIIIA Binding Affinities of 5' and 3' Deletion Mutants of *Xenopus* 5S RNA. The endpoint of each mutant is indicated by the solid line.

ever, it appears that deletion past helix II from the 5' terminus has a greater effect on TFIIIA binding than the removal of the entire helix IV/loop E/helix V domain, suggesting that TFIIIA makes stronger interactions with the 5' domain of the binding site. In contrast, similar experiments showed that it is the 3' end of the ICR which is more important for the binding of TFIIIA to DNA (16). Creating deletions in RNA in many cases has structural consequences: for example, structural studies on the 5S RNA deletion mutants demonstrated that the 5'Δ39 mutant undergoes significant structural rearrangement of nucleotides 30-66, which might also contribute to the decrease in TFIIIA binding observed for this mutant (15).

Differential Affinity of the Somatic and Oocyte 5S RNAs for TFIIIA

In *Xenopus* there are two types of 5S RNA genes: the oocyte-specific gene which is expressed only in oocytes, and the somatic-specific gene which is expressed exclusively in somatic tissues. The two genes differ in sequence at six

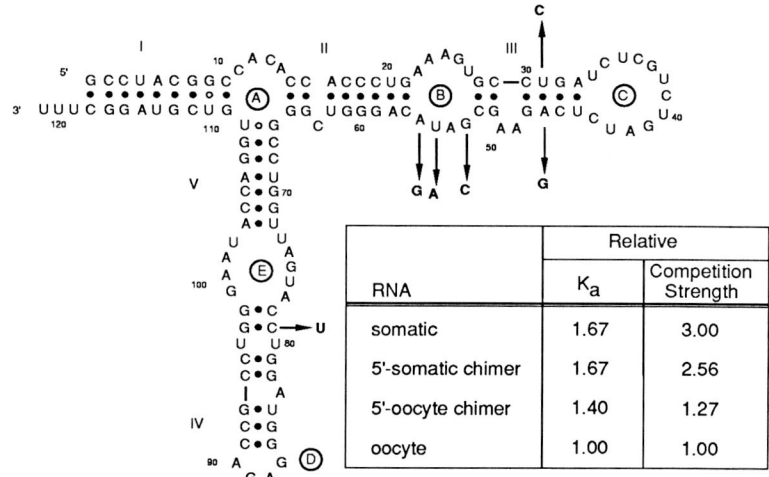

RNA	K_a	Relative Competition Strength
somatic	1.67	3.00
5'-somatic chimer	1.67	2.56
5'-oocyte chimer	1.40	1.27
oocyte	1.00	1.00

Figure 3. *Xenopus* oocyte 5S RNA, showing the six somatic-specific substitutions. Binding data for the two wild type and two chimeric 5S RNAs is also shown.

nucleotide locations (Figure 3), and there had been some evidence to suggest that TFIIIA had a higher affinity for the somatic vs. the oocyte gene (16). In order to investigate the affinity of the protein for the two 5S RNA species, we constructed the two wild type genes, and two chimeric genes: a 5'-somatic chimer carrying the 5 somatic specific substitutions in helix III and loop B, and a 5'-oocyte chimer which has only the somatic U substitution at position 79 (15). The results of two binding assays for these four 5S RNA species are shown in Figure 3. The somatic 5S RNA has a three fold higher competition strength compared to the oocyte 5S RNA, and this higher affinity for TFIIIA is conferred by one or more of the substitutions in either helix III or loop B. It seems most likely that it is the somatic-specific substitutions in loop B and not helix III, that enhance the affinity for TFIIIA. This proposal is based upon two observations: first, that the oocyte 5S RNA diverges from the eukaryotic consensus 5S RNA sequence in loop B, and second, that other eukaryotic 5S RNAs that have the consensus sequence in this loop, such as wheat germ 5S RNA, have an affinity for TFIIIA that is similar to the *Xenopus* somatic 5S RNA (8).

The Role of Highly Conserved Single Stranded Residues of 5S RNA in the Binding of TFIIIA.

   TFIIIA has approximately equal affinity for a variety of eukaryotic 5S RNA molecules (8-10). If one compares the sequences of those 5S RNAs known to bind to TFIIIA, the most highly conserved sequence elements are present in the single stranded loops of the 5S RNA. Other evidence suggests that TFIIIA may form sequence-specific contacts with the single stranded loops: very strong protection from chemical modification in 7S RNP particles has been observed in loops B and E of the 5S RNA (14). In addition, comprehensive studies on the conformation of *Xenopus* 5S RNA indicate that loop E may form an extended, hydrogen bonded structure consisting of a number of non-canonical base pairs (17). Such an unusual structure may be essential for the interaction with TFIIIA. In order to test the importance of single stranded nucleotides for the binding of TFIIIA, seven of the mutants shown in Figure 4 were constructed. The eighth mutant, 96-101 was constructed to convert the unusual loop E conformation into a Watson-Crick base paired A-type RNA double helix, in order to test whether in the absence of sequence-

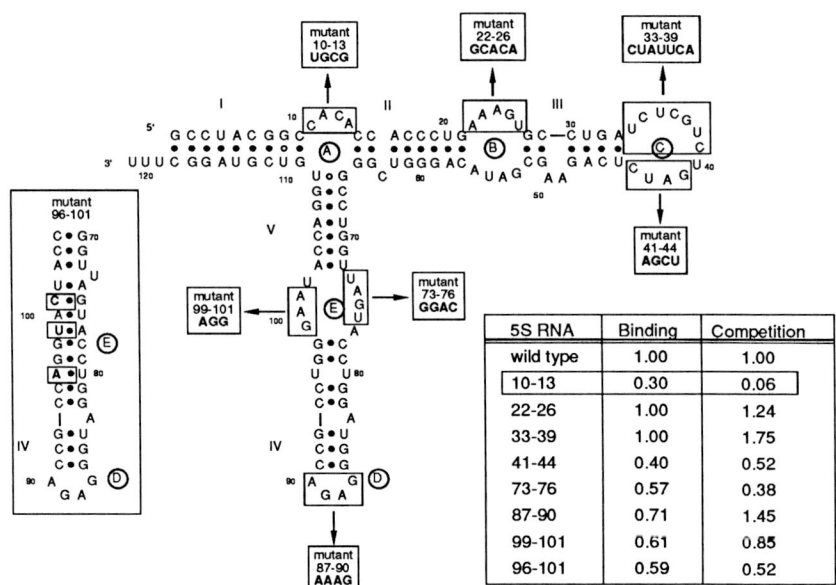

Figure 4. Relative Binding and Competition Strengths for single stranded mutants of *Xenopus* 5S RNA.

specific contacts, a special conformation in this region is recognized by TFIIIA.

The relative association constants and competition strengths for the mutants are summarized in Figure 4. From these data, it is obvious that with a single exception, the highly conserved single stranded nucleotides in *Xenopus* 5S RNA can be substituted with little effect on TFIIIA binding. Indeed, complete substitution of the sequence in loops B, C, D and E affect both the association constant and competition strength by a factor of 2 or less. Only substitution of the nucleotides in loop A has a dramatic effect on the interaction of the 5S RNA with TFIIIA, by decreasing the binding affinity by over three fold, and by decreasing the competition strength by almost twenty fold. Normally, such a result would be interpreted as indicating that substitution of the loop A nucleotides had resulted in the loss of specific RNA-protein contacts. However, five independent determinations of the TFIIIA footprint on *Xenopus* 5S RNA have failed to detect any protection of loop A nucleotides from modification when TFIIIA is bound (8,11-14). Both results can be rationalized if the nucleotides in loop A are critical to the conformation of the 5S RNA, the mutation thus altering the structure of the RNA molecule. In fact, it has been proposed that loop A forms a hinge that controls co-axial stacking of the helical domains of the 5S RNA (14). Further work with point mutants in loop A will be required in order to understand how these nucleotides are involved in the tertiary folding of the 5S RNA, and exactly how this folding might affect the binding of TFIIIA .

The Interaction of TFIIIA with the Base Paired Stems of 5S RNA

The lack of strong sequence-specific contacts between TFIIIA and the single stranded nucleotides of 5S RNA, coupled with the fact that the protein also binds specifically to the ICR in 5S DNA, suggests that TFIIIA may be a double stranded RNA binding protein. In order to assess the role of the double helical stems of *Xenopus* 5S RNA in the binding of TFIIIA, we have begun to construct a series of mutants like the ones shown in Figure 5. To date we have

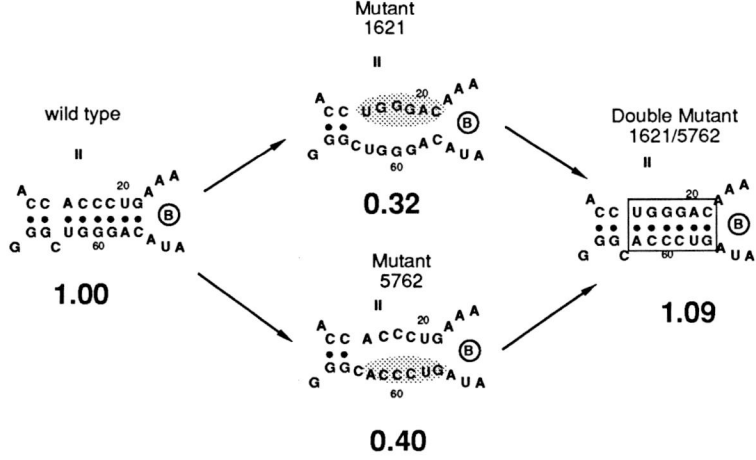

Figure 5. Effect of mutations in helix II on TFIIIA binding affinity.

created such mutants in helix II (nucleotides 16-21:57-62), helix IV (nucleotides 78-81:95-98) and helix V (nucleotides 67-70:105-108), and measured their association constants relative to the wild type 5S RNA. The data obtained indicate that breaking the base pairs in helices II and V reduces TFIIIA binding affinity by approximately 2.5 fold, while restoring base pairing by "flipping" the base pairs restores full binding activity. These results show that for those base pairs tested in helices II and V, the integrity of base pairing is important for TFIIIA binding, but the sequence of the base pairs is inconsequential. In contrast, disrupting the base pairs 78-81:95-98 in helix IV has little effect on TFIIIA binding. We are currently continuing this study, concentrating particularly on the base pairs which close loops, since these may represent special "hinges" where both base pairing and sequence may be important for TFIIIA binding.

DISCUSSION

When we began this work three years ago, relatively little was known about the RNA binding activity of TFIIIA, certainly in comparison to what was known about the DNA binding activity of the protein. The binding site for TFIIIA on 5S RNA had been determined by footprinting, but functionally important regions within this site had not been identified. Our studies with 5S RNA mutants have been designed to identify those nucleotides which are essential for TFIIIA binding.

With the various mutants that have been constructed, we have tested over 50% of the nucleotides in 5S RNA and so far have not found any evidence that TFIIIA forms strong sequence-specific contacts with the 5S RNA. This result is in contrast to the interaction of TFIIIA with DNA, where a number of essential nucleotides have been identified within the ICR, and is also unlike other RNA-protein interactions, for example the binding of R17 coat protein to its translational operator in R17 RNA (18). Although tests of all of the nucleotides within the TFIIIA binding site on 5S RNA have not been completed, our data suggest a model whereby TFIIIA uses a combination of the unique tertiary structure of 5S RNA with many weak sequence-specific contacts in order to achieve the required specificity in its RNA binding activity. That TFIIIA should display different modes of binding in interacting with RNA vs. DNA is not surprising. The nine nucleic acid fingers of TFIIIA differ in sequence, having evolved in an independent fashion. While all of the fingers presumably retained a general ability to bind to nucleic acids, it seems likely that some fingers evolved specifically to interact with 5S DNA, and others evolved specifically to interact with 5S RNA.

REFERENCES

1. Sakonju, S., Brown, D. D., Engelke, D., Ng, S.-Y., Shastry, B. S., Roeder, R. G. (1981). The binding of a transcription factor to deletion mutants of a 5S rRNA gene. Cell 23: 665.
2. Picard, B., Wegnez, M. (1979). Isolation of a 7S particle from Xenopus laevis oocytes: a 5S RNA-protein complex. Proc. Natl. Acad. Sci. USA 76: 241.
3. Pelham, H. R. B., Brown, D. D. (1980). A specific transcription factor that can bind either the 5S RNA gene or 5S RNA. Proc. Natl. Acad. Sci. USA 77: 4170.
4. Brown, R.S., Sander, C., Argos, P. (1985). The primary structure of transcription factor TFIIIA has 12 consecutive repeats. FEBS Lett. 186: 271.
5. Miller, J., McLachlan, A.D., Klug, A. (1985). Repetitive zinc-binding domains in the protein transcription factor IIIA from Xenopus oocytes. EMBO J. 4: 1609.
6. Berg, J.M. (1986). Potential metal-binding domains in nucleic acid binding proteins. Science 232: 485.
7. Evans, R.M., Hollenberg, S.M. (1988). Zinc fingers: Gilt by association. Cell 52: 1.
8. Romaniuk, P.J. (1985). Characterization of the RNA binding properties of transcription factor IIIA of Xenopus laevis oocytes. Nucl. Acids Res. 13: 5369.
9. Pieler, T., Erdmann, V. A., Appel, B. (1984). Structural requirements for the interaction of 5S rRNA with the eukaryotic transcription factor IIIA. Nucl. Acids Res. 12: 8393.
10. Andersen, J., Delihas, N. (1986). Characterization of RNA-protein interactions in 7S RNP from Xenopus laevis oocytes. J. Biol. Chem. 261: 2912.
11. Pieler, T., Erdmann, V. A. (1983). Isolation, characterization of a 7S RNP particle from mature Xenopus laevis oocytes. FEBS Letters 157: 283.
12. Andersen, J., Delihas, N., Hanas, J. S., Wu, C.-W. (1984). 5S RNA structure and interaction with

transcription factor A. 2. Ribonuclease probe of the 7S RNP from *Xenopus laevis* immature oocytes and RNA exchange properties of the 7S particle. Biochemistry 23: 5759.
13. Huber, P.W., Wool, I.G. (1986). Identification of the binding site on 5S rRNA for the transcription factor IIIA: Proposed structure of a common binding site on 5S rRNA and on the gene. Proc. Natl. Acad. Sci. U.S.A. 83: 1593.
14. Christiansen, J., Brown, R.S., Sproat, B.S., Garrett, R.A. (1987). *Xenopus* transcription factor IIIA binds primarily at junctions between double helical stems and internal loops in oocyte 5S RNA. EMBO J. 6: 453.
15. Romaniuk, P.J., Stevenson, I.L., Wong, H.-H.A. (1987). Defining the binding site of *Xenopus* transcription factor IIIA on 5S RNA using truncated and chimeric 5S RNA molecules. Nucl. Acids Res. 15: 2737.
16. Wormington, W. M., Bogenhagen, D. F., Jordan, E., Brown, D. D. (1981). A quantitative assay for *Xenopus* 5S RNA gene transcription *in vitro*. Cell 24: 809.
17. Romaniuk, P.J., Stevenson, I.L., Ehresmann, C., Romby, P., Ehresmann, B. (1988). A comparison of the solution structures and conformational properties of the somatic and oocyte 5S rRNAs of *Xenopus laevis*. Nucl. Acids Res. 16: 2295.
18. Romaniuk, P.J., Lowary, P., Wu, H.-N., Stormo, G., Uhlenbeck, O.C. (1987). The RNA binding site of R17 coat protein. Biochemistry 26: 1563.

ULTRASTRUCTURAL ANALYSIS OF RNA SPLICING IN VIVO[1]

Ann Beyer[2], Milan Jamrich[3], and Yvonne Osheim

Department of Microbiology, Cancer Center,
and Department of Biology
University of Virginia
Charlottesville, Virginia 22908

ABSTRACT We have analyzed the ribonucleoprotein structure and RNA processing patterns of nascent premessenger RNA transcripts on actively transcribing insect genes visualized in the electron microscope. Evidence is presented that the process of loop formation and removal on nascent transcripts represents splicing, and some details of the in vivo splicing process are described. Specifically, one pair of genes illustrates that there is a preferred, but non-obligatory, order for intron removal and another gene shows that the bulk of the spliceosome particle is removed with the intron upon splicing.

INTRODUCTION

When the transcripts of RNA polymerase II genes are dispersed using hypotonic spreading conditions and visualized in the electron microscope, it can be seen that they become complexed with protein as they are synthesized (1,2). In general, the ribonucleoprotein (RNP) structures formed on nascent transcripts are one of two types: RNP particles of ~25 nm diameter, which are superimposed on an RNP fibril of ~5 nm width. We find that the RNP particles are positioned at specific sites on the transcripts (1),

[1]This work was supported by grants from the ACS (NP-493B) and NIH (GM39271) to AB.
[2]Recipient of an ACS Faculty Research Award.
[3]Present address: Laboratory of Molecular Genetics, NICHD, National Institutes of Health, Bethesda, MD 20892

which correspond to 5' and 3' splice junction sequences on the genes that we have been able to identify in chromatin spreads (3). This direct evidence that 25 nm RNP particles correspond to spliceosome intermediates on the identified chorion genes (3) is supported by the observation that they behave exactly as predicted for spliceosome intermediates on other unidentified genes (4) based on results from in vitro splicing studies (reviewed, 5-7). The particles are first deposited at two separate sites on the transcripts, presumably corresponding to 5' and 3' splice junction regions (4). In some cases, the two separate particles appear to coalesce, forming a larger ~40 nm (spliceosome) particle and looping out the intervening sequence (2-4). Within one minute of formation (4), the intron loop is removed from the transcript, resulting in genes whose transcripts do not increase regularly in length from the initiation to the termination site as expected, but rather are shortened in discrete steps corresponding to intron loop length (2,4).

In our initial report on the phenomena of loop formation and removal, we referred to the process as "cleavage" of nascent transcripts (2) rather than splicing. We discussed, however, that it was difficult to distinguish between cleavage and splicing because the RNA loops that are most amenable to analysis, (i.e., those that clearly show transcript shortening upon removal), are generally greater than 3 kb in length and significantly longer than a typical exon of ~0.3 kb or less (8). When loop removal occurs, it is difficult to determine whether the relatively short 5' exons are spliced onto the remaining transcript or cleaved off with the loop, especially given the variation in RNA compaction between different transcripts when they are dispersed in this manner (±10%, ref. 2). In addition, because the spliceosome particle is removed with the intron loop (as will be reported herein) there is no indication in the ligated exons of the former position of the intron loop. Direct evidence for splicing (i.e., cleavage and ligation) would be the demonstration that an internal loop can be removed prior to a 5' terminal loop. Such an example is presented in this report. Our observations are also consistent with many biochemical studies of splicing which indicate that there are preferred but non-obligatory orders for intron removal (e.g. 9). The data presented here confirm and extend a recent report from our lab on the ultrastructural analysis of splicing in vivo (4).

METHODS

The following methods have been previously described: the Miller chromatin spreading procedure as applied to Drosophila embryos (1) and Calliphora fat bodies (2), and RNP fibril map construction (1,2).

RESULTS

We use an RNP fibril mapping technique to display the transcripts of a gene graphically such that similar sequences are aligned and RNP structural features are shown. These maps result in different patterns of RNP particles and loops that are characteristic for the transcripts of a given gene and identical for sister copies of the same gene (1-4). The particle/loop patterns can be used to derive possible exon-intron maps for the genes (3,4). The two genes shown in Figures 1 and 2 undoubtedly represent sister copies of the same gene based on their identical RNP fibril mapping patterns, as seen in Figure 3. The micrographs in Figures 1a and 2a are from a chromatin spread of Calliphora erythrocephala (blow fly) larval fat body, which is a polyploid tissue. Perhaps coincidently, the DNA templates of both genes have been broken, as evidenced by free ends of DNA at one end of the transcribed region on both genes. The visualized portions represent the 3' end of the gene in both cases, while the initiation ends are broken off. The longest transcripts on these genes, of 4-5 μm, correspond to ~30 kb of RNA, as determined by a previously-described relationship between μm and kb in these chromatin preparations (2). These long transcripts include two longer introns of ~17.7 and 4.8 kb, which are obvious as looped structures with a particle at their base on various transcripts (e.g. transcripts 2 and 4 on Figure 1 display the 17.7 kb intron loop and transcripts 8 and 10 show the 4.8 kb intron loop). In Figures 3a and c, the RNP fibril maps for these genes display the sequence-specific location of these loops, (which are shown in a linear form with a dotted line connecting the 2 loop-base sites). Figure 3b presents our proposed exon-intron structure for these transcripts based on their RNP morphology. In addition to the two large introns, four small introns are shown at the sites of the additional closely-spaced particles at two positions on the transcripts. Evidence for the existence of the small introns is the occurrence of closely-spaced 25 nm

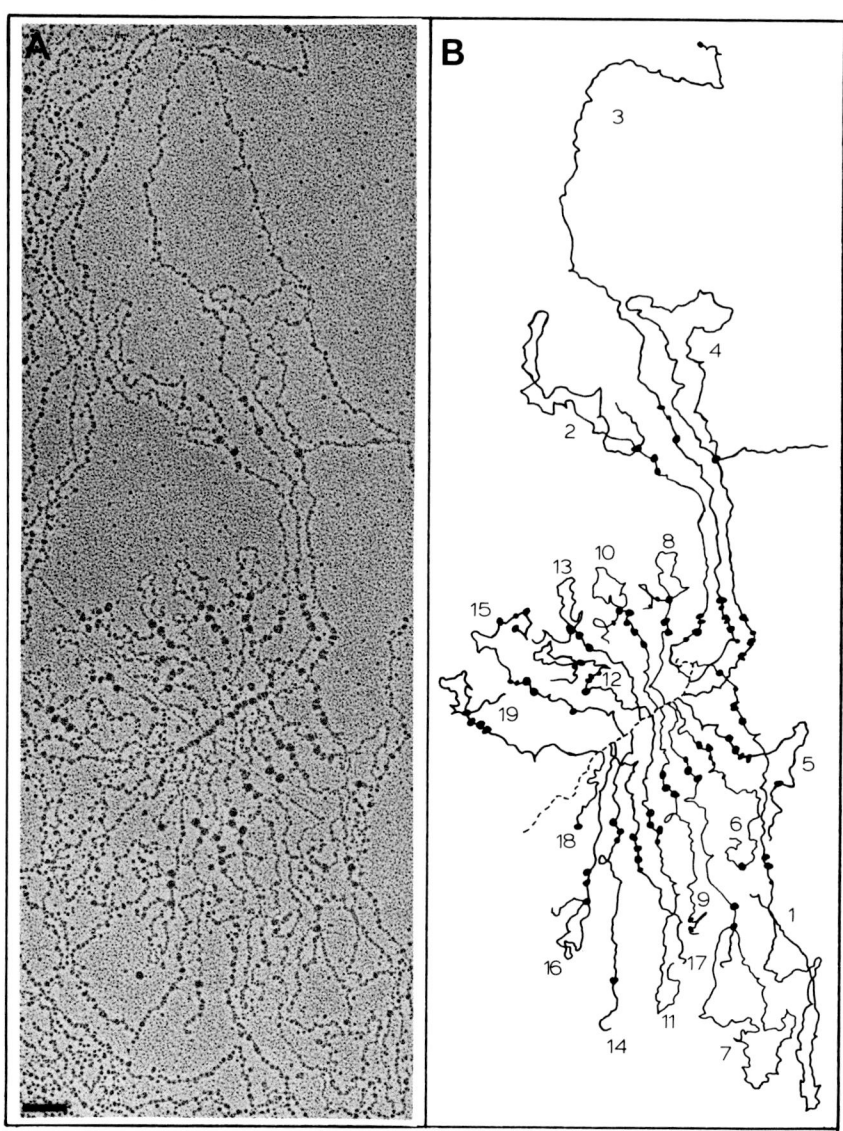

FIGURE 1.(a) Electron micrograph of a <u>Calliphora erythrocephala</u> gene displaying RNP particles, loops and splicing on nascent transcripts. Bar = 0.2 μm.
(b) Interpretive tracing of EM; (---) DNA. RNP fibril map shown in Figure 3.

Ultrastructure Analysis of RNA Splicing   137

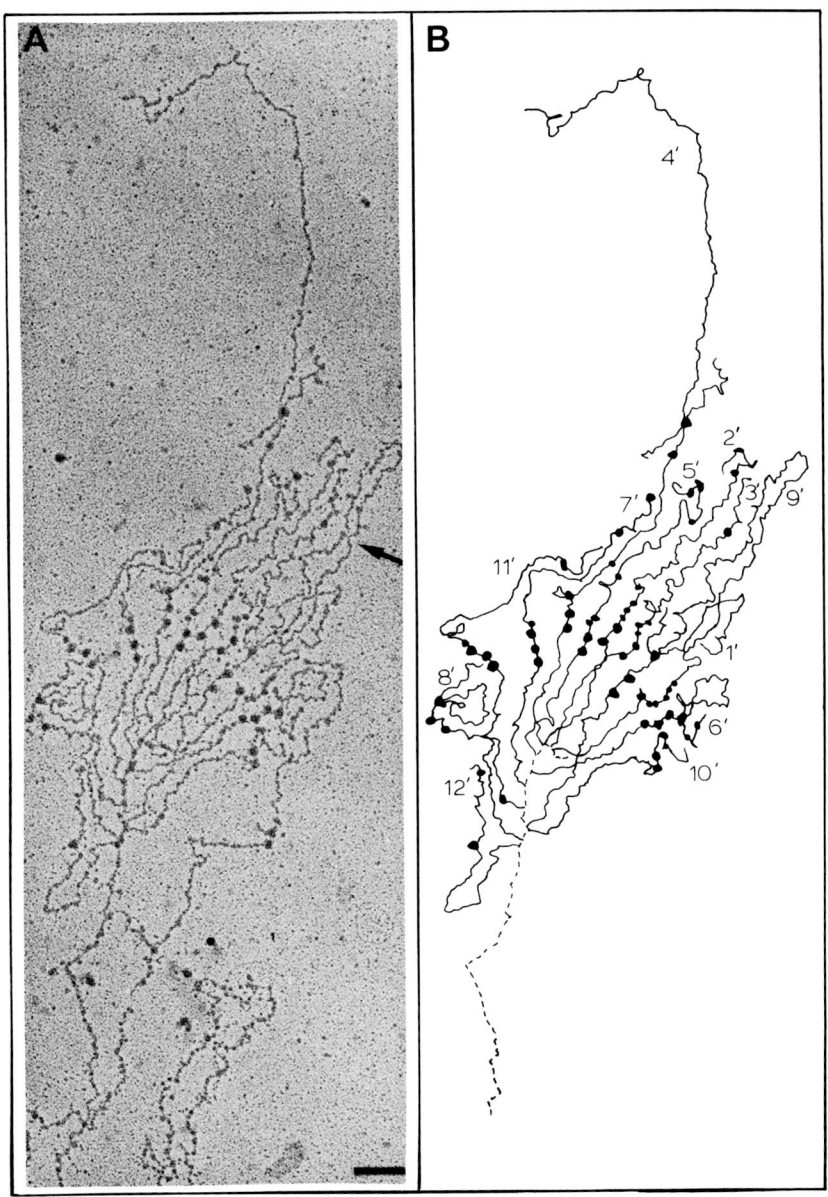

FIGURE 2. Same as Figure 1. Arrow indicates transcript 9' which has lost the internal ~4.8 kb intron.

particles (e.g. transcript 1'), which are replaced by half as many 40 nm particles (e.g. transcript 6'), which have a tendency to disappear from the transcripts. This is exactly the behavior of the particles on the Drosophila chorion gene transcripts, which have short introns that are not visible as loops (3). As will be shown in Figure 4, splicing (loop

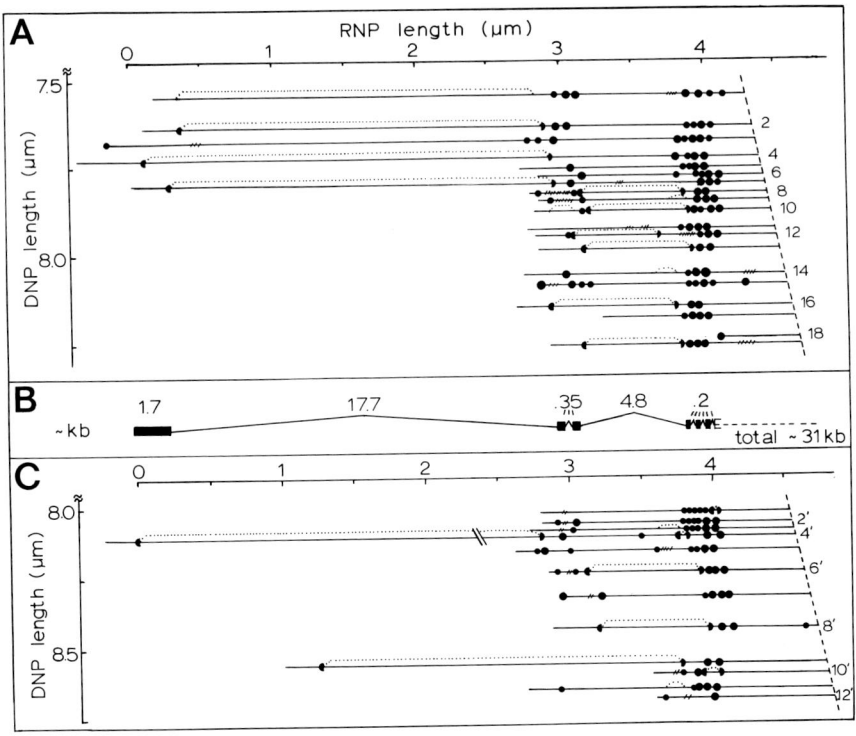

FIGURE 3.(a) RNP fibril map for gene shown in Figure 1. The spacing of the transcripts on the vertical axis corresponds to their position on the DNA. Transcripts are shown as straight lines of the appropriate length, with corresponding sequences approximately aligned (5' ends to left) (4). (●) RNP particles of 2 sizes: ~25 and 40 nm; (///) hairpin loops; (····) open loops. (b) Proposed exon-intron structure of transcripts. (■) Exons; (———) intron; (----) difficult to assign as intron or exon. (c) RNP fibril map for gene shown in Figure 2. The large loop on transcript 4' is broken.

removal) is accompanied by the loss of the 40 nm spliceosome particle from the nascent transcripts. Assuming that the additional particles occur at sites of short introns, and that their absence indicates that splicing has occurred, there are 10 different splicing intermediates that can be distinguished on these nascent transcripts, ranging from all introns present (transcripts 1-3 of Figure 1) to only one small intron remaining (transcript 12' of Figure 2). Although there is a general 5' to 3' progression, it appears that there is no obligatory order for intron removal in these transcripts, consistent with observations from biochemical approaches (9). Transcript 9' of Figures 2 (arrow) and 3c is particularly interesting because it has apparently lost the internal 4.8 kb intron while retaining the 17.7 kb intron at the 5' end of the RNA. This is a clear example of loop removal representing excision of an internal segment (splicing) rather than cleavage of a 5' terminal fragment.

Figure 4 represents a portion of a Drosophila embryo gene that clearly shows the loss of two spliceosome particles upon loop removal. Transcript 5 has two short RNA loops with large loop-base particles, which are preceded, on transcript 4, by smaller particles at three of the four splice junction regions for the two introns. Transcript 6 has lost the 5'-most loop and is appropriately shorter than expected for its position on the template. There is a small particle at the position of the excised intron on transcript 6 (arrow), but the bulk of the spliceosome particle is gone. Transcript 7 has lost both intron loops and displays no particles in the position of intron excision. These findings are general; that is, we typically see no particles or a small and, presumably, unstable 10-15 nm particle at the site of exon ligation. These _in vivo_ observations agree with results from _in vitro_ splicing systems, demonstrating that the excised intron is released in a complex that is comparable in mass to a spliceosome and contains U5, U6 and probably U2 snRNPs, while the ligated exons are released from the spliceosome in a smaller, more heterogeneous complex with unidentified factors (10).

DISCUSSION

In this and another recent article (4), we present evidence that splicing is the process represented by loop formation and removal on nascent transcripts, and that

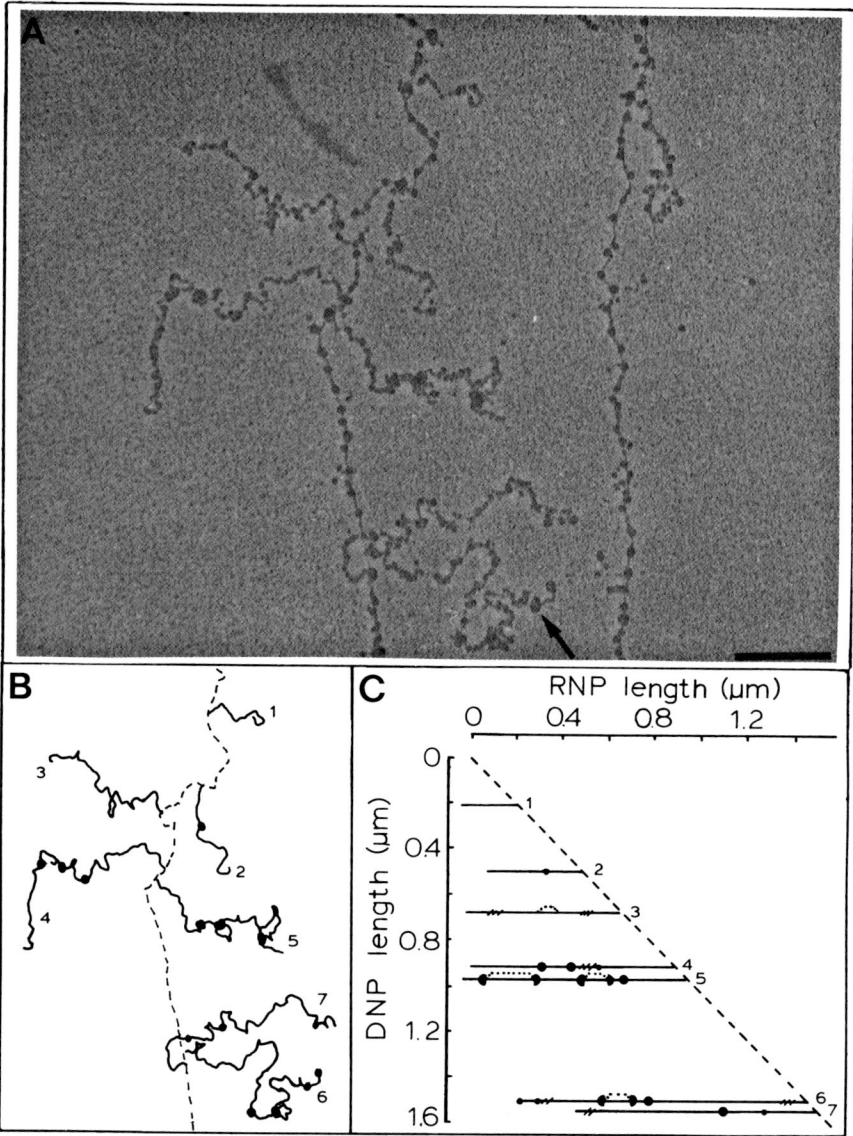

FIGURE 4.(a) Electron micrograph of a D. melanogaster gene. Bar = 0.2 μm. Arrow indicates the small particle left at the site of intron excision on transcript 6.
(b) Interpretive tracing of EM. (c) RNP fibril map.

co-transcriptional splicing occurs with a reasonable frequency (at least 14% of Drosophila embryo genes) (4). The direct evidence for splicing is the observation of this ultrastructure at known intron sites on chorion genes from Drosophila follicle cells (3), and the indirect evidence is the correspondence of what we see to what is expected for splicing based on biochemical studies in vivo and in vitro. For example, particles assemble at both ends of the intron but not in between (cf. 11) and particle assembly precedes loop formation, which is probably the ultrastructural representation of covalent branchpoint formation (4). We also find (4) that the 3' splice region is important in initiating spliceosome assembly (cf. 11,12), although we cannot distinguish ultrastructurally between events at the branchpoint, pyrimidine-rich region and 3' splice site, which are typically within 40 nucleotides of each other (6,7), since each 25 nm particle covers ~100 nucleotides of RNA (3,4). As shown herein, there is a preferred but non-obligatory order for intron removal (cf. 9), the bulk of the spliceosome particle is removed with the exon (cf. 10), and two spliceosomes can co-exist on the same RNA molecule (cf. 13). Our observations are consistent with the possibility that splice site selection may generally precede polyadenylation and that splicing itself is not rare on nascent transcripts. Our findings are in agreement with the first-come-first-served principle of splice site selection as proposed by Aebi et. al. (14) and supported by others (15,16).

Our observations also suggest a simple model for the RNP packaging of nascent transcripts. We suspect that hnRNP proteins (reviewed, 17) completely coat the transcript in a nonspecific manner as it is synthesized and are responsible for the 5 nm wide RNP fibrillar structure. The snRNPs (18) and other required splicing proteins (19) would then form the bulk of the 25 and 40 nm particles associated with splice junction sites. Immuno-EM studies are in progress to test these hypotheses.

REFERENCES

1. Beyer AL, Miller OL Jr, McKnight SL (1980). Ribonucleoprotein structure in nascent hnRNA is nonrandom and sequence-dependent. Cell 20:75.

2. Beyer AL, Bouton AH, Miller OL Jr (1981). Correlation of hnRNP structure and nascent transcript cleavage. Cell 26:155.
3. Osheim YN, Miller OL Jr, Beyer AL (1985). RNP particles at splice junction sequences on Drosophila chorion transcripts. Cell 43:143.
4. Beyer AL, Osheim YN (1988). Splice site selection, rate of splicing, and alternative splicing on nascent transcripts. Genes Dev 2:in press.
5. Padgett RA, Grabowski PJ, Konarska MM, Seiler S, Sharp P (1986). Splicing of messenger RNA precursors. Annu Rev Biochem 55:1119.
6. Green MR (1986). Pre-mRNA splicing. Annu Rev Genet 20:671.
7. Maniatis T, Reed R (1987). The role of small nuclear ribonucleoprotein particles in pre-mRNA splicing. Nature 325:673.
8. Naora H, Deacon NJ (1982). Relationship between the total size of exons and introns in protein-coding genes of higher eukaryotes. Proc Natl Acad Sci USA 79:6196.
9. Zeitlin S, Efstratiadis A (1984). In vivo splicing products of the rabbit beta-globin pre-mRNA. Cell 39:589.
10. Konarska MM, Sharp PA (1987). Interactions between small nuclear ribonucleoprotein particles in formation of spliceosomes. Cell 49:763.
11. Ruskin B, Green MR (1985). Specific and stable intron-factor interactions are established early during in vitro mRNA splicing. Cell 43:131.
12. Frendewey D, Keller W (1985). The stepwise assembly of a pre-mRNA splicing complex requires U-snRNPs and specific intron sequences. Cell 42:355.
13. Christofori G, Frendewey D, Keller W (1987). Two spliceosomes can form simultaneously and independently on synthetic double-intron messenger RNA precursors. EMBO J 6:1747.
14. Aebi M, Hornig H, Padgett RA, Reiser J, Weissman C (1986). Sequence requirements for splicing of higher eukaryotic nuclear pre-mRNA. Cell 47:555.
15. Kedes DH, Steitz JA (1987). Accurate 5' splice-site selection in mouse kappa immunoglobulin light chain premessenger RNA is not cell-type-specific. Proc Natl Acad Sci USA 84:7928.
16. Lowery DE, Van Ness BG (1988). Comparison of in vitro and in vivo splice site selection in kappa-immunoglobulin precursor mRNA. Mol Cell Biol 8:2610.

17. Chung SY, Wooley J (1986). Set of novel, conserved proteins fold pre-messenger RNA into ribonucleosomes. Proteins 1:195.
18. Steitz JA, Black DL, Gerke V, Parker KA, Kramer A, Frendewey D, Keller W (1988). Functions of the abundant U-snRNPs. In Birnstiel ML (ed): "Structure and Functions of Small Nuclear Ribonucleoprotein Particles," Berlin Heidelberg New York: Springer Verlag, p 115.
19. Ruskin B, Zamore PD, Green MR (1988). A factor, U2AF, is required for U2 snRNP binding and splicing complex assembly. Cell 52:207.

STRUCTURAL CHARACTERIZATION OF THE INTRON CONTAINING S25 RIBOSOMAL PROTEIN GENES FROM TWO TETRAHYMENA SPECIES[1]

Jan Engberg, Kirsten Bojsen and Henrik Nielsen.

Department of Biochemistry B, University of Copenhagen, DK-2200 Copenhagen, Denmark

ABSTRACT Tetrahymena thermophila contains a single genomic copy of the S25 ribosomal protein gene whereas two genomic copies were found in T. pigmentosa. Molecular characterization of these genes indicate that they are functional as well as interupted by an intron of about 1 kb in size located 68 bp downstream of the translation start site. In the coding region, only silent substitutions were found so that the amino acid composition is totally conserved among all three gene products. The TAA codon (normally used as stop codon in other organisms) is used once for Gln in the T. pigmentosa genes whereas CAA is used at the corresponding position in the T. thermophila gene. Apart from the consensus dinucleotides GT/AG at the splice site junctions very little similarity was observed among the intron sequences. The polypyrimidine stretch located close to the 3' splice site in introns of most higher eukaryotes was not observed in the Tetrahymena introns which may indicate that the sequence requirements for pre-mRNA splicing in Tetrahymena differ from those of higher eukaryotes studied thus far. Based on comparative sequence analyses it is suggested that the highly conserved sequence regions found at either side of the coding region are implicated in transcription initiation and polyadenylation, respectively.

[1]This work was supported by the Carlsberg Foundation and the Danish Natural Science and Medical Science Research Councils.

INTRODUCTION

The expression of the majority of the ribosomal proteins (r-proteins) is regulated coordinatively in Tetrahymena (1,2). In an attempt to investigate the molecular mechanisms involved in this regulation we have initiated the molecular cloning and nucleotide sequencing of several of the r-protein genes of Tetrahymena (cf. 3).

Ribosomal protein genes are well conserved in evolution and prokaryotic equivalents of the eukaryotic r-proteins can often be identified by sequence comparisons. These similarities are strictly confined to the coding regions of the genes. The ciliated protozoans, including Tetrahymena consist of an evolutionary ancient group of organisms (4,5,6) and it has been rather difficult to define, by means of sequence comparisons to higher eukaryotes, those genomic regions in Tetrahymena that are involved in transcriptional regulation such as the TATA-box and the CAAT-box (cf. 3, 7, 8). In the present study we took advantage of our previously aquired knowledge of the Tetrahymena phylogeny (6) and chose two highly divergent Tetrahymena species for the sequence comparisons assuming that functionally important regions would stand out as highly conserved sequence elements. The T. pigmentosa species chosen to be compared with the one already investigated (T. thermophila, 3), was found to harbor two genomic copies of the S25 r-protein gene. This situation enabled us to make inter- as well as intra-species sequence comparisons.

MATERIALS AND METHODS

Cultures of T. thermophila B 1868 (VII) and T. pigmentosa HG2 (II) were maintained as described previously (9). The molecular cloning and characterization of the T. thermophila S25 cDNA and the corresponding genomic sequence have been published recently (3). The cloning, the dideoxynucleotide sequencing and the primer extention analysis performed in the present study all followed standard recombinant DNA procedures as outlined in the following text. The details of the experiments will be published elsewhere.

RESULTS AND DISCUSSION

Cloning the T. pigmentosa genes.

The strategy used for cloning of the two S25 r-protein genes from T. pigmentosa was based on genomic restriction maps of the genes. These maps were derived by digesting total T. pigmentosa DNA with a variety of restriction enzymes, either alone or in pair-wise combinations, followed by probing the Southern-blots with radioactively labeled restriction fragments prepared from the T. thermophila S25 cDNA clone. Fig. 1 compares the genomic restriction maps around the coding region of the T. thermophila and T. pigmentosa S25 r-protein genes.

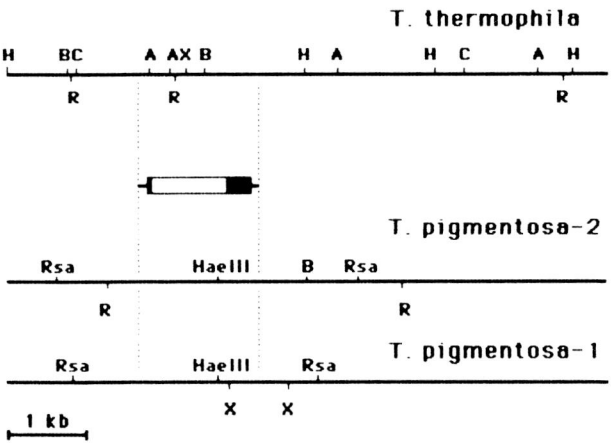

FIGURE 1. Alignment of the genomic restriction enzyme maps around the coding regions for the ribosomal protein S25 in T. thermophila and T. pigmentosa. The dotted lines delimit the transcribed regions including exons (filled boxes), intron (open box) and non-translated regions.
A: AvaII, B: BglII, C: ClaI, H: HindIII, R: EcoRI and X: XbaI.

Based on these data, total T. pigmentosa DNA was digested with either a single restriction enzyme or a combination of enzymes followed by isolation of size fractionated restriction fragments which were subsequently ligated to appropriately digested plasmid vectors and transformed into competent E. coli cells. These enriched plasmid libraries were screened using radioactively labeled T. thermophila cDNA fragments as probes. In the case of gene-1 the following restriction fragments were cloned: the 3.3 kb

RsaI-RsaI, the 2.1 kb RsaI-XbaI and the 0.8 kb XbaI-XbaI fragments. In the case of gene-2 the following fragments were cloned: the 4.1 kb RsaI-RsaI, the 2.7 kb EcoRI-BglII, the 1.5 kb EcoRI-HaeIII and the 1.2 kb HaeIII-BglII fragments. The ExoIII/Mung bean nuclease procedure (10) was then used to generate series of overlapping sub-clones which were subsequently used for the sequencing studies. These studies soon revealed the presence of an intron in the coding region of the T. pigmentosa genes at a position identical to that found in the T. thermophila coding sequence. These results are depicted in Fig. 2 which compares the nucleotide sequence of the coding region of the S25 gene of T. thermophila with that of the T. pigmentosa genes. As of this moment of writing we have not finished the sequencing of a few small internal regions of the T. pigmentosa introns so that the sizes of these can not be described at the nucleotide level, but both of them appear to be very similar in size to the T. thermophila intron. As shown in

Ribosomal protein S25 gene - coding region

```
              Met Gly Val Gly Lys Pro Arg Gly Ile Arg
T.th          ATG GGT GTT GGT AAA CCT AGA GGT ATT AGA
T.pi-1
T.pi-2

              Ala Gly Arg Lys Leu Ala Arg His Arg Lys
T.th          GCC GGT AGA AAA TTG GCT AGA CAC AGA AAG
T.pi-1
T.pi-2

              Asp Glu Arg                    Trp Ala
T.th          GAC GAA AG -  intron  - A TGG GCT
T.pi-1               -  intron  -      C
T.pi-2        T      -  intron  -      C

              Asp Asp Asp Phe Asn Lys Arg Leu Leu Gly
T.th          GAT AAC GAT TTC AAC AAG AGA CTT TTG GGC
T.pi-1
T.pi-2                  C

              Ser Arg Trp Arg Asn Pro Phe Met Gly Ala
T.th          TCC AGA TGG AGA AAT CCT TTT ATG GGT GCT
T.pi-1        T                 C
T.pi-2                          C

              Ser His Ala Lys Gly Leu Val Thr Gly Lys
T.th          TCT CAC GCT AAG GGT TTA GTC ACT GAA AAG
T.pi-1                                          G
T.pi-2            C                             G

              Ile Gly Ile Gly Ser Lys Gln Pro Asn Ser
T.th          ATC GGT ATT GAA TCT AAG CAA CCT AAC TCT
T.pi-1        T       C             T             C
T.pi-2        T       C             T             C

              Ala Val Arg Lys Cys Val Arg Val Leu Leu
T.th          GCC GTC AGA AAG TGT GTT AGA GTT TTG CTC
T.pi-1                                          T G
T.pi-2                                          A T G

              Arg Lys Asn Ser Lys Lys Ile Ala Ala Phe
T.th          AGA AAG AAC TCC AAG AAG ATC GCT GCT TTC
T.pi-1                                      C   C
T.pi-2                                      C   C

              Val Pro Met Asp Gly Cys Leu Asn Phe Leu
T.th          GTT CCT ATG GAT GGT TGC CTT AAC TTC TTA
T.pi-1                C                   T G
T.pi-2                            T T G   T

              Ala Glu Asn Asp Glu Val Leu Val Ala Gly
T.th          GCT GAA AAC GAC GAA GTC TTG GTT GCT GGT
T.pi-1
T.pi-2                              A

              Leu Gly Arg Gln Gly His Ala Val Gly Asp
T.th          CTT GGT AGA CAA GGT CAC GCC GTC GGT GAT
T.pi-1        C
T.pi-2

              Ile Pro Gly Val Arg Phe Lys Val Val Cys
T.th          ATT CCG GGT GTT AGA TTC AAA GTT GTT TGT
T.pi-1                A           T             C   C
T.pi-2                A   C                     C   C

              Val Lys Gly Ile Ser Leu Leu Ala Leu Phe
T.th          GTC AAG GGT ATT TCT CTC TTG GCT CTT TTC
T.pi-1        T           C   C               C
T.pi-2        T           C   C               C

              Lys Gly Lys Gly Lys Arg Ter
T.th          AAG GGT AAG AAG GAA AAG CGT TGA
T.pi-1                                  A A
T.pi-2                                  A A
```

FIGURE 2. Nucleotide sequence and deduced amino acid sequence in the coding region of the ribosomal protein S25 from two <u>Tetrahymena</u> species.

Fig. 2, the coding regions of T. pigmentosa-1 and -2 have 26 and 31 single nucleotide differences, respectively, relative to the T. thermophila gene (23 of which are common to both). All of these changes are silent so that the respective gene products are 100% identical at the amino acid level. Of particular interest is the use of the "termination-codon" TAA in amino acid position 62 in the two T. pigmentosa genes. The use of this codon for Gln has been observed in several protozoans (11-17). Furthermore, the use of the rare arginine codon CGU as the C-terminal codon in T. thermophila was noted previously (3). This observation may not be so significant in view of the fact that T. pigmentosa uses the frequently used arginine codon AGA arginine at this position.

Functionality of the T. pigmentosa genes.

In an initial attempt to determine whether both T. pigmentosa genes were transcribed in vivo, we used a synthetic 18-mer complementary to the region between + 16 and + 33 in a primer extension regiment using total T. pigmentosa mRNA and reverse transcriptase (cf. ref. 3). This experimental set-up resulted in a major extension product which mapped as indicated by the arrow in Fig. 4 (see later). Based on the observation that the two T. pigmentosa genes differ in nucleotide sequence in parts of the 5' non-coding region (see later) we are currently repeating the primer extension experiments using oligonucleotide primers which are specific for either of the two genes. Preliminary results show that both primers give rise to extended products of distinct sizes.

Intron sequence comparisons.

In sharp contrast to the sequence conservation observed in the coding regions, the corresponding intron regions showed an almost complete lack of sequence similarity. Fig. 3 shows a selected region of the intron sequences around the 5' and 3' splice sites, but the impression of an extensive sequence divergence is maintained when larger intron regions are considered. Underlined in Fig. 3 are the dinucleotides GT/AG which appear to be conserved at the splice sites of all introns (cf. 18). At the 5' intron junction in T. thermophila and T. pigmentosa-1, a stretch of 10 bp are identical. However, this similarity does not extend to the T. pigmentosa-2 gene. Apart from the GT nucleotides at the 5' splice site, the Tetrahymena sequences also show identity to the eukaryotic 5' splice consensus sequence (18) at positions -2, -1 and +6. At the 3' splice site, the Tetrahymena sequences show little similarity to the consensus sequence of higher eukaryotes: (Py)$_n$ N C/T AG↓G (18). Notably the stretch of pyrimidines is not present in the Tetrahymena introns. These sequence differences between Tetrahymena and higher eukaryotes make it tempting to investigate the precise sequence requirements of the pre-mRNA splicing reaction in Tetrahymena.

5' - splice junction

```
T.th.    AAG GAC CAA AG GTGAATATTTTTTTAATAACAATCA-
T.pi-1   AAG GAC CAA AG GTGAATATTTCTACATTCTTATTCC-
T.pi-2   AAG GAT CAA AG GTTCATTCTTCTTTTTCATTCTTCA-
```

3' - splice junction

```
T.th.    -TGCCATTATTCTCTTCCTCTCAAAATAAATTTATTAAATT
T.pi-1   -AAAATCATTTATCGAAGATAAATTAATATTAAAAATATCT
T.pi-2   -ATATAAAAAAATAAAAAAATGTTTTAATAAAATTAAAAAT

T.th.    CAATTTTAAATAAAAAG A TGG GCT GAT AAC GAT
T.pi-1   TAAAAATAAAAAAACAG A TGG GCC GAT AAC GAT
T.pi-2   ATATTATTATAATATAG A TGG GCC GAT AAC GAC
```

FIGURE 3. Comparison of three Tetrahymena S25 genomic sequences in the splice junction regions.

Regions of transcription initiation.

We have previously mapped the transcription initiation site for the T. thermophila S25 gene by primer extension and S_1 nuclease experiments (3) and a major start site at position -70 was observed. In recent experiments a minor start site at position -95 was also observed. These sites have been indicated by means of arrows in Fig. 4. Analogous primer extension analyses were performed in the case of T. pigmentosa and the preliminary results indicate the presence of a single major start site at the position indicated by the upward pointing arrow in Fig. 4. It may be noted that the major start sites are located at an identical distance (70 bp) from the translation start site in both species. In Fig. 4, small insertions have been introduced in the sequences in order to maximize the impression of the remarkable similarity which exists within 4 regions located close to the transcription start sites. The number of identical

```
                        BOX 1
T.th.     AAGAAAAGAAATATATAAAGGGGAAATTTAATGAAG-TCT
T.pi-1    ATTAAAATATAAAGATAAAGGGGAAATTTCGTGTGGTTCT
T.pi-2    GAAGAGAAAGAAGAATAAAGGGGAAATTTAATGCGC-TCT


                           BOX 2
T.th.     AATTAAATAAATGTATGTGTATATGGTTTGAATATGCATT
T.pi-1    AATTAAATTAATGTATATATGGATGTTTTCAATATCCATT
T.pi-2    AATTAAATAAGTGTATGTGTGTGTGGTTTGAATATGCATG


              ↓                      ↓
T.th.     TGTATTTAAGATACTTAAAAGCGGTGTGGAAGATTAAAAT
T.pi-1    CAAACTTAAAAGCTAA--------GA-CAAAGATTAAAAT
T.pi-2    GAGAGTTAAAAGCGAA--------GATCGAAGATTAAAAT
                                       ↑

             BOX 3
T.th.     TAAAAGAAAACAAAATAATTAAA--GTAAA-AAATATA--
T.pi-1    TAAAAGAAAACAAAATAAT-AAATAGCAAACAATTATT--
T.pi-2    TAAAAGAAAACAAAATAAT-AAATAGAAAACAAATATAAA


             BOX 4
T.th.     AAAAAGAAAATAAAAATCAAA    ATG
T.pi-1    AAAAAATAAAAAAAAATGAAA    ATG
T.pi-2    AAATAATCAATAAAAATCAAA    ATG
```

FIGURE 4. Comparison of three Tetrahymena S25 genomic sequences in the region of transcription initiation. Arrows indicate major transcription initiation sites as mapped by primer extension analysis. Boxes 1 to 4 demarcate regions of strong similarity.

positions in these sequences are 17 out of 19 (box 1), 32 out of 42 (box 2), 30 out of 30 (box 3) and 27 out of 36 (box 4). Very similar sequences upstream of the translation start site were also observed in a recently completed analysis of the r-protein L1 of T. thermophila (unpublished results). However, similar sequence regions appear not to be present in front of the Tetrahymena actin or histone genes (7,8) which may indicate that they are specific for r-protein genes.

Polyadenylation signals.

Fig. 5 shows the sequences of the 3' portion of the three S25 genes starting from the fifth codon of the terminal coding region. It appears that the thre sequences diverge completely downstream of the TGA stop codon. Such sudden sequence divergence was also noted in the 3' nontranslated portion of the two genomic copies of the histone H4 genes of T. thermophila (7). Closer inspection reveals a region located about 100 bp downstream of the translation stop codon in which the T. thermophila and T. pigmentosa sequences show identity in 26 out of 30 positions (the bracketed region in Fig. 5). Shortly downstream of this

```
T.th.    AAG GAA AAG CGT TGA GCTGCATATTAATAATAATA
T.pi-1   AAG GAA AAG AGA TGA TCTATGTATGTATGTGATAT
T.pi-2   AAG GAA AAG AGA TGA ACAAATTAATTATGTCTTTT

T.th.    CTTCCAATAATTAAATAAAAAGAGTATTTGCCTTAATATA
T.pi-1   CAAATTTANANNAATTATAAGTATCCATATCTAATTTTTA
T.pi-2   AATATTACAACCAAAATAAAACAACTATAACTTCTCTTTA

T.th.    TATTATTAACTGTAAATGAAAATAT----ATAGTAATTAT
T.pi-1   TTACTATATGTTATTGTCGATTTATATGCATAGTCATTAT
T.pi-2   TAATTTCAACAACAAAACAAATATAAATTAAACTGATCAA

T.th.    TTAAATTTACTCTCTCAAACATCTTTTTAAATTAATTAAT
T.pi-1   TTAAATATACTCCTTCAAA----TCTATTCATTCATTCTA
T.pi-2   TCTCTGTATTCGAAATCTAATTTATATCCTTCTAATTGTA

                 ↓
T.th.    CAATCAATCAATCAATT
T.pi-1   TCCTTCATTCTACTACT
T.pi-2   AATCGAAATATATAGTC
```

FIGURE 5. Comparison of three Tetrahymena S25 genomic sequences in the region of polyadenylation.

region, a series of short repeats is observed in both sequences (underlined in Fig. 5). Unfortunately the site of polyadenylation has been determined only in the case of T. thermophila (3, arrow in Fig. 5), so that the evidence for suggesting a functional relationship between the conserved region, the short repeats and the site of polyadenylation is weak at present. For the sake of completeness it should be mentioned that a sequence element similar to the bracketed region of Fig. 5 is observed in the T. pigmentosa-2 gene at a position about 60 bp further downstream of that of the T. pigmentosa-1 gene (not indicated in Fig. 5). In addition, we have recently completed the sequencing of the 3' ends of the cDNA's of three new r-protein genes of T. thermophila (L1, L21 and L37, unpublished results). In these sequences, an element corresponding to the bracketed region of Fig. 5 can not be observed in its full size, but a sub-set element: TATTTAAA (cf. Fig. 5) was present in all cases at approximately the same distance from the respective translation stop codon and the site of polyadenylation. This sequence element, therefore, may be the Tetrahymena equivalent of the polyadenylation signal (AATAAA) found in higher eukaryotes (cf. 19).

REFERENCES

1. Dreisig H, Andreasen PH, Kristiansen K (1984). Regulation of ribosome synthesis in Tetrahymena pyriformis. Eur J Biochem 140:469.
2. Dreisig H, Andreasen PH, Kristiansen K (1984). Regulation of ribosome synthesis in Tetrahymena pyriformis. Eur J Biochem 140.477.
3. Nielsen H, Andreasen PH, Dreisig H, Kristiansen K, Engberg J (1986). An intron in a ribosomal protein gene from Tetrahymena. EMBO J 5:2711.
4. Nanney DL (1986). In The molecular biology of ciliated protozoa. Gall JG (ed) Academic Press, New York. p. 1.
5. Sogin ML, Elwood HJ, Gunderson JH (1986). Evolutionary diversity of eukaryotic small-subunit rRNA genes. Proc Natl Acad Sci USA 83:1383.
6. Sogin ML, Ingold A, Karlok M, Nielsen H, Engberg J (1986). Phylogenetic evidence for the acquisition of ribosomal RNA introns subsequently to the divergence of some of the major Tetrahymena groups. EMBO J 5: 3625.
7. Horowitz S, Bowen JK, Bannon GA, Gorovsky MA (1987). Unusual features of transcribed and translated regions of the histone H4 gene family of Tetrahymena thermophila. Nucl Acids Res 15:141.
8. Cupples CG, Pearlman RE (1986). Isolation and characterization of the actin gene from Tetrahymena thermophila. Proc Natl Acad Sci USA 83:5160.
9. Nielsen H, Simon E, Engberg J (1985). Updating rDNA restriction enzyme maps of Tetrahymena reveals four new intron-containing species. J Protozool 32(3):480.
10. Instruction manual. Bluscript Exo/Mung DNA sequencing system. Stratagene. San Diego CA.

11. Horowitz S, Gorovsky MA (1985). An unusual genetic code in nuclear genes of Tetrahymena. Proc Natl Acad Sci USA 82:2452.
12. Helftenbein E (1985). Nucleotide sequence of a macronuclear DNA molecule coding for α-tubulin from the ciliate Stylonychia lemnae. Special codon usage: TAA is not a termination codon. Nucl Acids Res 13:415.
13. Preer JR, Preer LB, Rudman BM, Barnett AJ (1985). Deviation from the universal code shown by the gene for surface protein 51A in Paramecium. Nature (London) 314-:188.
14. Nomoto M, Imai N, Saiga H, Matsui T, Mita T (1987). Characterization of two types of histone H2B genes from macronuclei of Tetrahymena thermophila. Nucl Acids Res 15(14):5681.
15. Caron F, Meyer E (1985. Does Paramecium primaurelia use a different genetic code in its macronucleus. Nature 314:185.
16. Hanyu N, Kuchino Y, Nishimura S (1986) Dramatic events in ciliate evolution: alteration of UAA and UAG termination codons to glutamine codons due to anticodon mutations in two Tetrahymena tRNA^{Gln}. EMBO J 5(6):1307.
17. Andreasen PH, Dreisig H, Kristiansen K (1987). Unusual ciliate-specific codons in Tetrahymena mRNA are translated correctly in a rabbit reticulocyte lysate supplemented with a subcellular fraction from Tetrahymena. Biochem J 244:331.
18. Padgett RA, Grabowski PJ, Konarska MM, Seiler S, Sharp PA (1986). Splicing of messenger RNA precursors. Ann Rev Biochem 55:1119.
19. Gil A, Proudfoot NJ (1984). A sequence downstream of AAUAAA is required for rabbit β-globin mRNA 3'-end formation. Nature 312:473.

INTRON RECOGNITION IN PLANTS

Greg Goodall, Karin Wiebauer and Witold Filipowicz

Friedrich Miescher Institute
Postbox 2543, Basel CH4002, Switzerland

ABSTRACT

We have compared the splicing of pre-mRNA in plants and animals by transfecting the soybean leghemoglobin and human β-globin genes into both plant protoplasts and HeLa cells. Analysis of the splicing pattens obtained indicates that plants and animals differ in the mechanism of 3' splice site selection. A survey of animal and plant gene sequences indicated that the polypyrimidine tracts present at the 3' end of animal introns are not present in plant introns. Many plant introns, however, are AU-rich throughout the entire intron. To test the significance of the UA bias in plant introns we have chemically synthesized a hypothetical plant gene of mostly random sequence, but incorporating 5' and 3' splice site consensus sequences flanking a AU-rich "intron", embedded in a GC-rich sequence. This gene is efficiently spliced at the predicted sites in plant protoplasts. Using a series of unique restriction sites within the synthetic sequence, we have made a series of modifications to test the contribution of various sequence elements to intron recognition.

INTRODUCTION

In a compilation of plant intron splice site and putative branch point sequences Brown (1) has pointed out that the plant 5' and 3' splice site consensus sequences are almost identical to those of animal introns. Yet when animal genes are expressed in plant cells, the RNA products are usually not processed by the plant splicing aparatus. Barta et al. (2) have shown that none of the four introns of human growth hormone are removed when the gene is expressed in transgenic tobacco or sunflower and van Santen and Spritz (3) found that IVS1 of human α-globin is not removed in tobacco cells. Presumably some feature of plant introns is lacking in animal introns so that they are not processed by the plant splicing system. The problem does not seem to be associated with the branch point. Although no plant branch point has been determined to date, we can predict that plants and animals have similar requirements in this regard since, firstly, plant introns do not contain a highly conserved sequence such as the UACUAAC branch point sequence found in yeast introns, secondly the sequence in the region of U2 snRNA that is proposed to base pair with pre-mRNA is identical in animal and plant U2 snRNAs (4 and references therein), and thirdly, the survey of plant intron sequences by Brown indicated that the consensus of potential plant branch site sequences is essentially the same as in animals.

Conversely, the splicing of several plant introns has been analysed in HeLa nuclear extracts: the introns of wheat amylase, pea legumin (5), maize bronze locus, oat phytochrome type 3, bean phaseolin (3) and both introns of rubisco small subunit (6). All these introns were faithfully processed except that of phaseolin, which was however processed in HeLa cells in vivo, leading to the suggestion that the sequence requirements for splicing in animals are more relaxed than in plants.

We have analyzed the splicing of the three introns of soybean leghemaglobin in HeLa cells and extracts and of the two introns of human β-globin in plant cells (7). Our results point to a difference in the mechanism of 3′ splice site selection. We suggest that the elevated U and A content of plant introns is important for their processing and have devised a model system to test this proposal.

RESULTS

Processing of Soybean Leghemoglobin in a HeLa Cell Nuclear Extract.

Various segments of the soybean leghemoglobin c3 (Lb) gene (8), containing three introns of 120, 100 and 226 nt, were subcloned into pGEM vectors containing the SP6 and T7 polymerase promotors. The recombinant plasmids, linearized at the appropriate restriction sites, served as templates for the in vitro synthesis of splicing substrates containing either one or two introns. A time dependent accumulation of shorter RNA species, corresponding to spliced products, was observed only with RNA substrates containing IVS2. There was no indication of the excision of either IVS1 or IVS3.

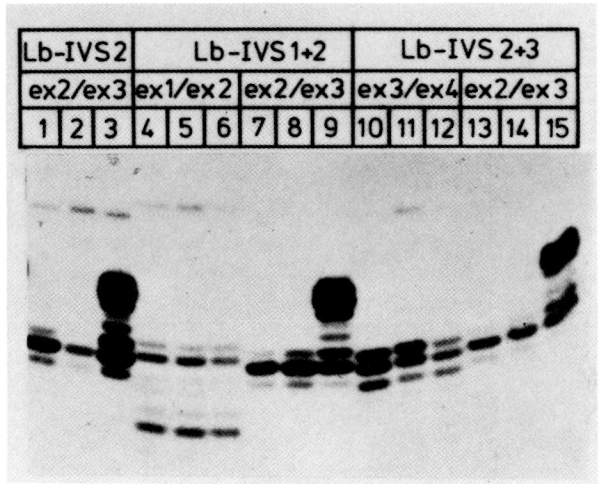

FIGURE 1. Analysis of in vitro processed gene transcripts by the oligonucleotide/nuclease S1 protection assay. The products of in vitro splicing reactions carried out in the presence (lanes 3,6,9,12,15) or absence (lanes 2,5,8,11,14) of ATP were analyzed with 5'-end labelled 30 nt oligonucleotide probes complementary to each of the three exon/exon junctions in correctly spliced Lb mRNA. Control analyses were carried out with untreated in vitro transcripts (lanes 1,4,7,10,13). The nuclease S1-resistant products were separated on a 20% gel. The Lb gene transcripts and oligonucleotide probes used for each assay are indicated above the lanes.

The products of the in vitro splicing reactions were further analyzed by the nuclease S1 protection assay, using 5'-end-labelled oligonucleotides as hybridization probes (Fig. 1). Each of the 30 nt probes was complementary to sequences spanning one of the three exon/exon junctions in Lb mRNA. Only RNAs from which the introns had been removed precisely would be expected to protect the probes against S1 digestion. Processing of Lb-RNAs containing either one or two introns was analyzed. Only the exon2-exon3 probe, diagnostic of a correct IVS2 excision, was protected by in vitro processed Lb RNAs (lanes 3, 9 and 15); RNAs incubated with the extract in the absence of ATP did not protect this probe (lanes 2, 8 and 14). These data confirm the results of direct gel analysis; only the second intron of Lb pre-mRNA is excised in the HeLa cell nuclear extract. The accuracy of IVS2 processing was additionally verified by sequence analysis. The sequence at the splice junction was found to correspond to the predicted cDNA sequence TTTTGGATTG/GTACGTGATT (8), confirming that the intron was removed precisely.

Processing of Leghemoglobin pre-mRNA in Transfected HeLa Cells.

In order to verify that the failure of the mammalian system to process IVS1 and IVS3 of the soybean gene was not an artifact of an in vitro assay, the splicing of Lb pre-mRNA was also analyzed in HeLa cells. The Lb gene was inserted into a mammalian expression vector which was transfected into HeLa cells. RNA was isolated 50 hr after transfection and its processing analyzed by RNase A/T1 mapping. The results of the mapping experiments, which are described in detail in Wiebauer et al. (7), are summarized schematically in Fig. 2. Of the three introns of leghemoglobin pre-mRNA only IVS2 was correctly and efficiently processed in HeLa cells. The 5' splice sites of the remaining two introns were faithfully recognized but correct processing of the 3' sites took place only rarely (IVS1) or not at all (IVS3); cryptic 3' splice sites were used instead.

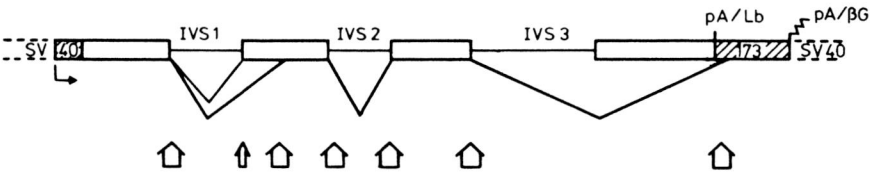

FIGURE 2. Schematic summary of the results of RNAse A/T1 mapping of the leghemoglobin pre-mRNA expressed in HeLa cells. Arrow thickness indicates the approximate efficiency of splice site usage. The cryptic 3' splice site used during processing of IVS3 is located in adjacent vector sequences.

Processing of Human β-globin pre-mRNA in Transfected Plant Protoplasts.

To study the fate of a mammalian pre-mRNA in plant cells the human β-globin (hβ) gene was inserted into the plant expression vector pDH51 and the resulting plasmid introduced into protoplasts of O. violaceus or N. tabacum by electroporation (7). RNA was isolated 24 hr after transfection and the processing of the two hβ gene introns (IVS1 of 130 nt and IVS2 of 850 nt) analyzed by RNase mapping (Fig. 3). RNA isolated from HeLa cells transfected with the gene inserted into a mammalian expression vector served as a positive control for a correctly spliced hβ pre-mRNA. IVS1 was not spliced in the plant protoplasts. IVS2 processing occured at a low level, but involved the authentic 5' splice site and a cryptic 3' splice site. The exact location of this site was confirmed by nuclease S1 analysis (see Wiebauer et al. (7) for details). This splicing of IVS2, although inaccurate, is the only established example of heterologous intron processing in plant cells.

Surveys of Plant and Animal Intron Sequences and Construction of a Model Plant Gene.

The failure to select the authentic 3' splice sites in the Lb introns 1 and 3 in HeLa cells as well as the pattern of β-globin IVS2 splicing in plant cells, which involves the authentic 5' but a cryptic 3' splice site, indicate that it is the process of 3' acceptor site recognition which differs in plant and animal mRNA splicing reactions. What structural features distinguish plant and animal introns, and in particular their 3' acceptor regions? We have compared the occurence of pyrimidine and purine nucleotides at positions -50 to +20 (3'-terminal G of IVS corresponding to position -1) in 179 plant and 800 vertebrate intron/exon borders (Fig. 4). In vertebrate introns the frequency of pyrimidines increases gradually from about position -30, reaching 75-90% at positions -15 to -5 (Fig. 4A). This 3'-proximal region represents the so-called polypyrimidine tract. The high AU/GC ratio of plant introns has been noted by many investigators. In most plant introns, the AU content is about 15-20% higher than that of the neighbouring exons, (for some exceptions see, e.g., ref. 9) with U residues contributing more to this AU excess than A residues. The transition from the AU-rich intron sequence (average AU content in positions -50 to -5 in 179 plant introns is 70.1%) to the downstream exon sequence

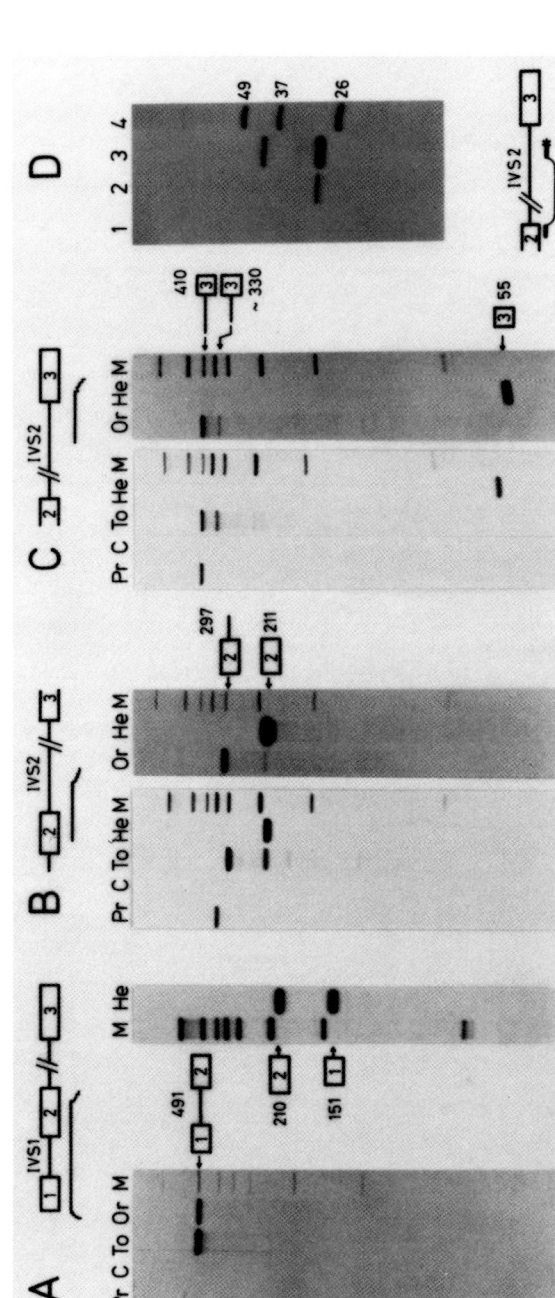

FIGURE 3. **(A – C)** RNase A/T1 mapping of the human β-globin pre-mRNA expressed in plant protoplasts and **(D)** analysis of the IVS2 processing by nuclease S1 assay. Analyses were carried out on RNA from non-transfected (**lanes C**) or transfected (**lanes To**) tobacco protoplasts, transfected *Orychophragmus* protoplasts (**lanes Or**) and transfected HeLa cells (**lanes He**). **Lanes Pr:** undigested probe. The probes, described in detail in Wiebauer et al. (7), are shown schematically above the autoradiographs. **(D)** RNA from control (**lane 1**) or transfected (**lane 3**) tobacco protoplasts and the unspliced SP6 RNA (**lane 2**) were hybridized to a 49 nt "cDNA" oligonucleotide probe specific for the exon 2/cryptic-3'-splice-site junction (see lower diagram) and nuclease S1-resistant fragments analyzed on a 20% gel. Lane 4 contains oligonucleotide size markers.

(average AU content in positions +2 to +20 is 51.4%) is shown in Fig. 4D. A similar transition is not observed when vertebrate intron/exon borders are inspected (Fig. 4B).

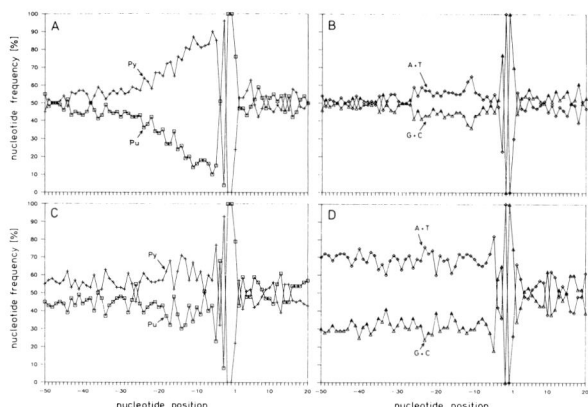

FIGURE 4. Distribution of pyrimidine (Py) and purine (Pu) (A and C) and A+T and G+C (B and D) nucleotides in the vicinity of the 3' splice sites in 800 vertebrate (A and B) and 179 plant (C and D) introns. The 3'-terminal G residue of IVS corresponds to position -1.

That the bias for U and A in plant introns is significant for their splicing is suggested by the fact that the animal introns which do not undergo splicing in plant cells (all four growth hormone gene introns, and IVS1 of the human α- and β-globin genes) have a low AU content (26.5 - 54.6%) whereas the β-globin IVS2 which is partially processed in plant cells (splicing between an authentic 5' site and a cryptic 3' site located within the intron) contains 68.9% AU. Furthermore it is the more AU-rich part of the β-globin IVS2 which undergoes excision in plant cells.

We are attempting to test the hypothesis that the AU bias in plant introns is important for their splicing and in addition to test whether there is some as yet unrecognized structural component that is required for recognition. To do this we have synthesized a model gene, of mostly arbitrary sequence, but incorporating the features we believe to be necessary for splicing of its hypothetical intron. The structure of the model gene is shown schematically in Fig. 5. An efficient plant promotor (the 35S promotor from

Cauliflower Mosaic Virus) lies upstream of the test sequence, which is fused to part of the soybean leghemoglobin gene including intron 1 and flanking exon sequences, followed by the polyadenylation region of Cauliflower Mosaic Virus. The leghemoglobin sequences serve as an internal control for splicing efficiency and mRNA stability. The test sequence consists of 186 nucleotides of chemically synthesized DNA of a sequence that is arbitrary but incorporates 5' and 3' splice site consensus sequences flanking an AU-rich region, embedded in a slightly GC-rich context. The RNA that is expressed from this gene in tobacco protoplasts is quite efficiently spliced. We are currently testing several variations of the sequence to test the contribution of the AU bias of the intron to the splicing efficiency.

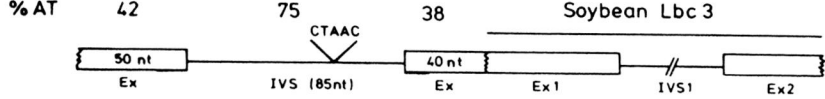

FIGURE 5. Schematic representation of the model gene.

CONCLUSIONS

Although not all plant introns are processed by the HeLa nuclear extract, all introns tested so far are recognized by HeLa cells in vivo, the location of the 3' splice site that is selected being dependent on the presence of a polypyrimidine tract. However, of the seven animal introns that have been tested for splicing in plants, only human β-globin IVS2 is processed, and inaccurately at that. Thus plants require in their introns a sequence element that is not usually present in animal introns. Examination of intron sequences, together with our results with the model intron, suggest that the plant-specific element is contributed by a U- and A-rich sequence.

REFERENCES

1. Brown J (1986). A catalogue of splice junction and putative branch point sequences from plant introns. Nucl. Acids Res. 14:9549.
2. Barta A, Sommergruber K, Thompson D, Hartmuth K, Matzke MA, Matzke AJM (1986). The expression of a nopaline

synthase-human growth hormone chimaeric gene in transformed tobacco and sunflower callus tissue. Plant Mol Biol 6:347.
3. van Santen VL, Spritz RA (1987) Splicing of plant pre-mRNAs in animal systems and vice versa. Gene 56:253
4. Vankan P, Filipowicz W (1988) Structure of U2 snRNA genes of Arabidopsis thaliana and their expression in electroporated plant protoplasts. EMBO J 7:791.
5. Brown JWS, Feix G, Frendewey D (1986). Accurate in vitro splicing of two pre-mRNA plant introns in a HeLa cell nuclear extract. EMBO J 5:2749.
6. Hartmuth K, Barta A (1986). In vitro processing of a plant pre-mRNA in HeLa cell nuclear extract. Nucl Acids Res 14:7513.
7. Wiebauer K, Herrero JJ, Filipowicz W (1988) Nuclear pre-mRNA processing in plants: Distinct modes of 3'-splice-site selection in plants and animals. Mol Cell Biol vol 8, in press
8. Brisson N, Verma DPS (1982). Soybean leghemoglobin gene family: Normal, pseudo, and truncated genes. Proc Natl Acad Sci USA 79:4055.
9. Klösgen WB, Gierl A, Schwarz-Sommer S, Saedler H.(1986). Molecular analysis of the waxy locus of Zea mays. Mol Gen Genet 203:237.

DIFFERENTIAL ACCUMULATION OF U4 SMALL NUCLEAR RNAS DURING CHICKEN DEVELOPMENT[1]

Gina M. Korf and William E. Stumph

Department of Chemistry and Molecular Biology Institute, San Diego State University
San Diego, California 92182

ABSTRACT There are only two U4 small nuclear RNA genes per haploid genome in the chicken. One gene, U4B, encodes the previously known chicken U4B RNA. The second gene, U4X, codes for a sequence variant of U4 RNA that was unknown prior to the cloning of the gene. By using U4B-specific and U4X-specific hybridization probes, U4B RNA and U4X RNA have been detected in each of seven adult and three embryonic chicken tissues examined. Notably, the relative accumulation of U4B and U4X RNA varies in both a developmental and tissue-specific manner. Since U4 RNA is a component of the spliceosome, it is conceivable that U4B and/or U4X RNA accumulation patterns may be relevant to mechanisms involved in the alternative splicing of pre-mRNAs.

INTRODUCTION

The genome of the domestic chicken contains two distinct genes, U4B and U4X, that code for U4 RNA (1). These genes are closely linked and each is present in only a single copy per haploid genome (1). This contrasts with the situation in other metazoans which have between 5 and 1000

[1]This work was supported by NSF DCB-8615964.

members within each gene family coding for each of the major snRNAs (2). The chicken U4B RNA gene codes for the chicken homolog of U4B RNA, a major sequence variant of U4 RNA previously identified in the chicken and in other vertebrates (3). The U4X RNA gene, on the other hand, codes for a novel U4 RNA sequence variant that was unknown prior to the cloning of the gene. This U4X RNA gene differs from the U4B RNA gene at seven nucleotide positions in the RNA coding region, and also at several nucleotide positions in the 5'-flanking DNA within transcriptional regulatory sequences that are otherwise conserved among the chicken U1, U2, and U4B snRNA genes (1). To begin to investigate the potential significance of the U4B and U4X RNA sequence heterogeneities, we have examined the in vivo accumulation of the U4B and U4X RNA gene products during chicken development. We report that U4B RNA and U4X RNA are both present in each of three embryonic and seven adult tissues examined. Interestingly, the relative accumulation of the two RNA species was found to vary in a temporal and tissue-specific manner.

MATERIALS AND METHODS

Isolation of total cellular RNA was by the method described in Maniatas et al. pg. 194 (4). RNA was electrophoresed in 6% polyacrylamide, 7M urea gels and transferred to Gene Screen Plus membrane filters by electroblotting for 1h at 60V, then 45 min at 30V. The chamber buffer consisted of 12mM Tris, 6mM sodium acetate, and 0.3mM EDTA (pH 7.5). Filters were prehybridized at 45°C for 4h in 0.9M NaCl, 1% SDS, 0.5mM EDTA, 10mM Tris (pH 7.6), 8 µg/ml E. coli DNA, and 20 µg/ml ATP. Oligonucleotide hybridization probes were labeled with [γ-^{32}P] ATP and T4 polynucleotide kinase. Hybridization was in prehybridization buffer containing at least 20 x 10^6 cpm of labeled probe. Filters were then rinsed twice, each time for 1h in 2 x SSC, 1% SDS, at 45°-50°C and autoradiographed. For the experiments shown in Figure 2 and 3, hybridization was first with the U4B probe. The filters were then autoradiographed; the U4B probe was eluted (by washing the filters two times for 1h in distilled water at 75°C); and the filters were then hybridized to the U4X probe. Synthetic U4B-like (SP6:U4B) and U4X-like (SP6:U4X) transcripts, approximately 233 and 178 nucleotides in length respectively, were synthesized from the cloned genes using SP6 RNA polymerase. The portions of the gene templates cloned into the pSP64 vector are diagrammed in Fig. 1.

RESULTS

The genomic organization of the closely linked chicken U4B and U4X RNA genes is diagrammed in Fig. 1. The two genes differ in sequence at seven nucleotide positions (1). The sequences of the encoded U4B and U4X RNA transcripts are shown at the top of Fig. 1. Five of the nucleotide differences are clustered at the 3' ends of the RNAs. To study the differential accumulation of the two gene products, two DNA 19-mer oligonucleotides were employed (U4B probe and U4X probe, Fig. 1). These are specifically complementary to U4B RNA and U4X RNA respectively.

U4B and U4X RNA Accumulation in Adult Tissues.

To determine the pattern of U4B and U4X RNA accumulation in different tissues, RNA was isolated from seven tissues of a single adult hen and examined by northern blot analysis. The amount of RNA loaded in each lane was adjusted to give a signal of uniform intensity when hybridized to the ^{32}P-labeled U4B probe (Fig. 2, upper panel). After autoradiography, the U4B probe was rinsed from the filter, and that same filter was then rehybridized to the radiolabeled U4X probe (Fig. 2, lower panel). Synthetic RNA transcripts (SP6:U4B and SP6:U4X) served as internal controls to ensure the hybridization specificity of the U4B and U4X probes.

```
     1         11         21         31         41         51         61         71         81         91        101        111        121        131        141
U4B  AGCUUGGCCAGUGGCAGUAUCGUAGCCAAUGGUAAUCCAGGCCGCCAAUAUUGCUAAAUGCUAAAUUGCUAUUUGACAUUGGCCAUUGGCAUUGGCCAUUGCCCAUGACAUUGAAAUAAUGCGGCAAUUUUGACAGUCUCGGAGA(CUG)
U4X  ---------------------------------U-------------G--------------------------------------------------------------------G--C----CC----G(---)
                                                                                                                        U4B probe: 3' AACTGTCAGAGATGCCTCT 5'
                                                                                                                        U4X probe: 3' AACTCTCGGAGGGGCCTCC 5'
```

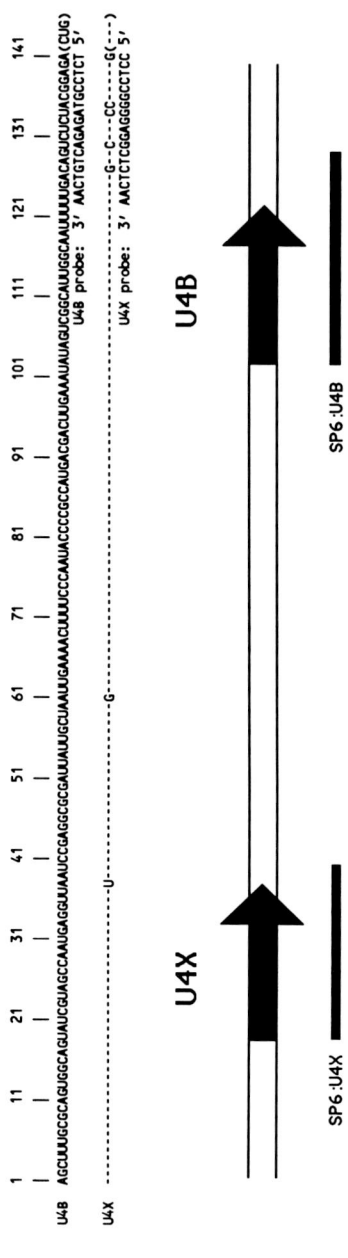

Figure 1. The chicken U4B and U4X RNA sequences, and organization of the genes that encode them. The U4B RNA and U4X RNA sequences encoded by the two genes are shown at the top of the figure (1). The dashes indicate sequence identity between the two U4 RNAs, whereas base differences are shown explicitly. Possible length heterogeneity at the 3' end of U4 RNA is indicated by the parentheses surrounding the last three nucleotides (3). Also shown are the sequences of two synthetic DNA oligonucleotides (U4B probe and U4X probe) that are specifically complementary to either U4B RNA or U4X RNA. The U4X and U4B RNA genes are closely linked in the chicken genome and are depicted in the center of the figure by bold arrows pointing in the direction of transcription (1). At the bottom of the figure, two restriction fragments are illustrated that were cloned into the pSP64 vector and transcribed with SP6 RNA polymerase to synthesize artificial U4X and U4B transcripts (SP6:U4X and SP6:U4B). These artificial transcripts were used as internal controls on the northern blots to demonstrate the specificity of hybridization.

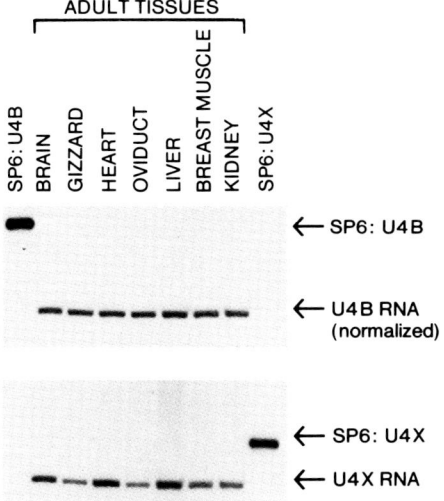

Figure 2. Northern blot analysis of chicken U4B RNA and U4X RNA: relative expression levels in adult tissues. RNA samples from seven tissues of an adult hen were electrophoresed in a polyacrylamide gel and electroblotted to a Gene Screen Plus membrane filter. The northern filter was hybridized first to the radiolabeled U4B probe (upper panel) and then to the radiolabeled U4X probe (lower panel). The amount of RNA loaded in each lane was adjusted so that the signals would be of uniform intensity when hybridized with the U4B probe. SP6:U4B and SP6:U4X denote artificial transcripts of the cloned genes (generated using SP6 RNA polymerase) that acted as internal controls to monitor the hybridization specificity of the U4B and U4X 19-mer probes.

U4B RNA and U4X RNA were both detected in each of the seven adult tissues examined. However, their relative level of accumulation varied from tissue to tissue. This is indicated by the different intensities of the U4X signal in the lower panel. Of the seven tissues examined, the U4X to U4B RNA ratio was highest in heart and liver, and lowest in gizzard and oviduct. Breast

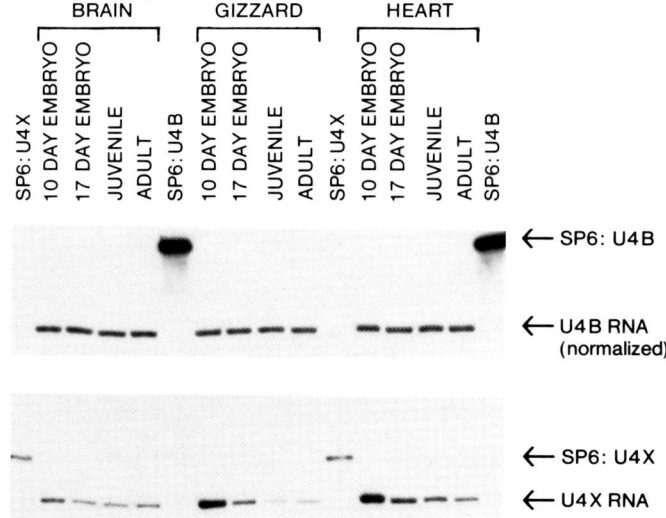

Figure 3. Northern blot analysis of U4B RNA and U4X RNA: relative expression levels during chicken development. Northern analysis as in Fig. 2, except using RNA from the brain, gizzard, and heart of 10 day embryos, 17 day embryos, a juvenile chick, and an adult hen. Upper panel: hybridization to the U4B probe; lower panel: hybridization of the same filter to the U4X probe.

muscle, kidney, and brain exhibited intermediate levels of U4X:U4B RNA accumulation.

U4B and U4X RNA Accumulation During Chicken Development.

Next we examined whether the relative accumulation of U4B and U4X RNA is regulated differentially as a function of chicken development. Total cellular RNA was isolated from the brain, gizzard, and heart of 10 day embryos, 17 day embryos, a chick 16 days post-hatching, and an adult hen. The upper panel of Fig.3 shows the results of hybridizing a northern blot of these samples to the U4B probe. After eluting the U4B probe, the filter was hybridized to the U4X probe (Fig. 3, lower panel).

In each of the three tissues, there is a decline in U4X expression (relative to the normalized U4B signal) as development proceeds from the 10 day embryo to the adult. Notably, however, each tissue exhibits a unique pattern of developmental regulation. In brain the 10 day embryonic U4X signal is relatively weak, and it undergoes only a minor reduction in intensity at later developmental stages. In contrast, in gizzard the U4X signal initially is relatively intense, and then it undergoes a striking decrease, relative to U4B, as the 10 day embryo develops to an adult. During these same stages, the heart exhibits a pattern of U4X to U4B expression that is different from the gizzard or brain. Overall, the U4X:U4B RNA ratio generally decreases as development proceeds. However, the timing of the decrease and the final U4X:U4B RNA ratio in a particular adult organ is governed in a tissue-specific manner.

Phylogenetic Distribution of U4X RNA.

Our cloning of chicken U4 RNA genes resulted in the identification of the novel U4 RNA which we have termed U4X RNA. Next, we investigated whether a U4X RNA homolog might also exist in other species of the vertebrate subphylum. To do this, we tested whether the chicken U4X oligonucleotide probe would hybridize to U4 RNAs from other species. RNA samples from HeLa cells and from tissues of five vertebrate animals were analyzed on northern blots (Fig. 4). As in Fig. 2 and 3, the amount of RNA loaded in each lane was normalized to give signals of equivalent intensity when hybridized with the U4B probe (Fig. 4, top panel). However, the U4B 19-mer probe is complementary to both U4B RNA and U4A RNA which are major sequence variants of U4 RNA in mammals (3). Therefore, in some species, the signals in the upper panel reflect normalization to combined levels of U4A RNA and U4B RNA, as well as any other closely related homologs that

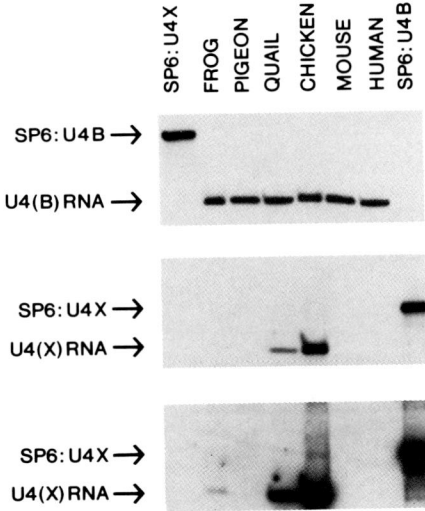

Figure 4. Northern blot analysis of U4B and U4X RNA in six vertebrate animals. Total cellular RNA was isolated from HeLa cells, from frog (X. laevis) kidney, and from homing pigeon, leg-horn chicken, bob white quail, and balb/c mouse liver. The top panel shows the normalized signals resulting from hybridization with the U4B probe. The middle panel shows the results of hybridizing another filter (but identical blot) to the U4X probe. The bottom panel is an autoradiogram of the same experiment shown in the middle panel, but exposed longer to reveal bands of weaker intensity.

might exist in a particular species. When another filter (but otherwise identical blot) was hybridized to the U4X probe, the results shown in the middle panel of Fig. 4 were obtained. The bottom panel in Fig. 4 is an autoradiogram of the same filter shown in the middle panel, but a longer exposure to reveal bands of weaker intensity.

The results reveal that the strongest U4X signal occurs using RNA from the homologous species (chicken). However, a clearly defined signal is also seen with the quail RNA. This suggests the existence of a U4X homolog in the quail, a member of the same avian family (Phasianidae) as the chicken. In contrast, the chicken U4X probe fails to hybridize with RNA from a member of a different avian family and order (pigeon) or with RNA from the two mammalian species (mouse and man). Somewhat surprisingly, the U4X probe hybridizes very weakly to a frog U4 RNA. This weak signal may represent hybridization to an evolutionarily related U4X RNA homolog in the frog (which either is expressed at a low level in the frog or which hybridizes poorly due to sequence divergence). Alternatively, the signal may be due to a U4X-like RNA that evolved separately in the frog and simply shares the property of GC-richness at its 3' end.

In summary, U4X RNA (operationally defined as being capable of hybridizing with the chicken U4X probe) is selectively distributed within the vertebrate subphylum. It appears to have a homolog in a species closely related to the chicken (quail), and potentially in a more distantly related species (frog).

DISCUSSION

We have found that the relative accumulation of the chicken U4B and U4X RNAs is regulated in a stage-dependent and tissue-specific manner. In each of the tissues examined, the U4X:U4B RNA ratio is greater in the embryonic tissue than in the corresponding adult tissue. However, each tissue exhibits a unique pattern of U4X:U4B expression both as a function of development and in comparing levels among adult tissues.

Whether this differential accumulation of the U4X and U4B RNAs is physiologically important to the organism is not yet known. However, it is conceivable that the sequence differences between the two chicken U4 RNAs may provide them with unique functional specificities. For example, the concentration of U4X RNA, or changes in the U4X:U4B RNA ratio, may play a role in regulating alternative splice site selection during development. Indeed, it has been suggested that changes in the concentrations or ratios of "constitutive" splicing factors (such as the snRNPs) may play a role in alternative splicing (2,5-10). Furthermore, regulated expression of genes encoding multiple snRNA sequence variants has also been observed during mouse and frog development (8-10). Thus, the differential expression of specific snRNA variants may provide a potential mechanism for regulating mRNA splicing patterns throughout development.

REFERENCES

1. Hoffman ML, Korf GM, McNamara KJ, Stumph WE (1986). Structural and functional analysis of chicken U4 small nuclear RNA genes. Mol Cell Biol 6:3910-3919.
2. Dahlberg JE, Lund E (1988). The genes and transcription of the major small nuclear RNAs. In Birnsteil M (ed): "Structure and Function of Major and Minor Small Nuclear Ribonucleoprotein Particles," Berlin: Springer-Verlag, p 38-70.
3. Reddy R, Busch H (1988). Small nuclear RNAs: RNA sequences, structure, and modifications. In Birnsteil M (ed): "Structure and Function of Major and Minor Small Nuclear Ribonucleoprotein Particles," Berlin: Springer-Verlag, p 1-37.
4. Maniatas T, Fritsch EF, Sambrook J (1982). "Molecular Cloning: A Laboratory Manual.", New York: Cold Spring Harbor Lab., p 194.
5. Laski FA, Rio DC, Rubin GM (1986). Tissue specificity of Drosophila P element transposition is regulated at the level of mRNA splicing. Cell 44:7-19.
6. Breitbart RE, Nadal-Ginard B (1987). Developmentally induced, muscle-specific trans factors control the differential splicing of alternative and constitutive troponin T exons. Cell 49:793-803.
7. Grabowski PJ, Sharp PA (1986). Affinity chromotography of splicing complexes: U2, U5, and U4/U6 small nuclear ribonucleoprotein particles in the spliceosome. Science 233:1294-1299.

8. Forbes DJ, Kirschner MW, Caput D, Dahlberg JE, Lund E (1984). Differential expression of multiple U1 small nuclear RNAs in oocytes and embryos of Xenopus laevis. Cell 38:681-689.
9. Lund E, Kahan B, Dahlberg JE (1985). Differential control of U1 small nuclear RNA expression during mouse development. Science 229:1271-1274.
10. Lund E, Dahlberg JE (1987). Differential accumulation of U1 and U4 small nuclear RNAs during Xenopus development. Genes and Dev 1:39-46.

DEVELOPMENTAL STAGE SPECIFIC RNA-EDITING OF A MITOCHONDRIAL TRANSCRIPT OF TRYPANOSOMA BRUCEI

Paul Sloof, Hans van der Spek, Janny van den Burg and Rob Benne

Laboratory of Biochemistry, University of Amsterdam, Academic Medical Center, Meibergdreef 15, 1105 AZ Amsterdam, The Netherlands

ABSTRACT Expression of some mitochondrial genes in trypanosomes involves insertion and deletion of uridines to produce functional mRNAs: RNA-editing. Here we report that a transcript, derived from the Trypanosoma brucei URF 2 gene, is edited at the 3'-terminal region: 11 uridines are inserted and one uridine is deleted. This form of editing is restricted to the bloodstream lifecycle stage in which the mitochondrion does not have a functional respiratory chain.

INTRODUCTION

Kinetoplast DNA (kDNA) of trypanosomes is one of the most unusual mitochondrial (mt) DNAs known at present, not only in terms of structural organisation but also in the way some of its genes are expressed.

kDNA consists of a single network made up from thousands of catenated circular DNA molecules. Two types of circular DNA occur: minicircles of 1-3 kilobasepairs (kb) in length depending on the species, which constitute 90-95% of the network, and 20-40 Kb maxicircles of which about 50 copies are present per network (1,2). The function of the minicircles is unknown, but it is clear that the maxicircle contains the mt genes (1-5). Genes were found for the mt ribosomal RNAs and for respiratory chain proteins such as apocytochrome b (Cyb), cytochrome c oxidase subunits I, II and III (CoI, CoII, CoIII) and NADH dehydrogenase subunits 1, 4 and 5 (ND1, ND4,ND5). Furthermore, unassigned reading frames (URFs) have been identified, some of which are present at corresponding locations in the maxicircles of three trypanosome species. The common trypanosome URFs will be referred to as Maxicircle URFs (MURF 1, MURF 2 and MURF 3, see ref. 2) whereas species specific URFs are indicated by the prefixes T, C or L (for Trypanosoma brucei, Crithidia fasciculata and Leishmania tarentolae, respectively).

The nucleotide sequences of some trypanosomal maxicircle genes and MURFs revealed a number of extraordinary features including the occurrence of frameshifts in protein coding regions, and the absence of genomic ATG initiation codons. In our study of the (-1) frameshift containing CoII genes of T. brucei and C. fasciculata we discovered that this frameshift is repaired at the RNA level (6). CoII transcripts contain a continuous reading frame, since in both organisms four extra, non-DNA-encoded, uridine residues are found at a location corresponding to the frameshift position in the genes. The presence in total T. brucei DNA of a second CoII gene version encoding the extra nucleotides, could be excluded by hybridization experiments using oligonucleotides as probes, which are complementary to either the RNA sequence (with the extra uridines) or to the DNA sequence. We therefore concluded that the extra uridines are inserted into the mRNA during or after transcription of the frameshift containing CoII gene in a process that we have called "RNA-editing" (6).

Since the first demonstration of this novel type of mRNA processing more examples have been described. By sequencing cDNAs and by direct RNA sequencing we have found that the (+1) reading frameshift of the MURF 3 gene of C. fasciculata is corrected by insertion of 5, non-DNA-encoded, uridines into the transcript at a location corresponding to the frameshift position in the gene (7). Other transcripts are extensively edited at the 5'-terminus both by uridine insertion and deletion (8-11). In these cases AUG initiation codons are created at similar locations in corresponding transcripts of the three trypanosome species.

The most striking case of RNA-editing has recently been reported by Feagin et al. (12) who showed that CoIII RNA in T. brucei displays a high degree of identity to the CoIII genes of C. fasciculata and L. tarentolae but lacks homology with its corresponding gene on the T. brucei maxicircle. This gene was previously designated TURF 2 (13).

We will maintain this nomenclature in the present study for reasons outlined in the discussion. We have reported that the TURF 2 gene lacks an upstream ATG codon and would encode a highly unusual mt protein since it would be rich in charged aminoacids (13). The reason why the TURF 2 gene escaped identification as the T. brucei CoIII gene by hybridization experiments (see ref. 5) has now become clear. From the nucleotide sequence of the 3'- two-thirds of the CoIII RNA (12) it appeared that only the sequence of the G's, A's and C's is identical to that of the TURF 2 gene and that the differences arise from extensive uridine insertion and deletion. Some 58% of the nucleotides in the sequenced part of the CoIII RNA are uridines which are not DNA encoded (12). This extensive way of RNA-editing results in a nucleotide sequence that encodes a protein, entirely different from the one encoded by the non-edited transcript.

Although at present a number of RNA-editing examples have been described, it proved difficult to formulate a possible mechanism since each case is unique with respect to number, pattern and site of uridine insertion and deletion. One approach that can lead to clues to a possible editing mechanism is to analyse sequences of cDNAs representing the same transcript at various stages of editing. Such studies have been performed for edited transcripts of the C. fasciculata MURF 3 gene (7). In this paper we report that sequence differences between the TURF 2 gene and a cDNA, representing a TURF 2 transcript, are limited to the 3'-end: 11 extra uridine-residues are present in the transcript whereas one U-residue is missing. Our analysis shows that: (i) a second TURF 2 gene version encoding the altered sequence, does not exist in the T. brucei genome. We conclude therefore that the alterations are the result of U-insertion and deletion in a RNA-editing process. (ii) The U-insertion pattern of TURF 2 cDNA is different from that found in the corresponding area of CoIII RNA (12). (iii) This form of TURF 2 transcript editing is restricted to the bloodstream lifecycle stage of T. brucei in which the mitochondrion is not functional.

METHODS

Cell culture, RNA and DNA isolation, Cloning and Sequencing

T. brucei strain 427 cultivation, isolation of total cellular and kDNA, purification of total cellular RNA and of a poly(A^+) fraction are described in ref. 3. The cloning and sequencing of the TURF 2 gene has been described in ref 13, whereas the construction of a cDNA bank of bloodstream form poly(A^+)RNA and the isolation of a TURF 2 cDNA clone are described in ref. 15.

Nucleic acid hybridization

Southern and Nothern blot analysis was performed according to ref. 6. The following oligonucleotides were used: Nucleotide coordinates
DNA-oligo: 5'-CTACCTCTTCATTCCAACTA-3'   577-597   (13)
RNA-oligo: 5'-CCAATAAAACTTAACTATCA-3'   see below

The DNA-oligo represents the 3'-end of the TURF 2 gene. The RNA-oligo is complementary to the 3'-end of the TURF 2 transcript including the extra uridines and excluding the deleted U (see 6). The hybridization targets of the two oligonucleotides are indicated in Fig. 1. Both oligonucleotides are end-labelled and used as probes in hybridization experiments, performed as described (6) except that the wash temperatures for the DNA-oligo and RNA-oligo were 56 °C and 50 °C respectively.

RESULTS

RNA-editing at the 3'-end of a TURF 2 transcript

The TURF 2 cDNA has been obtained from a cDNA clone bank which was constructed from total poly(A^+) RNA of bloodstream T. brucei (15). The nucleotide sequence of this 450 bp cDNA matched that of the TURF 2 gene, except at the 3'-terminal region where 11 extra uridines are present at 5 different locations. In addition, one U was not present in the RNA. The cDNA sequence revealed also the presence of a poly(AU) tail in the transcript that served as template during cDNA synthesis (Fig. 1)

Absence of a second TURF 2 gene version in the T. brucei genome

In order to verify whether there is a second TURF 2 gene version, encoding the transcript from which the cDNA is derived, we have performed hybridization studies with total T. brucei DNA, using as probes oligonucleotides that represent either the RNA ('RNA'-oligo) or the DNA nucleotide sequence ('DNA'-oligo). The hybridization targets of these oligonucleotides are indicated in Fig. 1. In Fig. 2 hybridization analysis of Southern blots of restricted total T. brucei DNA is shown. Care was taken also to blot high molecular weight DNA such as network segments consisting of minicircles without sites for the respective restriction enzymes. The hybridizing bands obtained with the DNA-oligo as probe (panel A) represent the known maxicircle fragments containing the TURF 2 gene. The RNA-oligo fails to hybridize to this blot under conditions that permit a clear hybridization signal with the TURF 2 cDNA (panels B and C, respectively). To exclude the possibility that the TURF 2 transcript is encoded by a small subset of maxicircles, the intensity of the hybridization signal representing one maxicircle has been determined in

```
                          RNA-oligo
                        ─────────────
TURF 2  RNA  UUUAGUUGUUGA-UA GUUAA---GUUUUAUUGGUAGUUUGUAGGAAGUUU₇A₁₂U₄A₆U₃A₉

                          DNA-oligo
                        ─────────────
TURF 2  DNA  TTTAGTTG    GA  ATG  AA    G    A   GGTAGTTTGTAGGAAGTT


CoIII   RNA  UUUAGUUGUUGAUUA GUUAAUUUGU---AUUGGUAGUUUGUAGGAAGUUU₆AUA₁₀
```

FIGURE 1. Comparison of the 3'-terminal nucleotide sequences of the TURF 2 gene, the TURF 2 transcript and the CoIII RNA. Hybridization target sequences for the TURF 2 RNA specific oligonucleotide (RNA-oligo) and the TURF 2 DNA specific oligonucleotide (DNA-oligo) are indicated. Arrows point to sites of sequence differences between the TURF 2 RNA and the CoIII RNA.

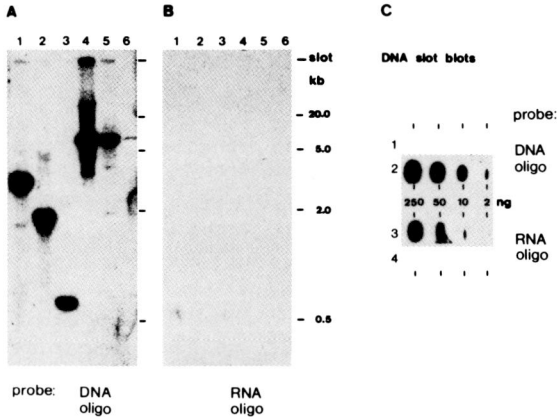

FIGURE 2. Search for additional TURF-2 gene versions. Southern blots of restricted T. brucei total DNA, 5 µg/lane, with TaqI(lanes 1), MboI(lanes 2), Alu(lanes 3), HindIII(lanes 4) and EcoRl(lanes 5; lanes 6 contain 0.1µg of the Eco digest) are hybridized with the DNA-oligo (panel A) and the RNA-oligo (panel B). Panel C shows DNA slot blots with in slots 1 and 3 TURF 2 cDNA and in slots 2 and 4 total T. brucei DNA.

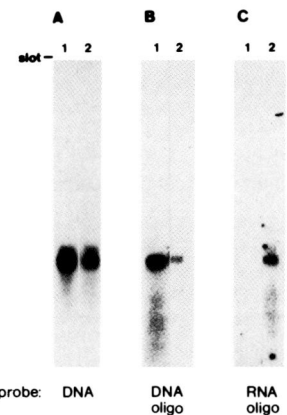

FIGURE 3. TURF 2 transcript analysis. Lanes 1 contain 2 µg procyclic poly (A$^+$) RNA; lanes 2 contain 2 µg poly(A$^+$) RNA from bloodstream T. brucei. Northern blots are hybridized with a cloned TURF 2 gene probe (panel A), the DNA-oligo (panel B) and the RNA-oligo (panel C).

panel A, lane 6 by applying 2% of the amount of DNA of the other lanes. Since the T. brucei network contains about 50 maxicircles, a single maxicircle encoding the altered sequence would have produced a similar signal in panel B. The absence of such a signal excludes the presence of divergent maxicircles encoding the TURF 2 transcript. In summary, the results presented in Fig. 2 are compatible with the presence of only one TURF 2 gene version, without the alterations found in the 3'-end of the transcript. We therefore conclude that the TURF 2 transcript has undergone editing at its 3'-terminus.

RNA editing of the TURF 2 transcript occurs exclusively in bloodstream T. brucei.

The involvement of RNA-editing in the expression of T. brucei maxicircle genes is mostly restricted to the procyclic lifecycle stage of T. brucei in which its mitochondrion contains a functional respiratory chain. The MURF 2 transcript, however, is edited not only in procyclic but also in bloodstream form T. brucei in which the mitochondrion is not functional (see for summary table in ref. 11). In Fig. 3 panel C is shown that the RNA-oligo hybridizes with a 600 nucleotide TURF 2 transcript, which is present in bloodstream form but not in procyclic RNA. Hybridization with the DNA-oligo lights up an unedited RNA species of similar length which is mainly found in RNA of the procyclic form of T. brucei (panel B). All transcripts hybridize to a DNA fragment containing the TURF 2 gene (panel A). These experiments show that RNA-editing in the 3'-terminal region of the TURF 2 transcript exclusively occurs in the bloodstream lifecycle stage of T. brucei.

DISCUSSION

Recently we reported that the nucleotide sequences of a 450 bp TURF 2 cDNA and the TURF 2 gene are identical, except at their extreme 3'-ends where in the cDNA 11 extra uridines are found and one genomic T-residue appears to be absent. Moreover, the cDNA sequence revealed the presence of a poly (AU) tail at the 3'-terminus of the transcript that served as template during cDNA synthesis (Fig. 1; 6). In this paper we show that a second TURF 2 gene version, containing the altered sequence, is absent both in nuclear and in mt DNA of T. brucei (Fig. 2). We therefore conclude that insertion and deletion of U-residues occurs during or after transcription, in a RNA-editing process (6). Furthermore we show that editing of this transcript is restricted to the bloodstream lifecycle stage of T. brucei and that the non-edited transcript mainly occurs in the procyclic form (Fig. 3). In fact this is the first demonstration of editing of a mitochondrial transcript that occurs in the bloodstream form only, a lifecycle stage in which the mitochondrion lacks a functional respiratory chain (14).

From recent studies it has been inferred that the TURF 2 gene gives rise to CoIII RNA after extensive editing of an initial, non-edited, TURF 2 transcript (12). With CoIII specific probes a smear of RNAs has been detected, ranging from about 600 nucleotides up to 960 nucleotides in length. These RNAs may represent CoIII specific RNAs in various stages of editing. Furthermore, these RNAs are found in both lifecycle stages of T. brucei (12). Apparently, CoIII RNA-editing occurs irrespective of the functional status of the mitochondrion.

In these studies, the TURF 2 RNA-oligonucleotide hybridizes to one distinct RNA species of about 600 nucleotides, only present in the bloodstream form RNA population (Fig. 3). This result is in agreement with the observation that the pattern of U-insertion at the 3'-end of procyclic CoIII RNA is different from the TURF 2 cDNA sequence (Fig. 1). Besides a divergent U-insertion pattern in the poly (AU) tail (arrow 3, Fig. 1), differences at 2 locations in the edited region are noteworthy (arrows 1 and 2, Fig. 1). These differences could indicate that the TURF 2 transcript is not a precursor to the CoIII RNA. The possibility arises, therefore, that the form of TURF 2 editing presented in this paper and the one that leads to CoIII RNA are two separate processes which are regulated in a developmental stage specific way. In this respect, it should be pointed out that in the non-edited TURF 2 transcript a 220 aminoacids long open reading frame is contained (13). It is not entirely clear why this reading frame is preserved, since statistically in DNA that has the nucleotide composition of the TURF 2 gene, 1 out of every 12 codons could be a stopcodon as long as it is removed from the transcript by RNA- editing. For this reason, future research will deal with the possible role of the TURF 2 open reading frame.

ACKNOWLEDGEMENTS

We thank Prof. H.F. Tabak for comments on the manuscript, Mr. P. Smits for initial experiments and Dr. J.E. Feagin for making available unpublished results.

REFERENCES

1. Benne, R. (1985) Mitochondrial genes in trypanosomes. Trends Genet. 1, 117
2. Simpson, L. (1988) The mitochondrial genome of kinetoplastid protozoa: Genomic organization, transcription, replication and evolution. Ann. Rev. Microbiol. 41, 363
3. DeVries, B.F., Mulder, E., Brakenhoff, J., Sloof, P., Benne, R. (1987) The variable region of the Trypanosoma brucei kinetoplast maxicircle: sequence and transcript analysis of a repetitive and a non-repetitive fragment. Mol. Biochem. Parasitol. 27, 71
4. Sloof, P., Van den Burg, J., Voogd, A., Benne, R., Agostinelli, M., Borst, P. Gutell, R, Noller, H.F. (1985) Further characterization of the extremely small mitochondrial ribosomal RNAs from trypanosomes: a detailed comparison of the 9S and 12S RNAs from Crithidia fasciculata and Trypanosoma brucei with rRNAs of other organisms. Nucl. Acids Res. 13, 4171
5. Sloof, P., Van den Burg, J., Voogd, A, Benne, R. (1987) The nucleotide sequence of a 3.2 kb segment of mitochondrial maxicircle DNA from Crithidia fasciculata containing the gene for cytochrome oxidase subunit III, the N-terminal part of the apocytochrome b gene and a possible frameshift gene; further evidence for the use of unusual initiator triplets in trypanosome mitochondria. Nucl. Acids Res. 15, 51
6. Benne, R., Van den Burg, J., Brakenhoff, J., Sloof, P., Van Boom, J.H., Tromp, M.C. (1986) Major transcript of the frameshifted CoxII gene from trypanosome mitochondria contains four nucleotides that are not encoded in the DNA. Cell 46, 819
7. Van der Spek, H., Van den Burg, J., Croiset, A., Van den Broek, M., Sloof, P., Benne , R. (1988) Transcripts form the frameshifted MURF 3 gene from Crithidia fasciculata are edited by U-insertion at multiple sites. EMBO J., submitted.
8. Feagin, J.E., Jasmer, D.P., Stuart, K. (1987) Develop-

mentally regulated addition of nucleotides within apocytochrome b transcripts in Trypanosoma brucei. Cell 49, 337
9. Feagin, J.E., Shaw, J.M., Simpson, L., Stuart, K. (1988) Creation of AUG initiation codons by addition of uridines within cytochrome b transcripts of kinetoplastids. Proc. Natl. Acad. Sci. USA 85, 539
10. Feagin, J.E., Stuart, K. (1988) developmental aspects of uridine addition to mitochondrial transcripts of Trypanosoma brucei. Mol. Cell. Biol. 8, 1259
11. Shaw, J.M., Feagin, J.E., Stuart, K., Simpson, L. (1988) RNA-editing of mitochondrial transcripts by addition and deletion of uridines generates conserved aminoacid sequences and methionine initiation codons in three kinetoplastids. Cell, in press.
12. Feagin, J.E., Abraham, J.M., Stuart, K. (1988) Extensive editing of the cytochrome c oxidase III transcript in Trypanosoma brucei. Cell, in press.
13. Benne, R., DeVries, B.F., Van den Burg, J., Klaver, B. (1983) The nucleotide sequence of a segment of Trypanosoma brucei mitochondrial maxicircle DNA that contains the gene for apocytochrome b and some unusual unassigned reading frames. Nucl. Acids Res. 11, 6926
14. Vickerman, K. (1985) developmental cycles and biology of pathogenic trypanosomes. Brit. Med. Bull. 41, 105
15. Hoeijmakers, J.H.J., Borst, P., Van den Burg, J., Weissmann, C., Cross, G.A.M. (1980) The isolation of plasmids containing DNA complementary to messenger RNA for variant surface glycoproteins of Trypanosoma brucei. Gene 8, 391

TRANSCRIPT ALTERATION BY mRNA EDITING
IN KINETOPLASTID MITOCHONDRIA[1]

Jean E. Feagin and Kenneth Stuart

Seattle Biomedical Research Institute
4 Nickerson St., Seattle, WA 98109-1651

ABSTRACT Several mitochondrial transcripts from kinetoplastid protozoan parasites are altered, probably posttranscriptionally, by the addition of uridines which are not encoded in their genomic sequences and the deletion of uridines which are encoded. The cytochrome oxidase III transcript of *Trypanosoma brucei* is extensively edited, with over 50% of the mRNA sequence resulting from editing. This phenomenon is called RNA editing. The consequences of RNA editing include extension of open reading frames, creation of translation start and stop codons, and correction of genomic frameshifts. The editing mechanism, which is as yet incompletely characterized, precisely determines the final nucleotide sequence of the mRNAs.

INTRODUCTION

The nucleotide sequences of several mitochondrial transcripts in *Trypanosoma brucei*, *Leishmania tarentolae*, and *Crithidia fasciculata* differ from their genomic sequences. Nucleotides which are not encoded in the DNA are added to transcripts (1-7) and in some cases, nucleotides which are encoded in the DNA are missing from the transcripts (4-6). All the transcript nucleotides which are affected are uridines. No DNA sequences which correspond

---

[1] This work was supported by NIH grant AI14102 to K.S. who is a Burroughs-Wellcome Scholar in Molecular Parasitology.

exactly to the altered transcript sequences have been
detected in either mitochondrial or nuclear DNA (1, 2, 6),
thus the addition and deletion of uridines must occur either
co- or posttranscriptionally. This process of transcript
sequence alteration is termed RNA editing.

RNA editing has a number of consequences. It extends
open reading frames (2-6), creates translation start (2-5)
and stop codons (6), and corrects genomic frameshifts (1,
J.M. Shaw and L. Simpson, unpublished results). Uridine
additions and deletions also occur in the 3' untranslated
sequence and poly(A) tails of some transcripts (1, 6, 7);
their role is uncertain but may affect transcript stability.
In *T. brucei*, where production of the mitochondrial
respiratory system is developmentally regulated, editing of
some transcripts is developmentally regulated, while others
are edited constitutively and yet others appear not to be
edited at all (2, 4, 5). This implies that the editing
machinery is always present but its activity is controlled
in a gene-specific fashion.

RESULTS AND DISCUSSION

Editing adds and deletes uridines

The sequences of portions of numerous transcripts from
kinetoplastid mitochondria have been determined and compared
to their respective DNA sequences. This has revealed that
the transcripts of approximately 50% of the genes differ
from their DNA sequences (Table 1). All edited transcripts
identified to date have uridine additions (1-7) and some
also have uridine deletions (4-6). Uridine deletions are
fewer in number than additions for any specific transcript.
The number of additions to a transcript ranges from 4
nucleotides for COII transcripts (1, J.M. Shaw and L.
Simpson, unpublished results) to at least 347 for the *T.
brucei* COIII transcript (6). Deletions range from 2
nucleotides removed from the *C. fasciculata* COIII transcript
(5) to at least 16 deleted from the T. brucei COIII
transcript (6). The *T. brucei* MURF2 transcript (5, 6)
provides an example of both addition and deletion (Fig. 1A).

Primer extension analyses show that for each edited
transcript, there also exists a shorter, less abundant
transcript (2-5). Several lines of evidence suggest that
the shorter transcripts are unedited. The primer extension

TABLE 1
EDITING EVENTS

genes	Tb +	Tb -	Lt +	Lt -	Cf +	Cf -
CYb	34	0	39	0	39	0
COI	0	0	0	0	0	0
COII	(4)	0	(4)	0	(4)	0
COIII	347*	16*	29	14	32	2
ND1	nd		0	0	nd	
ND4	0	0	0	0	nd	
ND5	0	0	0	0	nd	
MURF1	nd		nd		nd	
MURF2	26	4	28	4	28	0
MURF3	20*	7*	20	(5)	3*	0*

+ = uridines added
- = uridines deleted
* = incomplete data
() = frameshift correction
nd = not determined

products corresponding to the shorter transcripts are not seen when oligonucleotides specific for edited sequences are used in primer extensions (2, 3). In addition, the size difference between the products equals the net sum of editing events. For example, the *T. brucei* MURF2 transcript has 26 additions and 4 deletions (Table 1, Fig. 1A); its shorter product has, as expected, 22 nucleotides less than the principal product (5, 6).

Editing creates initiation codons.

Nearly half of kinetoplastid mitochondrial genes lack genomically encoded AUG initiation codons near the 5' end of their open reading frames. It was initially proposed that alternate initiation codons, coding for isoleucine or leucine, might be used (8-14). Determination of the 5' end sequences of the transcripts of such genes shows that in many cases, an AUG codon is created by addition of a uridine (2-5). The location of these created initiation codons is

A
```
        TAGAAAGGTATATAATCTATAATGA    AA G  GG G    ATTTTA   AG    A   TTG GCTTTGATTG AGTCTGTTTTTGATTG
NNNNNNNNGAAAGNANAUAAUCUAUAAUGAuuuuAAuuuGuuGuuuuA          AuuuAGuuuuAuuuUUGuGCUUUGAUUGuAGUCCGUGUUUUGAUUG
```

B Tb
```
    GTTAAGAATAATGGTTATAAATTTTATATAAA A G   CG G AGA       A A     A     A AGAAA   G G   GCTCTTTTAATGTCA
    NNNAANAAUAAUGNUAUAAAUUUUAUAUAAAuAuGuuuCGuuGuAGAuuuuuAuAuuuuuuuuAuuAuuuuuuuAuuAGAAuuuGuuGUCUUUUAAUGUCA
```
Lt
```
    ATAAATTTAATTTAAATTTTAAATAATTATAAA A G        CG G AGA      G A         A A   AGAAA    A G  G C TTTAACTTCA
    NNNNAUUUAAUUUAAAUUUUAAAUAAUUAUAAAuAuGuuuuuuCGuGuuAGuuuuuuGuAuuuuuuuuAuuAuuuuuAGAAuuuAuGuuGuCuUUUAACUUCA
```
Cf
```
    TCTATTTTTAATTTAAAGCTTAAAGTTAAA A G         CG G AGA     A A         G A   AGAAA   A G  G CTTTTAACGTCA
    NNNNUUUUAAUUUAAAGCUUAAAGUUAAAuAuGuuuuuuCGuCGuuAGAuuuuAuAuAnuuuuuuuGuAuuuuuuuGuAuuuAGAAuuuAuGuuGCUUUUAACGUCA
```

C
```
    TTTAGAATAAA       AA  ATAAG
    TTTAGAATAAAuuuuuuuuuuuAuuuuuATAAGA(26)
```

FIGURE 1. Comparison of DNA and RNA sequences for edited transcripts. DNA sequences are shown aligned above the corresponding RNA sequences for the 5' end of T. brucei MURF2 (A), the 5' ends of CYb (B) from T. brucei (Tb), L. tarentolae (Lt), and C. fasciculata (Cf), and the 3' end of T. brucei CYb (C). Gaps in the DNA sequence show sites of uridine addition and those in the RNA sequence show uridine deletions. Added uridines are in lower case. In frame AUGs are underlined. The 5'-most AUG in T. brucei MURF2 (A) has been shifted into frame by uridine addition. The 3' AUG in T. brucei CYb (B) is encoded in frame but homology with other kinetoplastids suggests the AUG created by uridine addition is probably the initiation codon. Sequence data is from references 2-4.

very similar among the different kinetoplastids for each gene. In the case of the apocytochrome *b* gene (3), uridines are added in similar numbers at similar sites within the 5' end of transcripts in all three kinetoplastids, extending the open reading frames and creating initiation codons at nearly the same site (Fig. 1B). The N-terminal amino acid sequence is also nearly the same between the three species. Comparison of editing of the 5' ends of COIII transcripts of *L. tarentolae* and *C. facsiculata*, on the other hand, show the sites and numbers of uridine addition and deletion to be rather different, although the final transcripts are very similar (5). The end result of editing, in this case, is to make two dissimilar DNA sequences quite similar at the transcript level and thereby to produce almost identical amino acid sequences for the protein products. It is noteworthy that in nearly every gene where an AUG codon is located at an appropriate site within the open reading frame to serve as an initiation codon, no 5' editing of the transcript of that gene is seen. Conversely, most genes which lack such potential AUG initiation codons have them created by uridine addition (5).

Editing corrects frameshifts.

The COII gene in all three kinetoplastids contains a frameshift; the 5' two-thirds of the transcript is in one reading frame and the 3' third is in another, based on comparisons with COII genes in other organisms (1, 9-11). Transcript sequence analysis has shown that the frameshift is corrected by the addition of four uridines to the transcript (1, J.M. Shaw and L. Simpson, unpublished results). They are added in the same sites in all three species and predict the same amino acid sequence. The MURF3 gene of *L. tarentolae* is also a frameshifted gene and the frameshift in MURF3 is corrected by addition of five uridines to the transcript (J.M. Shaw and L. Simpson, unpublished results).

Editing alters 3' sequences.

The 3' end sequences of several transcripts are edited, both in the poly(A) tail and the 3' untranslated sequence. These include the *T. brucei* CYb (Fig. 1C) and COIII transcripts (6) and the *L. major* NADH dehydrogenase (ND) 1 transcript (7). The 3' editing can be fairly extensive; 30 uridines are added in 12 sites in the poly(A) tail of the ND1 transcript. Differences in the 3' sequences of two COIII cDNAs (6) suggest that editing in this region may not be as closely controlled as within the open reading frame or, alternatively, the sequenced cDNAs may be derived from incompletely edited transcripts. The role that editing of the poly(A) tail plays is unclear but may relate to transcript stability since AU-rich sequences can increase mRNA stability (15).

Editing is gene-specific and developmentally regulated for some transcripts.

Production of the mitochondrial respiratory system is developmentally regulated in *T. brucei*. Energy is produced by glycolysis in mammalian bloodstream stages and by the mitochondrial respiratory system which is produced in procyclic (insect) but not bloodstream stages (16). With the exception of the ND5 gene, all the mitochondrial protein coding genes have two size classes of transcripts and the steady state abundance of many of these transcripts is also developmentally regulated (17). The larger size class of CYb, COI, COII, and COIII transcripts, for example, is more abundant in procyclic than bloodstream forms.

Comparison of RNA editing events in different life cycle stages reveals that the activity is always present but is regulated in a gene-specific manner, including developmental regulation for some transcripts. Edited CYb transcripts are most abundant in procyclic forms and are absent in bloodstream forms lacking the respiratory system. This is seen in both RNA primer extensions and sequencing ladders (2, 4) and RNA blots probed with oligonucleotides specific for edited and unedited sequences (Fig. 2). Interestingly, only the smaller class contains 5' unedited sequences, while the edited transcripts are heterogeneous in size. The editing which corrects the COII frameshift is similarly regulated, occurring principally in procyclic forms (4). In contrast, editing of the MURF2 (4) and COIII (6) transcripts is constitutive. There are also a number of transcripts which have no detected editing events (5). This implies an additional level of regulation which controls the editing in a gene-specific manner.

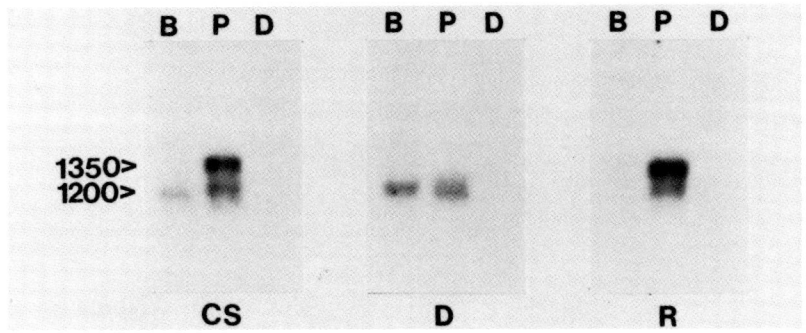

FIGURE 2. Northern blot analysis of CYb transcripts. Blots of total RNA from T. brucei bloodstream (B) and procyclic (P) forms and from mutants lacking mitochondrial DNA (D) were probed with oligonucleotides which detect all CYb transcripts (CS), unedited CYb transcripts (D), and edited CYb transcripts (R). Transcript sizes are given in nucleotides.

The T. brucei COIII transcript is extensively edited.

Nucleotide sequence analysis and heterologous hybridization using L. tarentolae and C. fasciculata COIII probes had failed to detect a COIII gene in mitochondrial or nuclear DNA of T. brucei (9, 14). However, sequencing T. brucei RNA using an oligonucleotide primer complementary to a COIII sequence conserved between the other two species revealed the presence of a COIII transcript very similar to those of L. tarentolae and C. fasciculata (6). Approximately 60% of the transcript sequence has been determined by a combination of RNA and cDNA sequencing. No genomic sequence exactly corresponds to the transcript sequence but a mitochondrial sequence just upstream of the CYb gene matches exactly except for the presence and absence of numerous thymidines (uridines in the RNA). This is the location of the COIII gene in L. tarentolae and C. fasciculata. Thus, the COIII gene of T. brucei is extensively altered by RNA editing. Of the 626 nucleotides known, 347 are added uridines, in 121 sites. There are also 16 deletions, in 7 sites. An example of the extent of the COIII editing is shown in Fig. 3.

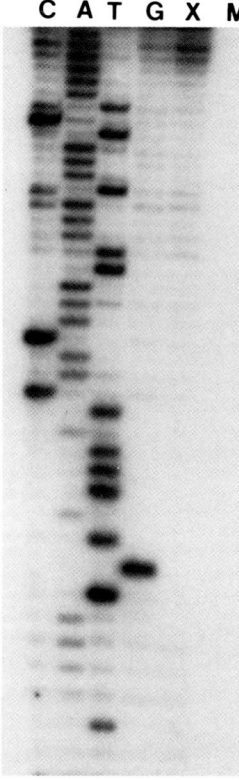

```
A
  C A T G X M

B
5'...
  DNA    A A G        G  G     A   A G GA
  RNA    AuAuGuuuuGuuGuuuAuuAuGuGA

         A GG      G       A    GGTTA
         uuAuGGuuuuGuuuuuuAuuGG UAuuuuuu

         AGA    A    AA    G   GA AAA  ACA
         AGAuuuAuuuAAuuuGuuGAuAAAuACAuuu

         ATTTG   TG  AG GG    A    G  AA
         uAUUUGuuUGuuAGuGGuuuAuuuGuuAAuu

              G     G GTTTTTGG    AGG
         uuuuuGuuuuGuGUUUUUGGuuuAGGuuuuu

              G  GTTTTG    G    G A  A GA   GA
         uuGuuG    UUGuuGuuuuGuAuuAuGAuuGA

         G    G  G    GTTTTG       G       G
         GuuuGuuGuuuG      GuuuuuuGuuuuuGu
```

FIGURE 3. *T. brucei* COIII transcript sequence. Total RNA was sequenced by the dideoxy chain termination method, using an oligonucleotide primer complementary to COIII transcript sequence and reverse transcriptase. Panel A shows part of an RNA sequencing gel; lane X has no dideoxynucleotides and lane M has no primer. Panel B shows a portion of the COIII transcript sequence, aligned below the corresponding DNA sequence. Added uridines are in lower case; gaps are left in the DNA sequence at sites of addition and in the RNA sequence at sites of deletion. The sequence which is the complement of that from the gel in A is underlined.

Editing is probably posttranscriptional.

Despite rigorous searches, no DNA sequences corresponding to edited transcripts have been found (1, 2, 6, J.M. Shaw and L. Simpson, unpublished results). Thus editing presumably operates via a novel mechanism. Although cotranscriptional editing cannot yet be ruled out, several lines of evidence suggest that editing is posttranscriptional. These include the existence of full-size unedited transcripts for each edited transcript (2-5) and of partially edited transcripts (2, 6). Also, editing occurs within poly(A) tails (6, 7). Analysis of cDNAs from partially edited COIII transcripts shows that they are fully edited at their 3' ends and unedited at their 5' ends; that is, their sequence corresponds to that determined by RNA sequencing at the 3' end and to that determined by DNA sequencing at the 5' end (6, J.M. Abraham, J.E. Feagin, and K. Stuart, unpublished results). This implies a 3' to 5' direction for editing, which is inconsistent with cotranscriptional editing.

Conclusions

RNA editing alters transcript nucleotide sequences by the addition and deletion of uridines. It appears to occur posttranscriptionally. The process is precise and operates in a gene-specific manner. RNA editing thus provides a powerful mechanism for controlling gene expression by altering transcript function and possibly stability. Thus far, editing has been observed only for mitochondrial transcripts in kinetoplastid protozoans. However, while the attachment of a spliced leader sequence to nuclearly encoded mRNAs was also originally seen only in kinetoplastids (18), recently an analogous phenomenon has been described for *Caenorhabditis elegans* (19). Upon further investigation, RNA editing may also be found in other organsims.

ACKNOWLEDGMENTS

We thank Andrea Perrollaz for technical assistance and Janet Shaw and Larry Simpson for communicating results prior to publication.

REFERENCES

1. Benne, R, Van den Burg, J, Brakenhoff, JPJ, Sloof, P, Van Boom, JH, Tromp, MC (1986). Major transcript of the frameshifted *coxII* gene from trypanosome mitochondria contains four nucleotides that are not encoded in the DNA. Cell 46:819.
2. Feagin, JE, Jasmer, DJ, Stuart, K (1987). Developmentally regulated addition of nucleotides within apocytochrome b transcripts in Trypanosoma brucei. Cell 49:337.
3. Feagin, JE, Shaw, JM, Simpson, L, Stuart, K (1988). Creation of AUG initiation codons by addition of uridines within cytochrome *b* transcripts of kinetoplastids. Proc Natl Acad Sci USA 85:539.
4. Feagin, JE, Stuart, K (1988). Developmental aspects of uridine addition to mitochondrial transcripts of *Trypanosoma brucei*. Mol Cell Biol 8:1259.
5. Shaw, JM, Feagin, JE, Stuart, K, Simpson, L (1988). Editing of kinetoplastid mitochondrial mRNAs by uridine addition and deletion generates conserved amino acid sequences and AUG initiation codons. Cell 53:401.
6. Feagin, JE, Abraham, JM, Stuart, K (1988). Extensive editing of the cytochrome c oxidase transcript in Trypanosoma brucei. Cell 53:412.
7. Spithill, TW, Samaras, N, Campbell, DA, Simpson, AM, Simpson, L (1988). Differential expression of the maxicircle gene encoding NADH dehydrogenase subunit 1 in promastigotes of two species of the protozoan parasite Leishmania. Proc Natl Acad Sci USA (in press).
8. Benne, R, De Vries, BF, Van den Burg, J, Klaver, B (1983). The nucleotide sequence of a segment of *Trypanosoma brucei* mitochondrial maxi-circle DNA that contains the gene for apocytochrome *b* and some unusual unassigned reading frames. Nucl Acids Res 11:6925.
9. de la Cruz, VF, Neckelmann, N, Simpson, L (1984). Sequence of six genes and several open reading frames in the kinetoplast maxicircle DNA of *Leishmania tarentolae*. J Biol Chem 259:15136.
10. Hensgens, LAM, Brakenhoff, J, De Vries, BF, Sloof, P, Tromp, MC, Van Boom, JH, Benne, R (1984). The sequence of the gene for cytochrome *c* oxidase subunit I, a frameshift containing gene for cytochrome *c* oxidase subunit II and seven unassigned reading frames in *Trypanosoma brucei* mitochondrial maxi-circle DNA. Nucl Acids Res 12:7327.

11. Payne, M, Rothwell, V, Jasmer, DP, Feagin, JE, Stuart, K (1985). Identification of mitochondrial genes in Trypanosoma brucei and homology to cytochrome c oxidase II in two different reading frames. Mol Biochem Parasitol 15:159.
12. Jasmer, DP, Feagin, JE, Stuart, K (1987). Variation of G-rich mitochondrial transcripts among stocks of Trypanosoma brucei. Mol Biochem Parasitol 22:259.
13. Simpson, L, Neckelmann, N, de la Cruz, VF, Simpson, AM, Feagin, JE, Jasmer, DP, Stuart, K (1987). Comparison of the maxicircle (mitochondrial) genomes of Leishmania tarentolae and Trypanosoma brucei at the level of nucleotide sequence. J Biol Chem 262:6182.
14. Sloof, P, van den Burg, J, Voogd, A, Benne, R (1987). The nucleotide sequence of a 3.2 kb segment of mitochondrial maxicircle DNA from Crithidia fasciculata containing the gene for cytochrome oxidase subunit III, the N-terminal part of the apocytochrome b gene and a possible frameshift gene; further evidence for the use of unusual initiator triplets in trypanosome mitochondria. Nucl Acids Res 15:51.
15. Shaw, G, Kamen, R (1986). A conserved AU sequence from the 3' untranslated region of GM-CSF mRNA mediates selective mRNA degradation. Cell 46:659.
16. Vickerman, K (1985). Developmental cycles and biology of pathogenic trypanosomes. Brit Med Bull 41:105.
17. Feagin, JE, Jasmer, DP, Stuart, K (1986). Differential mitochondrial gene expression between slender and stumpy bloodforms of Trypanosoma brucei. Mol Biochem Parasitol 20:207.
18. Parsons, M, Nelson, RG, Watkins, KP, Agabian, N (1984). Trypanosome mRNAs share a common 5' spliced leader sequence. Cell 38:309.
19. Krause, M, Hirsh, D (1987). A trans-spliced leader sequence on actin mRNA in C. elegans. Cell 49:753.

THE GENERATION OF POLY(A) HEADS ON VACCINIA LATE mRNA: PROPOSAL OF A SLIPPAGE MECHANISM.

Hendrik G. Stunnenberg*, Luisa de Magistris, Beate Schwer

European Molecular Biology Laboratory, Meyerhofstrasse 1, P.O box 10.2209, 6900 Heidelberg, F.R.G.

ABSTRACT   Vaccinia late mRNAs are characterized by the presence of a leader sequence which is not encoded in the viral genome. We have shown that in vivo the leader sequence of transcripts from the 11k late promoter consists of a capped homopolymeric stretch of approximately 35 A-residues: the poly(A) head. Analysis of the transcripts generated in a vaccinia virus specific cell-free transcription system strongly suggests that the poly(A) head is synthesized de novo, and that its synthesis is directly coupled with transcription initiation at the late specific and essential TAAAT promoter motif. The data are consistent with a slippage (stuttering) mechanism during transcription initiation. Analysis of the poly(A) head transcripts synthesized in vivo from 11k promoter mutants provides genetic data in favour of a slippage model.

INTRODUCTION

Vaccinia virus is exclusively located in the cytoplasm throughout its infectious cycle. Most (if not all) of the enzymes necessary for transcription and replication are encoded in the viral genome (reviewed in 1) although it cannot be excluded that host factors are recruited at some stage during the viral infection.

Gene expression can be divided into an early and late phase. Vaccinia promoters, early as well as late, do not share sequence homologies with other eukaryotic genes and are only recognized by the viral RNA polymerase. Early genes are transcribed (in a co-linear fashion) immediately following infection and prior to viral DNA replication (2-5 hours post-infection). Early transcription is dependent on enzymes and factors which are present in the infectious virus particle and de novo protein synthesis is not required (1). Early transcripts are terminated 20-40 nucleotides downstream of a T$_5$NT sequence motif (2). Late genes are activated post-replicationally; the transcripts have heterogeneous 3' termini because of the absence of discrete termination sites (2). A highly conserved and essential sequence motif (consensus T/A.T/A.TAAAT G/A.R.R) is characteristic of late promoters (3)

Several laboratories have reported that late transcripts are "discontinuously" synthesized obtaining a poly(A) leader sequence at their 5' termini (4,5,6). The junction between the poly(A) head and sequences co-linearly encoded in the genome is located within the TAAAT sequence motif (4,5). We have shown that in vivo the poly(A) head consists of a capped homopolymeric A-stretch of ~ 35 nucleotides (5).

We have studied poly(A) head formation using a cell-free transcription system derived from virus infected cells. This system shows a strong preference for vaccinia late promoters. We show that transcripts synthesized in this extract are indistinguishable from mature in vivo mRNAs (7). Furthermore, we show that cis-acting sequences flanking the late specific TAAAT motif affect the length of the poly(A) head formed in vivo (8).

RESULTS

Poly(A) head transcripts synthesized in vitro

Whole-cell or cytoplasmic S100 extracts were prepared from mock- or virus-infected (strain WR) HeLa S3 suspension cultures. The extracts were depleted of nucleic acids by DEAE-Sephacel column chromatography and tested for their ability to direct RNA synthesis from various promoters: the vaccinia 11k late promoter fused to a C-less cassette (11kΔC), the Adenovirus major late promoter and a Xenopus 5S maxigene. Vaccinia specific 11k transcripts are only synthesized in extracts from virus infected cells whereas HeLa RNA polymerase II and III activities are only detectable in extracts from mock-infected cells.

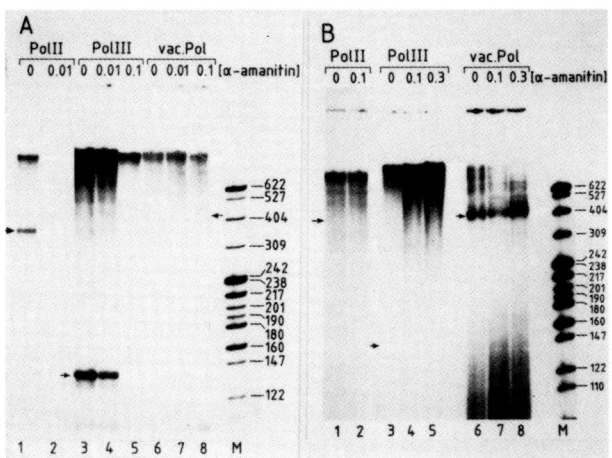

Fig.1: <u>RNA polymerase activities in extracts from mock-infected (**A**) or vaccinia virus infected HeLa cell suspensions (**B**)</u>. The extracts were prepared and run-off transcription assays performed as described (7). The HeLa RNA polymerase II and III activities were determined using pML(C2AT)19, truncated at position +390, in the absence of GTP (9) and the 5S maxigene of Xenopus (10), respectively. Vaccinia late specific RNA polymerase activity was determined using the 11KΔC template, truncated at position +390, in the absence of CTP. Different concentrations of the inhibitor α-amanitin were added to the reaction mixture as indicated (μg/ml). (M): ^{32}P-labeled HpaII digested pBR322 size markers are indicated. The arrows indicate the positions of the respective run-off transcripts (from EMBO J. 1988 7:1183).

We have shown that a poly(A) head of approximately 35 residues is present at the 5' end of the in vitro transcripts (7). Furthermore, it is evident that only run-off transcripts with a poly(A) head are detectable, even after very short incubation times, indicating that poly(A) head formation is either coupled with transcription or very rapidly added to nascent RNA chains (7) Similar results were obtained by Wright and Moss (11).

Poly(A) head formation is coupled to transcription initiation.

We have constructed a reporter gene lacking A-residues (ΔA) fused to the 11k late promoter. Transcription of the reporter gene is now independent of ATP substrate concentration, whereas ATP is required if the poly(A) heads are synthesized de novo. Run-off assays in the absence of exogenously added ATP did not result in the synthesis of poly(A) head transcripts nor in (longer) precursor transcripts lacking a poly(A) leader.
Addition of ATP to a final concentration of ≤ 100 µM appeared to result in the synthesis of transcripts with a short and heterogeneous A-head whereas full-length A-heads are.only generated upon addition of ATP to a final concentrations ≥ 250 µM (7). It was also evident that at least 50 µM ATP was required for transcription initiation. The residual endogenous ATP concentration (<10 µM) remaining in the extract after DEAE-column chromatography and extensive dialysis was not sufficient for transcription initiation. Transcription in the presence of dinucleotides (GpppA, ApA, UpA or GpA) without additional ATP added to the reaction results in the synthesis of run-off transcripts with short 5' A-stretches (Fig.2A). Only if the ATP concentration is increased to 250-500 µM, are full-length A-heads generated. Transcription assays performed in the presence of the cap-analogue m7GpppA result in higher levels of RNA transcripts. Again, transcripts have short and heterogeneous poly(A) heads in the absence of exogenous ATP. Full-length

A-heads are only synthesized if exogenous ATP is added (Fig. 2B). Addition of m7GpppG, ApG, ApU or GpppG did not result in the generation of specific run-off transcripts (data not shown).

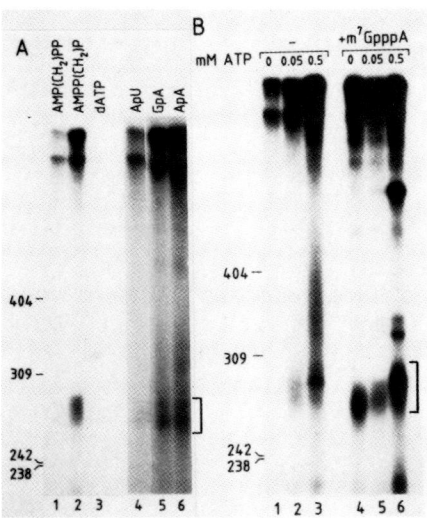

Fig. 2 <u>Use of ATP-analogues and dinucleotides</u> (**A**): Runoff transcripts from the 11KΔA template synthesized in the absence of ATP (7). The transcription reactions were supplemented by 0.5 mM of AMP(CH$_2$)PP (lane 1), 0.5 mM AMPP(CH$_2$)P (lane 2), 0.5 mM dATP (lane 3) and by 0.5 mM of each of the dinucleotides: ApU (lane 4), GpA (lane 5) and ApA (lane 5). (**B**): Transcripts synthesized from the 11KΔA construct truncated at position +265 in the absence and presence of 0.5 mM m7GpppA (lane 1-3 resp. 4-6). The transcription reactions were supplemented with different amounts of ATP as indicated. (M): ^{32}P-labeled HpaII digested pBR322 size markers. The bars indicate the position of the runoff transcripts (from EMBO J 1988 7:1183).

Several conclusions can be drawn from the <u>in vitro</u> transcription experiments. First, only dinucleotides with an A-residue at the 3' position can be used as primers for the generation of poly(A) head transcripts, indicating that free poly(A) polymerase activity is not involved in this reaction but that the synthesis of the A-stretch is template dependent. Second, shorter poly(A) stretches are generated at the 5' termini of "mature" run-off transcripts (with respect to the reporter gene) indicating that the Km for ATP during A-head formation is higher then that during transcription elongation. Third, the "linkage" of the poly(A) head to the reporter gene transcript does not require ATP as an energy source.

Cis-acting sequences affecting the length of the A-head in vivo.

The in vitro results strongly suggest that poly(A) head formation is directly coupled with transcription initiation at the A-residues of the TAAAT motif. We have constructed several mutants in the sequences flanking the TAAAT sequence motif and determined the effect of these mutations on the formation of poly(A) head transcripts in vivo. Insertion of a pyrimidine residue immediately downstream of the TAAAT motif resulted in a reduced level of transcription. More importantly, the length of the poly(A) head is significantly shorter (15-25 nt) as compared to wild-type A-heads (~35 nt). Substitution of the proximal part of the promoter **TAAAT**GAAT (wt) into **TAAAT**AAAGGAT (mutant Δ6) results in heterogeneous A-heads of 15-35 nt in length. We could show by RNA sequencing that the location of the junction between the A-head and genomic sequences is not affected by the mutations (8). These observations are not compatible with a splicing event because mutations affecting the splice acceptor site would either result in a reduced efficiency of splicing or alternatively in the use of "cryptic" acceptor sites, but would be unlikely to affect the location of the 5' donor splice site, i.e. the length of the poly(A) tract.

DISCUSSION
Possible mechanisms

The generation of a capped 5' terminal poly(A) leader which is not encoded in the viral chromosome, and the "joining" of this leader to sequences co-linearly encoded in the genome could be explained by several mechanisms, for instance (trans)splicing (12), primed initiation (13,14), slippage (15) or sequential addition of A-residues. Our biochemical and genetic data clearly rule out a splicing-like mechanism. Primed initiation and post-transcriptional addition appear very unlikely because transcripts lacking an A-head cannot be detected and the formation of the poly(A) head appears to be concomitant with transcription initiation at the TAAAT motif. Furthermore, neither the ATP dependent variation in the length of the poly(A) head nor the requirement of dinucleotides with an A-residue at the 3' position could be explained by these models. A primed initiation model as described for corona virus (14) would require a base complementarity between the leader and genomic sequences which is not apparent.

Proposal of a slippage mechanism

The alternative model for the generation of a 5' terminal poly(A) stretch would be a slippage (stuttering) mechanism. This model was suggested by Kassavetis and co-workers (14) for bacteriophage T4 transcription in the late phase of infection. They postulated that if transcription initiation starts on three consecutive A-residues (TAT<u>AAA</u>TA of the late T4 promoter), the short transcript pppApApA would be loosely bound and could slip in a 3'->5' direction. As a consequence, an additional A-residue could be incorporated and the transcript would thus contain four A-residues. Several rounds of slippage and A-incorporation would result in a 5' poly(A) stretch.

Our <u>in vitro</u> data are consistent with a slippage mechanism. The dinucleotide and cap-analogue experiments indicate that initiation of poly(A) head synthesis is dependent on an A-residue at the 3' position and therefore seems to be template restricted Based on these experiments, transcription initiation can be placed at the A-residues of the vaccinia promoter

(TAAAT) which closely resembles the start site of the bacteriophage T4 promoter. Furthermore, we observe an ATP concentration dependent variation in the length of the poly(A) head *in vitro* and transcripts lacking a poly(A) head are not detectable.

An intriguing question is how a relatively discrete length of poly(A) is normally generated, i.e. what triggers the transition from slippage/A-addition to productive transcription? This process is probably comparable with the phenomenon of abortive versus productive transcription as described for transcription by E. coli RNA polymerase (16). The stressed intermediate model postulated by Straney and Crothers (16) implies a conformational constraint on the initiation complex since the polymerase-promoter contacts are not released during the synthesis of short primary transcripts. This constraint can be released by either ejecting the short transcript (abortion) or release of upstream promoter-polymerase contacts, (simultaneous) loss of sigma factor and transition into an elongation complex (productive transcription). An implication of the stressed-intermediate model, which could be verified, is that mutations in sequences involved in polymerase-promoter binding would affect the abortive/productive transcription ratio. Likewise, one can predict from our slippage model, that mutations in sequences flanking the site of transcription initiation (TAAAT motif) could change the conformation of a "slippage committed" initiation complex thereby affecting the intrinsic poly(A) length "measurement". The analysis of *in vivo* RNA transcripts derived from promoter mutants confirmed that cis-acting DNA sequences affect the length of the poly(A) head without affecting the site of A-addition.

A question which we are now addressing, is the biological function of the 5' poly(A) head. Preliminary data suggest that the presence of this unusual leader sequence affects the stability and/or the translation efficiency of the late messenger.

REFERENCES

1. Moss B (1985). "Replication of poxviruses" In B.N. Fields (ed) "Virology" p 685. Raven Press, NY.
2. Yuen L, Moss B (1986). Multiple 3' ends of mRNA encoding Vaccinia virus growth factor occur within a series of repeated sequences downstream of T clusters. J Virol 60:320.
3. Hänggi M, Bannwarth W, Stunnenberg HG (1986) Conserved TAAAT motif in vaccinia virus promoters: overlapping TATA box and site of transcription initiation. EMBO J 5:1071.
4. Schwer B, Visca P, Vos JC, Stunnenberg, HG (1987). Discontinuous transcription or RNA processing of vaccinia virus late messengers results in a 5' poly(A) leader. Cell 50:163.
5. Bertholet C, Van Meir E, Heggeler-Bordier B, Wittek R (1987). Vaccinia virus produces late mRNAs by discontinuous synthesis Cell 50:153.
6. Patel DD, Pickup DJ (1987) Messenger RNAs of a strongly-expressed late gene of cowpox virus contain 5'-terminal poly(A) sequences. EMBO J 6:3787.
7. Schwer B, Stunnenberg HG (1988) Vaccinia virus late transcripts generated in vitro have a poly(A) head.EMBO J 7:1183.
8. De Magistris L, Stunnenberg HG (1988) Cis-acting sequences affecting the length of the poly(A) head of vaccinia virus late transcripts. Nucl Acid Res 16 (8): in press
9. Sakonju S, Bogenhagen DF, Brown, DD (1980) A control region in the center of the 5S RNA gene directs specific initiation of transcription: I. The 5' border of the region..Cell 19:13.
10. Sawadogo M, Roeder, RG. (1985) Factors involved in specific transcription by human RNA polymerase II: analysis by a rapid and quantitative in vitro assay. Proc Natl Acad Sci USA 82:4394.
11. Wright CF, Moss, B (1987) In vitro synthesis of vaccinia virus late mRNA containing a 5' poly(A) leader sequence. Proc Natl Acad Sci USA 84:8883.

12. Sharp PA (1987) Trans splicing: variation on a familiar theme? Cell 50:147.
13. Beaton AR, Krug RM (1981). Selected host-cell capped RNA fragments prime influenza viral RNA transcription in vivo. Nucl Acids Res 9:4423.
14. Makino S, Stohlman SA, Lai MMC (1986) Leader sequences of murine coronavirus mRNAs can be freely reassorted: Evidence for the role of free leader RNA in transcription. Proc Natl Acad Sci USA 83:4204.
15. Kassavetis GA, Zentner PG, Geiduschek EP (1986). Transcription at bacteriophage T4 variant late promoters. J Biol Chem 261:14256.
16. Straney DC, Crothers DM (1987) A stressed intermediate in the formation of stably initiated RNA chains at the Escherichia coli lac UV5 promoter J Mol Biol 193:267.

THE ROLE OF 16S RNA IN RIBOSOME FUNCTION: *IN VITRO* SYNTHESIS, ASSEMBLY AND FUNCTION OF 30S RIBOSOMES CONTAINING SINGLE BASE ALTERATIONS IN THE 16S RNA

D. Negre, P.R. Cunningham, C. Weitzmann, R. Denman, J. Colgan, K. Nurse, and J. Ofengand

Roche Institute of Molecular Biology
Roche Research Center
Nutley, New Jersey 07110 USA

An *in vitro* system for the site-specific mutagenesis of 16S RNA was used to make 10 single base changes around C1400, a residue known to be at the decoding site. C1400 was replaced by U, A, or G, 5 single base deletions from 1397-1404 were made, and C or U was inserted next to C1400. Another mutant had 7 added 3'-end nucleotides such that a stem and loop involving the anti Shine-Dalgarno sequence could form. The activity of these 11 mutants in A and P site binding, and in initiation-dependent and initiation-independent peptide synthesis was analyzed. None of the base substitutions of C1400 were markedly inhibitory although C1400 is almost completely conserved in 16S RNA. The insertions and deletions blocked initiation-dependent peptide synthesis but stimulated the initiation-independent reaction. The effects on tRNA binding were variable. The extra stem and loop at the 3' end blocked initiation-dependent peptide synthesis, but not the other assays. The only alteration to block all ribosomal function was the deletion of G1401. It appears that while the conserved and crosslinkable C1400 is not essential for function, the adjacent conserved G1401 is.

INTRODUCTION

The *Escherichia coli* ribosome consists of two unequally sized subunits, 30S and 50S, each containing a single large M_r RNA, the 16S and 23S RNAs, complexed with 21 or 32 proteins, respectively. The 50S subunit also contains one molecule of 5S RNA. Previously, workers in the field considered the proteins of the ribosome to be the functional entities by analogy with the prevailing belief that enzymes were exclusively proteins. It has become increasingly evident, however, that the role of the large M_r RNAs is not simply that of a scaffold on which to spatially organize functionally active proteins, but rather that the RNA itself is involved in the function of the ribosome. This view is supported by the direct role played by the 3'-end of 16S RNA in mRNA discrimination (1), by the existence of a number of antibiotic resistant mutants due to either a single base change in the 16S or 23S RNA or to methylation of a specific base (2,3), and by the unique spatial proximity of C1400 to the 5'-anticodon base of P site bound tRNA (4).

In addition, ribosomal RNA contains several long evolutionarily conserved sequences, and methylated or otherwise modified residues present at discrete sites (5). Interestingly, both of these features tend to be associated with functionally significant segments of ribosomal RNA. For example, the crosslinkable C1400 residue is located in the center of a highly conserved 14 nucleotide sequence which also contains two of the ten methylated residues of 16S RNA.

In order to learn how ribosomal RNA is involved in ribosome function, we developed a system for introducing single base alterations in 16S RNA in a way which allows the subsequent analysis of pure mutant ribosomes for their capacity to carry out all of the partial reactions normally involved in the process of protein synthesis (6,7). Other workers have previously reported making ribosomes mutant in their 16S RNA moiety (reviewed in 6), but assays of ribosome function in protein synthesis were limited by the presence of an unknown fraction of wild-type ribosomes in the mutant ribosome preparations. More recently, a system for directing a specific mRNA exclusively to the mutated ribosomes has been devised (8), and preliminary results on mutant ribosome function *in vivo* have been reported (9).

In this paper, we describe the effect of eleven mutations, ten of which involve or are nearby to the above-mentioned C1400 residue, on the protein synthesizing capacity of the ribosome.

RESULTS

In vitro generation of mutant *E. coli* 30S ribosomes. In order to obtain mutant 30S ribosomes for functional analysis which were uncontaminated by wild-type particles, both mutant RNA and mutant ribosomes were made *in vitro*. Transcription *in vitro* from suitably mutated DNA was used to obtain 16S RNA which was then combined with a complete set of isolated 30S proteins and reconstituted into an intact 30S particle. When combined with isolated 50S subunits, pure mutant 70S particles were obtained. The parent plasmid containing the wild-type 16S RNA sequence fused to the class III ϕ10 T7 polymerase promoter was inserted into pUC19 and mutated by the cassette technique (6,10). The set of mutants is listed in Table 1. Excellent transcription of full-length RNA was obtained in all cases, an average yield being 500 moles RNA per mole of template. Thus, milligram quantities of 16S RNA were available. The expected RNA product was confirmed in every case by RNA sequencing analysis (6). As shown in Table 1, there are two important differences between the synthetic RNA and natural RNA. First, there are three extra G residues at the 5'-end, and the second nucleotide of the natural sequence has been changed from A to G. Second, there are no methylated nucleotides in the synthetic RNA. In addition, in about 20% of the molecules, there is an extra nucleotide at the 3'-end (6), and presumably the 5'-terminal G is a 5'-triphosphate.

Despite these deviations from natural RNA, the synthetic product reconstituted into ribosomal particles which both looked like ribosomes (6) and were functional (see below). Compared to the classical reconstitution conditions (11), it was necessary to increase both ionic strength and temperature to obtain good reconstitution (6). Previously, we reported that G1400 did not assemble well (6). Subsequently, we found that by reducing the $Mg(OAc)_2$ from 20 mM to 16 mM, adding 3 mM spermidine, 5 mM mercaptoethanol, 100 units/ml of RNasin (Promega), and doubling the RNA, but not the protein, concentration, suitable reconstitution of the G1400 mutant could be obtained without seriously degrading the ability of the other mutants to reconstitute. For functional studies, the fractions cosedimenting with a marker 30S peak were pooled and recovered (6). Comparison by HPLC (12) of the protein content of each mutant ribosome with control 30S did not reveal the specific absence of any proteins.

TABLE 1

MUTANT RIBOSOMES CONSTRUCTED[a]

	1	1393	1400	1408	1542
NAT	A A A	U A C A C A C C G	m⁴Cm C C C G U m⁵C	A	A
SYN	G G G · G ·	· · · · · · · · ·	· · · · · · · C	·	·
U1400	G G G · G ·	· · · · · · · · ·	U · · · · · · C	·	·
A1400	G G G · G ·	· · · · · · · · ·	A · · · · · · C	·	·
G1400	G G G · G ·	· · · · · · · · ·	G · · · · · · C	·	·
Δ1397	G G G · G ·	· · · · · · · · ·	· · · · · · · C	·	·
Δ1398	G G G · G ·	· · · · · · · · ·	· · · · · · · C	·	·
Δ1400	G G G · G ·	· · · · · · · · ·	· · · · · · · C	·	·
Δ1401	G G G · G ·	· · · · · · · · ·	· · · · · · · C	·	·
Δ1402	G G G · G ·	· · · · · · · · ·	· · · · · · · C	·	·
U1400.1	G G G · G ·	· · · · · · · · ·	· U · · · · · C	·	·
C1400.1	G G G · G ·	· · · · · · · · ·	· C · · · · · ·	·	·
3'(+7)	G G G · G ·	· · · · · · · · ·	· · · · · · · C	·	· G G U C U A G

[a]Numbers refer to the sequence of natural 16S RNA (15). NAT, 16S RNA isolated from 30S subunits; SYN, 16S RNA made in vitro (6); Δ, a deletion of that residue also noted by a dotted rectangle; 1400.1, insertion of a single additional base 3' to the 1400 residue; 3'(+7), RNA made by linearizing the plasmid DNA with XbaI instead of MstII. Runoff transcription yielded the extra length RNA. Δ1400 is equivalent to Δ1399. Δ1402 is equivalent to Δ1403 and Δ1404.

Functional activity of synthetic 30S ribosomes. The functional activity of 30S reconstituted from natural and synthetic RNA is compared in Fig. 1 with isolated 30S particles. The main partial reactions of protein synthesis were studied, namely, P and A site binding, the initiation reaction, and peptide chain elongation. All assays were performed with 30S plus a 1.5 fold molar excess of 50S ribosomes. The results were corrected for reaction in the presence of 50S alone. P site binding and both peptide synthesis assays (6,7) were codon-dependent, A site binding (13) was EFTu-dependent, and all assays were proportional to the amount of 30S ribosomes added. Natural reconstituted 30S was almost fully active in all the assays confirming the results of Held et al. (11), albeit under different reconstitution and assay conditions. As reported previously (6), synthetic 30S was about half as active as isolated 30S in P site binding. Fig. 1 shows that this result can be extended to the other assays as well. The mutant bearing 7 extra nucleotides at the 3'-end of its RNA was as active as synthetic 30S in all assays except for initiation-dependent dipeptide formation which was strongly inhibited. This surprising result can be explained by the fact that, by chance, the sequence of the contiguous MstII and XbaI restriction sites (Table 1) created the possibility of intramolecular base-pairing between residues 1534-1537 and 1542-1545 of the 16S RNA. This placed

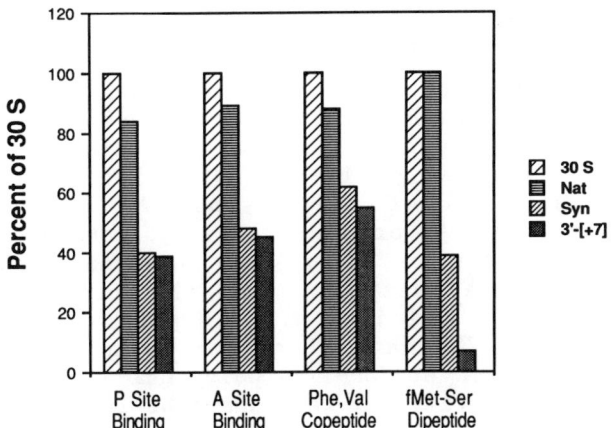

FIGURE 1. Functional activities of synthetic and natural reconstituted 30S ribosomes.

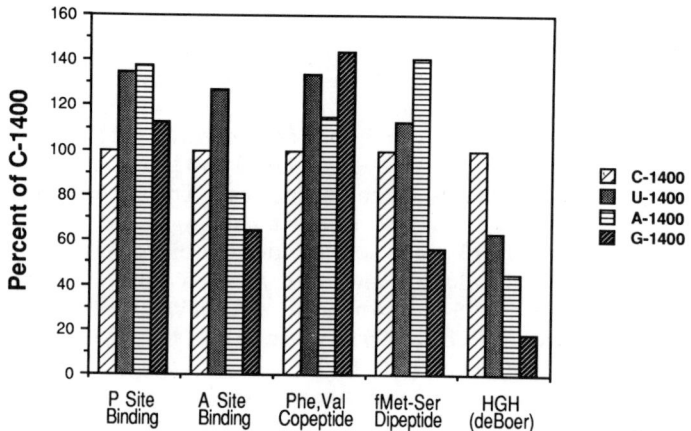

FIGURE 2. Functional activities of position 1400 mutants. HGH values are those for the *in vivo* synthesis of immunoprecipitable human growth hormone polypeptide using the specialized ribosome system described by de Boer (8), taken when the amount of protein produced was approximately proportional to time (9).

residues 1537-1541, the anti Shine-Dalgarno sequence for the β-lactamase mRNA used in the assay, in a stem-loop structure where it was probably unavailable for binding.

Base substitutions at C1400. The functional effect of base changes at C1400, the invariant C residue which is uniquely crosslinkable to tRNA, is shown in Fig. 2. In this Figure, the activity is given relative to the parent C1400 as 100 percent. Despite the almost complete conservation of C at position 1400 or its equivalent in small subunit RNA of all organisms examined (14), replacement by U, A, or G had only moderate effect. The largest change was produced by G1400 but even in this case, the inhibition was only two-fold and only for A site binding and chain initiation. The HGH set of values is calculated from the data of deBoer *et al.* (9) for the *in vivo* expression of HGH-like polypeptide. These results show a lower activity for the same substitutions, but are similar in that even G1400 possessed some activity. As the results were obtained *in vivo*, they not only reflect the ability to make a peptide large enough to be immuno-reactive, but also the ability of the mutant RNAs to be assembled into ribosomes. As we previously showed that the

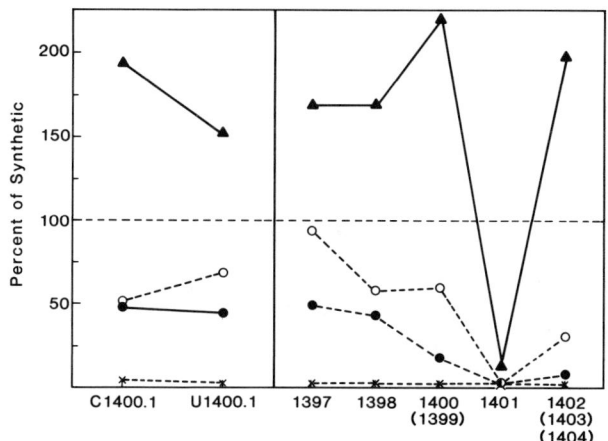

FIGURE 3. Functional activities of insertions and deletions between positions 1397 and 1404. Values are expressed as percent of synthetic, which were 40, 48, 62, and 39 percent of isolated 30S for P site binding, A site binding, Phe,Val copeptide synthesis, and fMet-Ser dipeptide synthesis, respectively. ●, P site binding; O, A site binding; ▲, Phe,Val copeptide synthesis; X, fMet-Ser dipeptide synthesis. Position numbers in () denote equivalent positions on the 16S RNA.

G1400 mutant assembled poorly (6), it is not surprising that our functional tests of G1400, which only examine pre-assembled ribosomes, should be higher than those observed by de Boer and coworkers.

<u>Insertion and deletion mutants</u>. Since base substitution at position 1400 had little effect, the position was deleted. This caused a satisfyingly large drop in P site binding, and completely inactivated fMet-Ser formation. However, Phe,Val copeptide synthesis was stimulated, reaching a value 137% of isolated 30S (Fig. 3). A series of single base deletions on either side of 1400 were then made to determine if this result was due to a general effect of one base shortening of this single-stranded sequence or to a more specific effect. A variety of changes were found. The most striking was the specific requirement for position 1401, the deletion of which was the only mutation to inhibit all functions.

The effect of insertions adjacent to 1400 was explored because we wished to know which C would be crosslinked if an additional one were placed in the CCG1401 sequence. Normally, only C1400 is crosslinked (7). In this mutant, C1400.1 was crosslinked (data not shown). That is, the C residue 5' to G1401 appears to be the one in the correct spatial position rather than the second C 3' to A1398. The functional effects of the two insertions are also shown in Fig. 3.

DISCUSSION

These results show that synthetic ribosomes are active in all of the partial reactions of protein synthesis. This was shown previously for P site binding (6) and for peptide synthesis (7), and is now shown for factor-dependent A-site binding. It is thus clear that the 10 methylated bases of 16S RNA are not essential for any of these partial reactions and that the 5' end of 16S RNA can be modified as well. While the 3' end modification inhibits initiation of protein synthesis, this probably is due to the specific sequence used and not to a general requirement for a normal 3' end.

Methylated bases may play a role in assembly. The synthetic RNA requirement for higher temperature and a higher salt concentration suggests that the methylated residues act to reduce the activation energy for folding under physiological conditions. In addition, methylation might help to direct folding along a productive, as opposed to a dead-end, path by creating a new lowest energy state at critical branch points of the assembly pathway. In this regard, it should be noted that when the standard conditions (11) were used with synthetic RNA, a particle of 25S was found which was functionally inactive (7).

The failure of C1400 substitutions to affect function more strongly was surprising. This residue is almost entirely conserved among small subunit rRNAs from widely diverse species and is even more highly conserved than its surrounding residues of the conserved 1394-1408 sequence (14). Moreover, it is uniquely positioned in the 30S ribosome since it is the only residue which can be crosslinked to the tRNA anticodon under the described conditions (7). Nevertheless, no functional effect of base substitution could be detected *in vitro*. As similar results were reported *in vivo* (9), there appears to be no strong rationale for the retention of C at this position.

Three conclusions can be drawn from the insertion-deletion studies. First, since all of the mutations inactivate peptide chain initiation, the initiation complex appears to be very finely tuned in this region. Second, all but one of the changes markedly stimulated initiation-independent peptide synthesis to levels even above that of isolated 30S, and there appeared to be an approximate inverse correlation with P site binding. If the relative affinity of tRNA for the P site controls the rate of translocation, then since translocation is the rate-limiting reaction in peptide formation (15), such a correlation might be expected. Since the P site assay is simply filter binding, a low affinity association would not be detected. This may also be the explanation for the high activity in Phe,Val synthesis for Δ1402 with an almost nonexistent P site binding value. Third, deletion of G1401 blocked all functional activity. Possibly, the 30S fails to associate with 50S. Whatever the explanation, it is striking that a single base deletion has such a marked effect. Base substitutions at G1401 are under study.

The ability of P site bound tRNA to be crosslinked is a measure of the orientation of the tRNA anticodon loop in the P site (7,16). For all mutants with measurable levels of P site binding, the crosslinking yield was like that of isolated 30S, except for A1400 and G1400 which, as purines, were not expected to be crosslinked (17). This result indicates that those tRNA molecules which did bind to the P site were bound correctly insofar as their anticodon was concerned.

Additional mutagenesis studies are underway, at these and other sites, in order to obtain new insight into the way ribosomal RNA participates directly in the protein synthetic process.

REFERENCES

1. Shine J, Dalgarno L (1974). The 3'-terminal sequence of *E. coli* 16S ribosomal RNA: Complementarity to nonsense triplets and ribosome binding sites. Proc Natl Acad Sci USA 71:1342.
2. Helser T, Davies JE, Dahlberg JE (1972). Mechanism of kasugamycin resistance in *Escherichia coli*. Nature New Biol 235:6.
3. Cundliffe E (1987). On the nature of antibiotic binding sites in ribosomes. Biochimie 69:863.

4. Ofengand J, Ciesiolka J, Denman R, Nurse K (1986). Structural and functional interactions of the tRNA-ribosome complex. In Hardesty B, Kramer G (eds): "Structure, Function and Genetics of Ribosomes", New York: Springer-Verlag, p. 473.
5. Noller HF (1984). Structure of ribosomal RNA. Ann Rev Biochem 53:119.
6. Krzyzosiak W, Denman R, Nurse K, Hellmann W, Boublik M, Gehrke CW, Agris PF, Ofengand J (1987). *In vitro* synthesis of 16S ribosomal RNA containing single base changes and assembly into a functional 30S ribosome. Biochemistry 26:2353.
7. Ofengand J, Denman R, Negre D, Krzyzosiak W, Nurse K, Colgan J (1988). Structural and functional relationships at the decoding site of the *Escherichia coli* ribosome. In Sarma RH, Sarma MH (eds): "Structure and Expression: I. From Proteins to Ribosomes", New York: Adenine Press, p. 209.
8. Hui AS, de Boer HA (1987). Specialized ribosome system: Preferential translation of a single mRNA species by a subpopulation of mutated ribosomes in *Escherichia coli*. Proc Natl Acad Sci USA 84:287.
9. de Boer HA, Eaton DH, Hui AS (1987). A novel approach for mutational analysis of the 16S rRNA molecule using the specialized ribosome system in *Escherichia coli*. In Tuite MF, Picard M, Bolotin-Fukuhara M (eds): "Genetics of Translation: New Approaches" NATO ASI Series H Cell Biology, New York: Springer-Verlag, Vol. 14, p. 343.
10. Krzyzosiak WJ, Denman R, Cunningham P, Ofengand J (1988). An efficiently mutagenizable recombinant plasmid for *in vitro* transcription of the *E. coli* 16S RNA gene. Mol Gen Genetics (submitted).
11. Held WA, Mizushima S, Nomura M (1973). Reconstitution of *Escherichia coli* 30S ribosomal subunits from purified molecular components. J Biol Chem 248:5720.
12. Kerlavage AR, Weitzmann C, Hasan T, Cooperman BS (1983). Reversed-phage high-performance liquid chromatography of *Escherichia coli* ribosomal proteins. Characteristics of the separation of a complex protein mixture. J Chromatogr 266:225.
13. Hsu L, Lin FL, Nurse K, Ofengand J (1983). Covalent crosslinking of *Escherichia coli* phenylalanyl-tRNA and valyl-tRNA to the ribosomal A site via photoaffinity probes attached to the 4-thiouridine residue. J Mol Biol 172:57.

14. Ofengand J, Ciesiolka J, Nurse K (1986). Ribosomal RNA at the decoding site of the tRNA-ribosome complex. In van Knippenberg P (ed): "NATO ASI Series", New York: Plenum Press, Vol. 110, p 273.
15. Gast FH, Peters F, Pingoud A (1987). The role of translocation in ribosomal accuracy. J Biol Chem 262:11920.
16. Ofengand J, Liou R (1981). Correct codon-anticodon base pairing at the 5' anticodon position blocks covalent crosslinking between transfer ribonucleic acid and 16S RNA at the ribosomal P site. Biochemistry 20:552.
17. Ofengand J, Liou R (1980). Evidence for pyrimidine-pyrimidine cyclobutane dimer formation in the covalent crosslinking between transfer ribonucleic acid and 16S RNA at the ribosomal P site. Biochemistry 19:4814.

RIBOSOMAL RNA AND UGA-DEPENDENT PEPTIDE CHAIN TERMINATION[1]

Emanuel J. Murgola,[2] H. Ulrich Göringer,[3]
Albert E. Dahlberg[3] and Kathryn A. Hijazi[2]

Department of Molecular Genetics, The University of Texas M. D. Anderson Cancer Center, Houston, Texas 77030[2] and Section of Biochemistry, Brown University, Providence, Rhode Island 02912[3]

ABSTRACT We have isolated and characterized a novel codon-specific translational suppressor in Escherichia coli: a mutant 16S ribosomal RNA. It is UGA-specific: on glucose minimal medium, it does not suppress UAA or UAG mutations or missense mutations related to UGA by a single base change. The suppressor is context-dependent, leading to suppression of UGA at only two of the four codon positions tested in trpA. The suppressor mutant resulted from the deletion of the C residue at nucleotide position 1054 in 16S rRNA. The existence of this suppressor has led us to propose that peptide chain termination at UGA codons normally involves base-pairing of the mRNA codon with a particular 5'-UCA-3' triplet in 16S ribosomal RNA. Preliminary results from site-directed mutagenesis experiments support this hypothesis. We suggest further that the helical region encompassing residue 1054 of 16S ribosomal RNA is a domain for termination at all three termination codons, UGA, UAA and UAG.

[1]This work was supported by American Cancer Society grant NP167 to E.J.M. and by grants from the National Institute of General Medical Sciences to E.J.M. (GM21499) and to A.E.D. (GM19756). H.U.G. was supported by an Otto Hahn Fellowship from the Max Planck Gesellschaft, Federal Republic of Germany.

INTRODUCTION

In the last few years it has become increasingly apparent that ribosomal RNA (rRNA) plays an important functional role in mRNA-dependent polypeptide synthesis. Considerable progress has been made in identifying structure/function relationships at specific sites in the 16S, 23S and 5S rRNAs of Escherichia coli (for review see ref. 1). Of particular interest are the examples of direct base pairing interactions between rRNA and mRNA at the different stages of translation. It was shown that, during the initiation of protein synthesis, the polypyrimidine sequence (CCUCC) near the 3' end of 16S rRNA pairs directly with the polypurine Shine-Dalgarno sequence 5' to the AUG initiation codon in mRNAs (2,3). More recently this same sequence in rRNA has been implicated in base pairing to a pentaguanylate sequence in release factor 2 (RF2) mRNA during elongation to produce a +1 frameshift (4). Here we report that rRNA may also base pair with mRNA during the termination step of translation.

RESULTS AND DISCUSSION

Suppressor Isolation.

Recently we isolated a novel codon-specific translational suppressor in E. coli: a mutant 16S rRNA that suppresses UGA mutations (5,6). This unusual UGA suppressor was obtained "the old fashion way", spontaneously and in the chromosome, without the assistance of plasmids and elegant mutagenic tricks. Using an appropriate UGA mutation in trpA (which codes for the alpha subunit of tryptophan synthetase), we constructed the parent E. coli strain so as to avoid turning up UGA suppressors previously isolated by us and others, such as mutant forms of tryptophan tRNA and of glycine tRNAs. Beginning with the tryptophan (Trp) auxotroph trpA(UGA211), we plated a culture on glucose minimal medium to select for Trp-independent colonies. The protoWhich were then screened to distinguish suppressed trpA mutants from trpA revertants. Of 10 suppressed mutants chosen for further analysis, one proved to be caused by a mutation in rrnB, one of seven rRNA operons in the E. coli chromosome. Cloning and DNA sequence analysis of the suppressor gene led to the identification of the suppressor mutation as a deletion of the C residue at nucleotide position 1054 of 16S rRNA. Details of the isolation, mapping, cloning and sequence analysis are presented elsewhere (5,6). Acceptable genotypic designations for the new suppressor are either rrsB(SuUGA) or rrsB(SuUGA-ΔC1054) where rrs stands for ribosomal RNA of the small subunit (hence 16S rRNA). [Phenotypically the mutant may be referred to simply as ΔC1054.] Clearly the existence of seven rrn operons in E. coli does not prevent the isolation of chromosomal rRNA suppressor mutants.

The Suppressor is UGA-Specific but Context-Dependent.

On glucose minimal medium, rrsB(SuUGA-ΔC1054) does not suppress UAA or UAG mutations, even at positions in a gene where it suppresses UGA. That was tested at four positions in trpA, and at several positions in the phage T4 genome. Furthermore, it does not correct missense mutations in which the mutant codon differs from UGA by a single base (AGA, CGA, UGU, UGC, UGG). The failure of rrsB(SuUGA) to suppress those mutations, particularly the missense ones, argues that general translational accuracy is not affected by the suppressor mutation, at least that it is not displaying greatly increased misreading. Rather, the specificity for UGA suggests a direct involvement of the mutant rRNA in UGA-dependent peptide chain termination.

The suppressor, however, did not correct all UGA mutations tested. In trpA, for example, where the mutations are well-characterized, four UGA mutations were tested, at positions 15, 211, 234 and 243 (there are 268 amino acids in the alpha subunit). But only two of those four UGA mutations are suppressible by rrsB(SuUGA-ΔC1054), those at positions 211 and 243. This pattern of context-dependent suppression differs significantly from that of UGA suppressors derived from each of the three glycine tRNAs (unpublished data). The differential suppression by the rRNA UGA suppressor could be due to the mutant 16S rRNA allowing only certain tRNAs to misread UGA at 15 and 234, namely those that insert amino acids that are not functional at those positions. But this "screening" function would have to be very specific since the amino acid requirements are different at those two positions. Alternatively, the ΔC1054 mutation may allow no readthrough at 15 and 234, while allowing readthrough (suppression) at 211 and 243. Experiments are in progress to distinguish between these two possibilities. Although either situation would be interesting, the second possibility might reveal some intrinsic property of 16S rRNA in the specificity of UGA "recognition" (see the hypothesis proposed below).

Effect of ΔC1054 on ribosomal protein S5.

Ribosomal proteins and rRNAs participate in the cooperative assembly of the ribosome. It is not unlikely, therefore, that an rRNA mutation, particulary a deletion, might affect the incorporation of a ribosomal protein into a mature subunit. Consequently, we examined the proteins of the small subunit from a strain containing the ΔC1054 mutation. The suppressor gene was cloned in the multicopy plasmid pKK3535 (pBR322 containing the rrnB operon inserted at the BamHI site) and the plasmid introduced into strain HB101. The 70S ribosomes were isolated from cells grown in L-broth (plus ampicillin) and the ribosomal proteins were separated by one-dimensional SDS-polyacrylamide gel electrophoresis. As a control, ribosomes were isolated from HB101 containing the wild-type pKK3535. Comparison of the two gels indicated that the proteins from the mutant subunits differed from wild-type only in that the mutant has about 50% of the amount of S5 found in wild-type ribosomes. Direct sequencing of the 16S rRNA showed that plasmid-coded mutant ribosomes constitute approximately 55% of the total ribosome pool. So the simplest interpretation of our result is that none of the ΔC1054 mutant 70S ribosomes has S5.

There are reasons to believe, however, that the S5 loss is not intimately involved in the UGA-suppressor phenotype of the ΔC1054 mutation but rather that it is most likely an unrelated consequence of the deletion. The hypothesis proposed in the next section leads to a testable prediction about S5.

mRNA-rRNA Base-Pairing: A Plausible Model for UGA-Dependent Peptide Chain Termination.

Attempts to explain the suppressor phenotype of the ΔC1054 mutation have led us to a model for normal UGA-dependent termination and the suggestion that a similar mechanism operates for termination at UAA and UAG codons. We propose that for UGA-dependent termination, anti-parallel base-pairing is required between the mRNA UGA codon and a particular complementary triplet in 16S rRNA. For the rRNA triplet we suggest specifically one of the two tandem 5'-UCA-3' sequences comprising nucleotides 1199 to 1201 or 1202 to 1204 (Fig. 1). Such base-pairing prevents any tRNA from misreading UGA and therefore allows for subsequent peptidyl-tRNA hydrolysis by the release factor RF2. We imagine, therefore, that the deletion of C1054 makes the 5'-UCA-3' triplet inaccessible for base-pairing with the UGA codon, possibly by altering the conformation of the 16S rRNA in that region, and so allows misreading of UGA and readthrough by some normal tRNA. The role of RF2 is not specified in this statement of the model.

This proposal is supported by several observations. First, the 1054 helix (Fig. 1) has several features that are highly conserved among 16S-like rRNAs, in particular the 1053-1054 "bulge" and one or two 5'-UCA-3' triplets, on the opposite side of the helix, in the 1200 region (8). It is

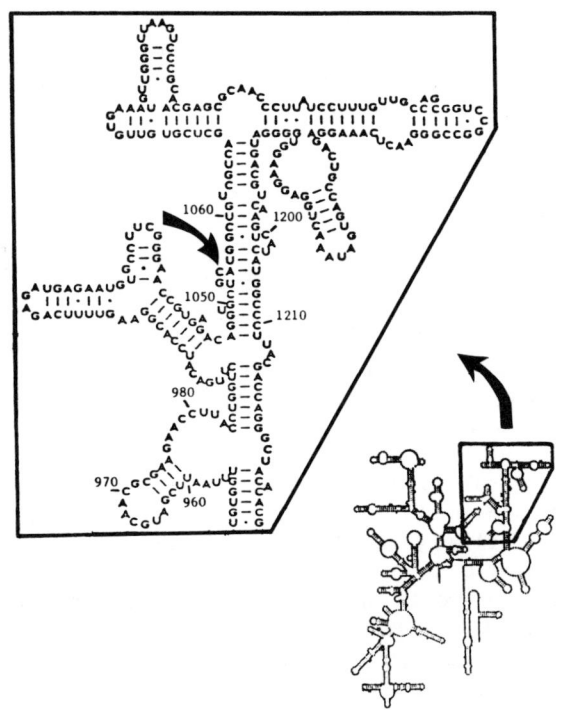

FIGURE 1. Sequence and secondary structure in the region of C1054 of E. coli 16S rRNA, enlarged from the model of Noller et al (7).

notable that <u>Mycoplasma capricolum</u> does not have two tandem 5'-UCA-3' triplets, or even one, in the 1200 region. Instead it has the hexamer 5'-CUACUA-3' (8). This seems significant since in that organism UGA is not a termination signal; it codes instead for Trp (9). Next, structural studies indicate that no ribosomal proteins bind to the 1054 helix, only "around" it (7,10). That could allow for "entry" or at least proximity of mRNA and RF2. This observation and the presence in that area of conserved triplets complementary to UAA and UAG lead us to suggest further that the helical region encompassing residue 1054 is a domain for termination, for peptide chain terminations signaled not only by UGA, but also by UAA or UAG. This suggestion implies that RF1 also functions in that region, which is reasonable from an evolutionary point of view when one considers that, for eucaryotic terminations, there is only one codon-dependent release factor (11). Finally, it is comforting that a recently published three-dimensional model of 16S rRNA (10) shows the 1054 helix located very close to the 1400 region, to which A- and P-site tRNAs have been crosslinked (12).

Predictions From the Base-Pairing Hypothesis.

Based on our model for peptide chain termination, several predictions can be made. First, base-substitution mutations in the 5'-UCA-3' triplet comprising nucleotides 1199-1201, 1202 to 1204, or both, will decrease the complementarity to UGA and should therefore lead to less or no termination at UGA codons and be manifested as UGA suppressors. Next, since we suggested that the loss of S5 in the ΔC1054 is a consequence of the deletion mutation rather than an indication of S5 being involved in UGA termination, the 1199-1204 base-substitution mutants that exhibit UGA suppression will likely not have lost S5. Finally, the "new" triplets produced by site-directed, base-substitution mutagenesis, will be complementary to some codon other than UGA. So when the new triplet is 5'-UUA-3', it will be important to ask whether UAA termination is increased. And similarly for 5'-CUA-3', we would ask whether UAG termination is increased. When the new triplet is complementary to a <u>sense</u> codon, one would like to know whether termination sometimes occurs at that codon. If the base-pairing hypothesis for termination is correct, it may be that in wild-type cells the rRNA sometimes "misreads" a sense codon and leads to premature peptide chain termination (by hydrolysis or drop-off of peptidyl-tRNA).

Site-directed Mutagenesis: Preliminary Results.

In preliminary experiments, we have made single base changes at positions 1199 and 1202 in 16S rRNA by oligonucleotide-directed mutagenesis. The U residues at 1199 and 1202 were independently changed to C residues, resulting in the two mutants C1199 and C1202. In each case, a 5'-CCA-3' triplet was produced. When these mutations are present in plasmid pKK3535, the strains grow more slowly than cells containing the wild-type plasmid. In one test for UGA suppression, C1199 ribosomes read through the UGA codon while C1202 ribosomes did not. In contrast to ΔC1054, both base-substitution mutants have the wild-type amount of small subunit ribosomal protein S5. These results support the base-pairing termination hypothesis and indicate that the loss of protein S5 is not directly responsible for UGA termination in mutant ΔC1054.

SUMMARY AND CONCLUSION

The characterization of rrsB(SuUGA-ΔC1054) and subsequent analyses support our hypothesis that peptide chain termination, at least at UGA codons, involves base-pairing between the UGA codon in mRNA and a specific 5'-UCA-3' triplet in 16S rRNA. Suppression of UGA mutations (due to decreased termination at UGA) can be achieved not only by deletion of C1054 but also by base-substitution at nucleotide 1199. The UGA suppression appears to be context-dependent, and the context pattern differs dramatically from that seen with the UGA suppressors derived from the three glycine tRNAs. We have suggested that the helical region encompassing residue 1054 in 16S rRNA is a domain for termination at all three termination codons. Current models of 16S rRNA folding are consistent with this suggestion.

The question of whether a termination codon is "recognized" by a protein-RNA interaction (release factor-mRNA) or by an RNA-RNA interaction (rRNA-mRNA) has never been resolved (13,14). Our results indicate that a major determinant in the recognition process is anti-parallel base-pairing between the termination codon and a complementary triplet in 16S rRNA.

ACKNOWLEDGMENTS

We are grateful to George Pennabble for occasional technical assistance and helpful discussions, to Walter J. Pagel for editorial consultation, and to Martha S. Trinkle for assistance in the preparation of the manuscript.

REFERENCES

1. De Stasio EA, Göringer HU, Tapprich WE, Dahlberg AE (1988). Probing ribosome function through mutagenesis of ribosomal RNA. In Tuite MF, Bolotin-Fukuhara M, Picard M (eds): "Genetics of Translation," NATO ASI Series, Vol. H14, Berlin: Springer, p 17.
2. Jacob WF, Santer M, Dahlberg AE (1987). A single base change in the Shine-Dalgarno region of 16S rRNA of Escherichia coli affects translation of many proteins. Proc Nat Acad Sci USA 84:4757.
3. Hui A, de Boer HA (1987). Specialized ribosome system: Preferential translation of a single mRNA species by a subpopulation of mutated ribosomes in Escherichia coli. Proc Nat Acad Sci USA 84:4762.
4. Weiss RB, Dunn DM, Dahlberg AE, Atkins JF, Gesteland RF (1988). Reading frame switch caused by base pair formation between the 3' end of 16S rRNA and the mRNA during elongation of protein synthesis in Escherichia coli. EMBO J. 7:1503.
5. Murgola EJ (1988). trpA and its protein: Something old, something new .... In Tuite MF, Bolotin-Fukuhara M, Picard M (eds): "Genetics of Translation," NATO ASI Series, Vol. H14, Berlin: Springer, p 195.
6. Murgola EJ, Hijazi KA, Göringer HU, Dahlberg AE (1988). Mutant 16S ribosomal RNA: A codon-specific translational suppressor. Proc Nat Acad Sci USA 85:4162.
7. Noller HF, Stern S, Moazed D, Powers T, Svensson P, Changchien L-M (1987). Studies on the architecture and functions of 16S rRNA. Cold Spring Harbor Symp Quant Biol 52:695.
8. Gutell RR, Weiser B, Woese CR, Noller HF (1985). Comparative anatomy of 16S-like ribosomal RNA. Prog Nucleic Acid Res Mol Biol 32:155.
9. Yamao F, Muto A, Kawauchi Y, Iwami M, Iwagami S, Azumi Y, Osawa S (1985). UGA is read as tryptophan in Mycoplasma capricolum. Proc Nat Acad Sci USA 82:2306.

10. Brimacombe R, Atmadja J, Stiege W, Schüler D (1988). A detailed model of the three-dimensional structure of Escherichia coli 16S ribosomal RNA in situ in the 30S subunit. J Mol Biol 199:115.
11. Caskey CT, Forrester WC, Tate W (1987). Peptide chain termination. In Clark BFC, Petersen HU (eds): "Gene Expression: The Translational Step and its Control," Copenhagen: Munksgaard, p. 149.
12. Ofengand J, Ciesiolka J, Denman R, Nurse K (1986). Structural and functional interactions of the tRNA-ribosome complex. In Hardesty B, Kramer G (eds): "Structure, Function, and Genetics of Ribosomes," New York: Springer, p. 473.
13. Steege DA, Söll DG (1979). Suppression. In Goldberger RF (ed): "Biological Regulation and Development," New York: Plenum, Vol. I, p. 433.
14. Craigen WJ, Caskey CT (1987). The function, structure and regulation of E. coli peptide chain release factors. Biochimie 69:1031.

TRANSLATION CONTROL OF THE S10 RIBOSOMAL PROTEIN OPERON OF ESCHERICHIA COLI

L. Lindahl, P. Shen, and J. M. Zengel

Department of Biology, University of Rochester
Rochester, New York 14627

ABSTRACT We present a model for the autogenous control of translation of the S10 ribosomal protein operon. The notion is that an unbasepaired region in the RNA immediately upstream of the Shine-Dalgarno sequence of the first gene in the operon is used as a ribosome entry site. We propose that translation inhibition is due to intramolecular basepairing of this ribosome entry site with upstream leader RNA.

INTRODUCTION

The translation of most ribosomal protein (r-protein) operons of Escherichia coli is regulated autogenously. That is, a specific protein encoded by each operon is not only a structural component of the ribosome, but also a regulatory protein. When this protein accumulates in excess of the available binding sites on nascent rRNA, it represses the translation of its own operon (for reviews, see refs.1,2). In the S10 r-protein operon, translation is regulated by L4, the product of the third gene of the operon (3,4). L4 also regulates the transcription of the S10 operon by stimulating transcription termination at a specific site in the S10 leader (4-6). However, this communication focuses only on the L4-mediated translation regulation.

Genetic experiments in our laboratory have shown that a specific hairpin structure in the mRNA, located in the leader region 50-67 bases upstream of the S10 initiation codon, is essential for L4 regulation of translation of the S10 gene (4). Thus, mutations on either side of the stem of this hairpin abolish L4-mediated translation regulation. Translation regulation is restored in a mutant with compensatory base changes which preserve the secondary structure of the RNA even though the primary sequence differs from wild-type in four positions. Base changes in the loop of the hairpin also eliminate translation control. Interestingly, analysis of the secondary structure of wild-type and mutant transcripts has shown that at least one of the base substitutions in the loop destabilizes the stem of the hairpin (P. Shen, J. M. Zengel and L. Lindahl, manuscript submitted). Thus the effect of the loop mutations may be indirect, resulting from their impact on the stem.

In the translation initiation complex, the ribosome appears to contact no more than 20 bases upstream of the initiation codon (7). The hairpin discussed above is therefore clearly outside the ribosome binding site. We suspect that the secondary structure of the leader region between the hairpin and the ribosome binding site could be important for the L4-mediated inhibition of translation. To begin exploring this possibility, we have determined the folding pattern of the S10 leader transcript (P. Shen, J. M. Zengel and L. Lindahl, manuscript submitted). Based on these structural studies we have devised a model for the translation control of the S10 operon which is presented here.

RESULTS AND DISCUSSION

We have subjected the S10 leader transcript to secondary structure mapping with RNases T2 and V1 and with dimethylsulfate modification (P. Shen, J. M. Zengel and L. Lindahl, manuscript submitted). The results of these experiments indicate that the leader of the S10 operon may assume two secondary structures which differ only with respect to the region immediately upstream of the Shine-Dalgarno region (Fig. 1). That is, this region of the S10 leader is accessible to both single- and double-strand specific RNases. We therefore believe that there is an equilibrium between the two structures shown in Fig. 1. The relative strength of the bands created by the single- and double-stranded nucleases indicates that form A is the predominant structure. The calculated free energy of this structure is also about 10% more favorable than the energy of the folding pattern of form B.

Since the Shine-Dalgarno sequence of the S10 gene is involved in the same intramolecular base-pairing in both of

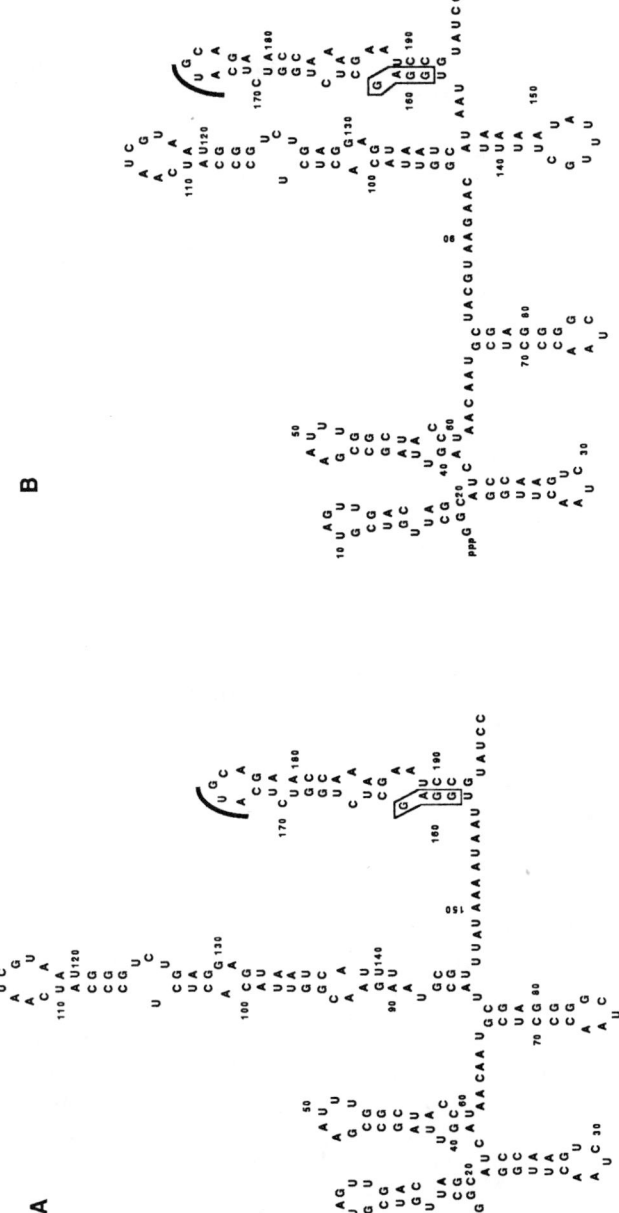

FIGURE 1. Alternate secondary structures for the S10 leader transcript.

the proposed models, it is likely that the Shine-Dalgarno region is not directly available for interaction with the 16S rRNA on an initiating 30S ribosomal subunit. It is therefore conceivable that the ribosome first must make contact with another region of the messenger before a contact between the Shine-Dalgarno sequence and the 16S rRNA can be established. The sequence between bases 146 and 157 in form A would offer the ribosome a stretch of single-stranded RNA with which it could make such an initial contact. However, in form B this region is base-paired and therefore presumably not available for ribosome binding. Thus, we speculate that form A is open for translation initiation, whereas form B is not. The regulation of translation of the S10 gene could then be explained if the equilibrium between forms A and B is affected by the concentration of free r-protein L4. The simplest model would be if L4 itself binds to the leader and thereby shifts the equilibrium from form A to form B. Binding of regulatory proteins to their specific messages has been demonstrated for the L10 (8,9) and alpha r-protein operons (10). However, we have not yet been able to demonstrate specific binding of L4 to the S10 leader. Also, it is not clear where on the S10 message L4 might bind. Our genetic analysis of the L4-mediated regulation of the S10 operon has not revealed any specific importance of the primary sequence homologies between the S10 leader and the 23S rRNA previously pointed out (11). At this point, we cannot rule out the possibility that other factor(s) are necessary to facilitate L4's binding, or that perhaps L4 acts indirectly via other factor(s).

Results obtained with several mutations in the S10 leader are consistent with, but do not prove, the hypothesis that the equilibrium between forms A and B of the leader (Fig. 1) could be the basis for L4-mediated regulation of translation. First, a deletion of bases 15 through 86 of the S10 leader has little or no effect on translation regulation, even though it eliminates transcription regulation (4). This is consistent with the proposed model because this deletion leaves essentially intact the structures involved in the proposed L4-mediated shift from the translationally open form A to the translationally closed form B (Fig. 1). Second, a 49 base deletion removing bases 94 through 142 eliminates translation control, but the translation efficiency of the mutated messenger is almost as high as that of wild-type mRNA (4). These effects are also consistent with the model because this deletion would prevent formation of the hairpin immediately upstream of the Shine-Dalgarno sequence in form B and would leave intact the single-stranded region proposed to be the initial binding site for the initiating ribosomes. Finally, an 8 base deletion removing bases 158 through 165 reduces the translation efficiency of the mutated message by about 10-fold (L. Lindahl, R.H. Archer and J.M. Zengel, unpublished results), in accordance with the proposal that this region is required for efficient translation.

Clearly, the model proposed here is speculative and needs to be tested. Mutations manipulating the relative stability of the two proposed folding patterns will of course be crucial for this. However, we must also find out why the hairpin between bases 106 and 123 is essential to the L4-mediated regulation. Some of the mutations in this region also affect the transcriptional regulation, suggesting that this may be the region where L4 binds. However, this hairpin is not involved in the shift between the two forms of the messenger shown in Fig. 1. Another challenge in the testing of the proposed model will therefore be to define the connection between the structure around base 110 with the structure of the region next to the translation initiation site.

ACKNOWLEDGMENTS

This work was supported by a Public Health Service Grant and a Research Career Development Award (to LL) from the National Institute of Allergy and Infectious Diseases.

REFERENCES

1. Nomura M, Gourse R, Baughman, G (1984). Regulation of the synthesis of ribosomes and ribosomal components. Ann Rev Biochem 53:75
2. Lindahl L, Zengel JM (1986). Ribosomal genes in Escherichia coli. Ann Rev Genet 20:297
3. Yates JL, Nomura M (1980). E. coli ribosomal protein L4 is a feedback regulatory protein. Cell 21:517.
4. Freedman LP, Zengel JM, Archer RH, Lindahl L (1987). Autogenous control of the S10 ribosomal protein operon in Escherichia coli: Genetic dissection of transcriptional and posttranscriptional regulation. Proc. Nat. Acad. Sci. USA 84:6516.
5. Lindahl L, Archer RH, Zengel JM (1983). Transcription of the S10 ribosomal protein operon is regulated by an

attenuator in the leader. Cell 33:241.
6. Lindahl L, Zengel JM (1988). Molecular mechanisms of the autogenous control of the S10 ribosomal protein operon of Escherichia coli. In Tuite M, Picard M, Bolotin-Fukuhara M (eds): "Genetics of Translation", New York: Springer-Verlag, p 105.
7. Steitz JA (1979). Genetic signals and nucleotide sequences in messenger RNA. In Goldberger RF (ed): "Biological Regulation and Development", vol 1, New York: Plenum Press, p 349.
8. Johnsen M, Christensen T, Dennis PP, Fiil NP (1982). Autogenous control: Ribosomal protein L10-L12 complex binds to the leader sequence of its mRNA. EMBO J 1:999.
9. Christensen T, Johnsen M, Fiil NP, Friesen JD (1984). RNA secondary structure and translation inhibition: analysis of mutants in the rplJ leader. EMBO J 3:1609
10. Deckman IC, Draper DE (1985). Specific interaction between ribosomal protein S4 and the alpha operon messenger RNA. Biochemistry 24:7860.
11. Olins P, Nomura M (1981) Regulation of the S10 ribosomal protein operon in E. coli. Cell 26:205.

EUKARYOTIC VIRAL 5'-LEADER SEQUENCES ACT AS TRANSLATIONAL ENHANCERS IN EUKARYOTES AND PROKARYOTES[1]

Daniel R. Gallie[2], Clarence I. Kado[3], John W.B. Hershey[4], Michael A. Wilson[5], and Virginia Walbot[2]

[2]Department of Biological Sciences, Stanford University
Stanford, California 93405-5020

ABSTRACT   Many plant and animal RNA viruses possess an extensive untranslated leader at their 5' end. In addition to their probable role in RNA-dependent replicase recognition, these leaders play an important part in determining the translation of reporter mRNAs in vivo. The leader from tobacco mosaic virus (TMV) was found to be the most stimulatory, enhancing translation up to several hundred fold. This enhancement has been observed in electroporated protoplasts from a variety of plants, in microinjected Xenopus laevis oocytes, and in prokaryotic species. From analysis of deletions of the 67 base TMV leader (Ω) we determined that the sequence involved in disome formation was responsible for the enhancement observed in X. laevis oocytes. In contrast, mutations within this subsequence had little effect on translation in protoplasts, suggesting that inherit differences do exist between plant and animal systems at the translational level. By isolating the ribosomal and initiation factor-containing fractions from wheat germ lysate and rabbit reticulocyte lysate and

[1]This work was supported by Cancer Biology Training Grant CA 2T32CA09302-11; NIH grants CA-11526 and GM22135; and Diatech Ltd.
   [3]Department of Plant Pathology, University of California, Davis, CA 95616
   [4]Department of Biological Chemistry, University of California, Davis, CA 95616
   [5]John Innes Institute, Norwich, United Kingdom

reconstituting them in various homologous or heterologous combinations, the components of each lysate responsible for the characteristic response to the presence of Ω in an mRNA were identified.

INTRODUCTION

The factors influencing the process of eukaryotic mRNA selection and translational initiation are poorly understood. Proposals, based on surveys to determine a sequence upstream of the start codon which would act as an eukaryotic equivalent of a Shine-Dalgarno region, a sequence essential for the expression of prokaryotic mRNAs in E. coli, have failed to account for the numerous exceptions that exist. The relaxed scanning model proposed by Kozak (1) has allowed some insight into the process of how eukaryotic ribosomes determine the proper start site of a particular monocistronic mRNA. Comparatively little has been done, however, to understand how other sequences within an mRNA might affect translation.

TMV RNA is one of the most efficient and competitive mRNAs in in vitro translation, and in vivo it must compete effectively with plant cellular mRNAs for available translational machinery. In some measure, this efficiency must be a consequence of the particular 5' leader sequence of TMV RNA. The untranslated leader (called Ω) is 67 bases in length and is devoid of guanosine residues (except for the cap structure, m^7GpppG). Ω forms disomic initiation complexes with 80S wheat ribosomes in vitro in the presence of sparsomycin (2). In addition to the legitimate initiation codon (AUG), an AUU sequence in the non-coding leader has been implicated in 80S ribosome binding (3, 4). There are 51 bases separating the AUU from the AUG codons, so that 80S ribosomes could bind to each site without steric hindrance. Cross-linking has been demonstrated between the 5' region of TMV RNA and the 3' terminus of wheat 18S rRNA (5), supporting the idea that disome binding could be an important mechanism in translational enhancement. Several untranslated viral leader sequences have been shown to bind two 80S ribosomes (TMV and turnip yellow mosaic virus (TYMV)(6); brome mosaic virus RNA3 (BMV)(7); Rous sarcoma virus (RSV) (8)); or even three ribosomes (alfalfa mosaic virus RNA3 (A1MV)(9). In addition to the AUG start codon causing 80S ribosomal formation, AUU codons have also been shown to cause the large ribosomal subunit to join the small subunit. Not all AUU codons upstream of the start codon possess this ability. Only those within certain positions in the leader, depending on the primary and/or secondary (and possibly tertiary) structure, enable these AUU codons to become 80S ribosomal binding sites.

Because of its lack of guanosines, Ω can be isolated from TMV RNA by RNase T_1 digestion. When subsequently purified and then circularized with T_4-RNA ligase, it failed to bind 80S wheat ribosomes but did bind one prokaryotic 70S ribosome per Ω fragment as efficiently as the linear form (2). More recently, E. coli ribosomes have been shown to interact with Ω to express the large 5' proximal open reading frame in TMV RNA and bring about cotranslational disassembly of TMV virions in vitro (10). In addition, eukaryotic ribosomes from two species, rabbit reticulocyte lysate and wheat germ, can also cause the disassembly of TMV virions through the process of translation (10). It may be fortuitous that Ω possesses the ability to interact with 70S as well as 80S ribosomes despite the absence of any consensus Shine-Dalgarno sequence (11) to base pair with the 3' end of 16S rRNA. Nevertheless, the enhancing nature of Ω on translation offers attractive opportunities to exploit this sequence as a translational enhancer in a wide variety of cell types.

RESULTS

The stimulatory nature of the TMV 67 base leader (Ω) has been observed in vitro and in vivo, in eukaryotic and prokaryotic translational systems, using reporter mRNAs of either prokaryotic or eukaryotic origin (12, 13, 14). To quantitate the effect of Ω in plant protoplasts we used the β-glucuronidase (GUS) reporter gene (15) with or without a 5'-proximal Ω sequence. In the SalI gene fragment used, the native GUS has 19 nucleotides upstream of the AUG start codon. The context of this initiator AUG codon (5'-CCCUUAUGU-3') was, according to Kozak (1), inefficient for eukaryotic translation (hereafter referred to as "bad context" GUS). To determine whether the effect of Ω on mRNA translation was influenced by the context of the initiation codon, a derivative of the GUS gene with an initiation codon context (5'-CGACCAUGG-3') close to the consensus sequence for optimal eukaryotic translational initiation was constructed (14). This derivative (hereafter referred to as "good context" GUS) has only 7 nucleotides upstream of the AUG in the SalI fragment used.

Eight micrograms of each transcript were electroporated into tobacco mesophyll protoplasts and incubated for 20 hours at 25°C and processed as described

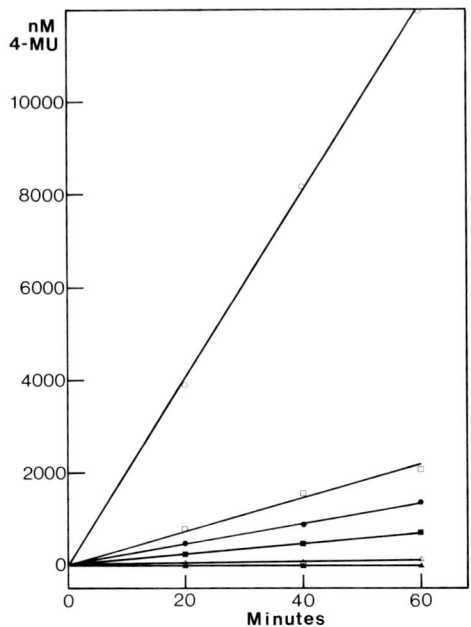

Fig. 1. Kinetic analysis of β-glucuronidase activity in extracts from electroporated tobacco mesophyll protoplasts. Graphical display of the rate of appearance of the reaction product (4-methyl-umbelliferone, (MUG). Extract volumes representing equivalent amounts of protein were added to each assay. ▲, "bad" or "good context" GUS mRNAs; △, 5'-capped "bad" or "good context" GUS mRNAs; ■, Ω-"bad context" GUS mRNA; □, Ω-"good context" GUS mRNA; ●, 5'-capped-Ω-"bad context" GUS mRNA; ○, 5'-capped-Ω-"good context" GUS mRNA.

(16). Fluorimetric quantitation of the kinetics of GUS activity (Fig. 1) and hence of GUS mRNA expression (Table 1), demonstrated that the levels of expression of uncapped "good" or "bad context" GUS mRNAs were below the limit of detection. GUS activity was detectable when the transcripts were capped and/or when Ω was present. In all cases, the presence of Ω enhanced expression markedly, stimulating the "bad context" GUS mRNA approximately 20-fold. Stimulation of "good context" GUS mRNA by Ω was even greater, showing an 80-fold increase with the capped form of the transcript (Table 1).

TABLE 1
TRANSLATIONAL ENHANCEMENT BY Ω ON GUS mRNAS ELECTROPORATED INTO TOBACCO PROTOPLASTS

mRNA	Initiation codon context	Specific activity (nmoles MUG hydrolysed/min/µg protein)	Relative fold stimulation
Uncapped			
GUS	bad	<0.01	1
Ω-GUS	bad	0.18	>18
GUS	good	<0.01	1
Ω-GUS	good	0.35	>35
5'-Capped			
GUS	bad	0.03	1
Ω-GUS	bad	0.61	20
GUS	good	0.04	1
Ω-GUS	good	3.20	80

As mentioned previously, the ability of an untranslated leader to form disome structures could be associated with the ability to enhance translation. To test this possibility, the disome-forming leader sequences from TYMV, BMV RNA3, and RSV were tested for their stimulatory ability on translation using the "bad context" form of the GUS gene (16). The monosome forming-leader from AlMV RNA4 was also tested. In marked contrast to the TMV leader which was stimulatory whether uncapped (>25 fold) or capped (>18 fold or >50 fold when compared to uncapped form without Ω) in electroporated tobacco mesophyll protoplasts, the leaders from AlMV RNA4, BMV RNA3, and RSV (only the 5' proximal 112 bases, containing the second ribosomal binding site, of the 380 base leader was tested) were stimulatory only when capped and at lower levels than observed with Ω (8 fold or >20 fold when compared to uncapped form without a viral leader; Table 2). This absolute requirement for a cap must reflect inherit differences in the manner in which these leaders act compared to the TMV leader. The ability to enhance translation is not dependent on a leader's capacity to bind more than one ribosome. AlMV RNA4, a monosome former, does enhance translation, whereas the TYMV leader failed to enhance translation even though it is known to form disomes; therefore, a consistent correlation between disome formation and translational enhancement does not exist.

TABLE 2
TRANSLATIONAL ENHANCEMENT BY VARIOUS VIRAL LEADERS ON GUS mRNAS ELECTROPORATED INTO TOBACCO PROTOPLASTS

mRNA	Specific activity (nmoles MUG hydrolysed/ min/μg protein)	Relative fold stimulation
Uncapped		
GUS	<0.01	1
Ω-GUS	0.25	>25
TYMV-GUS	<0.01	–
AlMV RNA4-GUS	<0.01	–
BMV RNA3-GUS	<0.01	–
RSV-GUS	<0.01	–
5'-Capped		
GUS	0.03	1
Ω-GUS	0.54	18
TYMV-GUS	<0.01	–
AlMV RNA4-GUS	0.23	8
BMV RNA3-GUS	0.23	8
RSV-GUS	0.23	8

Other viral leader sequences as translational enhancers in X. laevis oocytes

The ability of a particular leader to enhance translation in plant protoplasts does not predict its behavior in animal cells. Because Xenopus oocytes contain high levels of endogenous GUS activity, a new reporter gene was required. pSP64-based leader constructs, each containing the CAT gene, were transcribed and the uncapped mRNAs microinjected into oocytes.

Although Ω did significantly stimulate translation (10 fold) in microinjected Xenopus oocytes, the leaders from TYMV, AlMV RNA4, BMV RNA3, and RSV failed to do so (Table 3)(16), again reflecting the difference between Ω and the other viral leader sequences.

TABLE 3
THE EFFECT OF VIRAL LEADER SEQUENCES ON TRANSLATION IN X. LEAVIS OOCYTES

mRNA	% Conversion of chloramphenicol to acetylated products	Relative fold enhancement
no RNA	0.1	–
CAT	6.3	1.0
Ω–CAT	46.5	7.4
TYMV–CAT	1.3	0.2
AlMV–CAT	5.0	0.8
BMV–CAT	5.4	0.9
RSV–CAT	6.7	1.1

Mutational analysis of Ω for altered ability to enhance translation

In order to determine those sequences of Ω responsible for its enhancing ability, deletions were introduced throughout the 67 base region (Fig. 2). Ω contains 3 eight base direct repeats, the 5' copy of which is separated from the two 3' copies by a repetitive $(CAA)_n$ sequence. The 5' copy of the 8 base repeat is the sequence which contains the second ribosomal binding site. Although the 5' untranslated leader sequences from four different strains

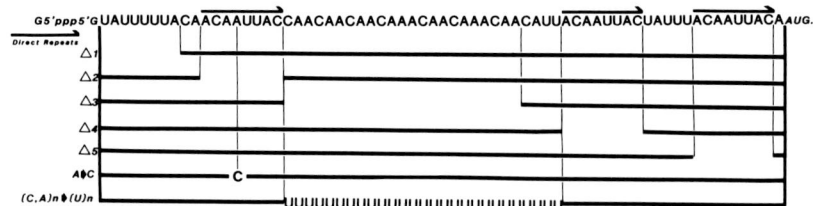

Fig. 2. Ω-derived mutants designed to test for alterations in translational enhancement. The 5' untranslated leader sequence of TMV is shown at the top. The AUG initiation codon of the first open reading frame (126 Kd protein) is shown at the right end. Ω is indicated by the sequence in large letters and the 8 base direct repeats by the arrows above. The sequence present in each mutant construct is designated by a solid line.

of TMV vary in length they all contain approximately equivalent repeats and a poly(CAA) sequence (17, 18). These conserved structural features of TMV leaders may reflect their role as information sources for the Ω-associated phenomena observed. In our mutational approach towards delineating those sequences necessary for translational enhancement, we constructed a series of Ω-based oligonucleotide sequences incorporating deletions of each structural feature (Fig. 2).

In tobacco mesophyll protoplasts, derivatives which deleted just one copy of the direct repeats did not significantly alter translational enhancement with the exception of $\Omega_{\Delta 3}$, which lost 48% of its activity (19)(Table 4). Replacement of the poly(CAA) with poly(U) resulted in an 87% reduction in activity. The single base substitution mutant, $\Omega_{A \rightarrow C}$, actually resulted in a small but reproducible increase (28%) in translational enhancement over that seen for Ω.

This same series of deletion derivatives suggested that the 5' direct repeat was responsible for the enhancement observed in Xenopus oocytes (19). $\Omega_{\Delta 2}$, the mutant in which the sequence for the second ribosome binding site has been deleted, was the most severely

TABLE 4
THE EFFECT OF MUTATIONS WITHIN Ω ON TRANSLATIONAL ENHANCEMENT OF GUS mRNAS ELECTROPORATED INTO TOBACCO PROTOPLASTS

mRNA	Specific activity (nmoles MUG hydrolysed/ min/μg protein)	Relative fold stimulation
GUS	0.054	1
Ω-GUS	2.5	46
$\Omega_{\Delta 1}$-GUS	2.3	43
$\Omega_{\Delta 2}$-GUS	2.7	50
$\Omega_{\Delta 3}$-GUS	1.3	24
$\Omega_{\Delta 4}$-GUS	2.1	39
$\Omega_{\Delta 5}$-GUS	2.5	46
$\Omega_{A \to C}$-GUS	2.8	52
$\Omega_{A, C \to U}$-GUS	0.33	6

affected: 71% of the enhancing ability was lost compared to the intact Ω sequence (Table 5). The single base substitution mutant, $\Omega_{A \to C}$, in which the AUU implicated in the binding of the second ribosome was changed to CUU, lost 55% of its enhancing ability. Although the deletion of the poly(CAA) resulted in a reduction of enhancement by 46%, substitution of this same region with poly(U) ($\Omega_{C, A \to U}$) surprisingly did not alter the ability of this sequence to enhance translation from that seen for Ω itself.

Mutational analysis of Ω in prokaryotes

There are many differences between translational initiation in prokaryotes and eukaryotes; Ω stimulates translation in both, though not necessarily by the same mechanism. Whether the Ω-associated enhancement seen in prokaryotes is artifactual or not, the fact that Ω can stimulate prokaryotic translation is of value in identifying those factors which influence translation in prokaryotes.

TABLE 5
THE EFFECT OF MUTATIONS WITHIN Ω ON TRANSLATION IN X. LAEVIS OOCYTES

mRNA	% Conversion of chloramphenicol to acetylated products
No mRNA	0.1
CAT	6.2
Ω-CAT	59.2
$Ω_{Δ1}$-CAT	29.2
$Ω_{Δ2}$-CAT	17.4
$Ω_{Δ3}$-CAT	31.6
$Ω_{Δ4}$-CAT	24.6
$Ω_{Δ5}$-CAT	31.9
$Ω_{A→C}$-CAT	26.7
$Ω_{A,C→U}$-CAT	58.6

A vector construct containing a derivative of the trp promoter allowing in vivo production of leader-GUS mRNAs has been described (16):

```
   -35                       -10
                                       *
CTGTTGACAATTAATCATCGAACTAGTTAACTAGTACGAAGCTTGTCGACGGATCC
                                      HindIII SalI BamHI
```

The * indicates the site of transcription initiation. GUS gene cassettes were introduced at the SalI site after the various HindIII/SalI leader sequence constructs had first been inserted into the vector.

As observed previously, Ω stimulated translation of GUS mRNA 8 fold (Table 6)(16). Of the five deletion mutants, three had little or no effect on Ω's enhancing ability (19). The replacement of the poly(CAA) region with poly(U) ($Ω_{C,A→U}$) resulted in only a small reduction in enhancement. Surprisingly, two mutants, $Ω_{Δ2}$ and $Ω_{Δ3}$, exhibited a nine fold increase in translation over that seen for Ω. In addition, $Ω_{A→C}$ possessed elevated enhancing potential as it was over six fold more active in stimulating translation that Ω itself.

Because it is possible the Ω-associated translational

enhancement may be a phenomenon unique to E. coli, which might suggest an interaction other than at the level of translational initiation, we tested $\Omega_{\Delta 3}$ in a variety of Gram-negative organisms and determined whether $\Omega_{\Delta 3}$ might replace the requirement for the Shine-Dalgarno (S-D) sequence in these species. A GUS gene derivative containing a more efficient initiation context for prokaryotes accounts for the generally higher levels of GUS specific activity in Table 7 than that seen in Table 6. Table 7 illustrates that the presence of $\Omega_{\Delta 3}$ at the 5' end of an mRNA can replace the requirement for a S-D sequence (GUS mRNA) in a variety of Gram-negative bacteria, but does not significantly stimulate the translation of an mRNA (CAT) that does contains an S-D region.

TABLE 6
THE EFFECT OF MUTATIONS WITHIN Ω ON TRANSLATIONAL ENHANCEMENT OF GUS mRNAS IN E. COLI

mRNA	Specific activity (nmoles MUG hydrolysed/ min/µg protein)	Relative fold stimulation
GUS	3.8	1
Ω-GUS	32.0	8
$\Omega_{\Delta 1}$-GUS	30.6	8
$\Omega_{\Delta 2}$-GUS	276.0	73
$\Omega_{\Delta 3}$-GUS	285.0	75
$\Omega_{\Delta 4}$-GUS	30.4	8
$\Omega_{\Delta 5}$-GUS	41.9	11
$\Omega_{A \to C}$-GUS	199.0	52
$\Omega_{A, C \to U}$-GUS	18.1	5

In vitro analysis of the mechanism by which Ω enhances translation

Ω-associated enhancement observed in rabbit reticulocyte lysate (MDL) was markedly less than that observed in wheat germ (WG) lysate (12). The differential between the two in vitro translation systems was exploited to identify that lysate component which was responsible for a lysate's characteristic response to Ω.

TABLE 7
THE EFFECT OF A S-D REGION ON Ω-ASSOCIATED ENHANCEMENT IN PROKARYOTES

Bacterial strain	Specific activity (nmol/min mg) GUS	Specific activity (nmol/min mg) Ω_{Δ_3}-GUS	Relative fold stimulation	Specific activity (nmol/min mg) CAT	Specific activity (nmol/min mg) Ω_{Δ_3}-CAT	Relative fold stimulation
E. coli HB101	9.4	2284	243	9626	10187	1.1
S. typhimurium 21D34	44.0	5597	127	30878	35471	1.2
E. amylovora 1D32R	7.0	393	57	12900	9429	0.7
X. vitians 7D51R	<1.0	<1	—	277	935	3.4
A. tumefaciens LBA4301	<1.0	94	94	3458	9823	2.8
A. rhizogenes 3D12R	17.0	581	34	4453	10420	2.3
R. meliloti 1U45	5.3	544	103	2875	4543	1.6

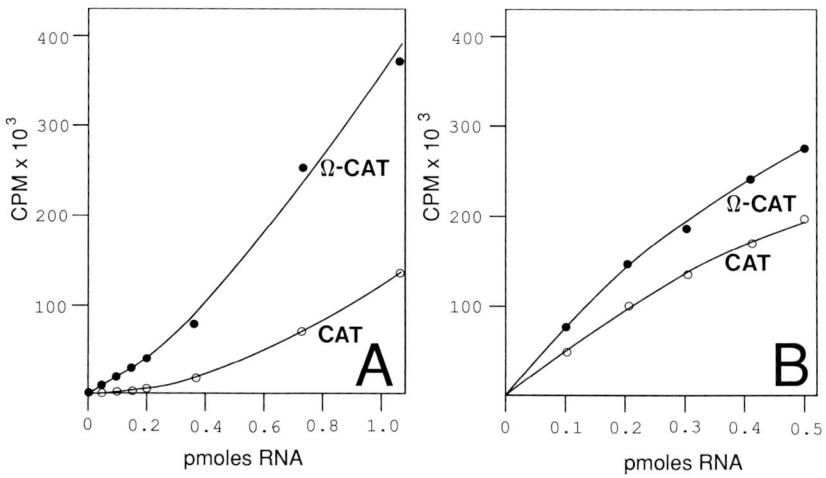

Fig. 3. Ω's effect on CAT mRNA translation in the two in vitro systems.

Because mRNAs which contain Ω may interact with ribosomes more efficiently, it was of interest to determine the effect that a range of ribosome concentrations would have on Ω's ability to stimulate translation. Each translation system was fractionated into a ribosome/initiation factor-cleared supernatant (S-100) and high-salt-washed ribosomes free of initiation factors (HSW Rib), and then the HSW Rib fraction was added back to the S-100 component. When the HSW Rib of WG were present in the homologous S-100 at 0.01X to 0.05X normal concentration, the presence of Ω at the 5' end of the CAT mRNA afforded no translational advantage to the mRNA (Fig. 4A). Above 0.05X normal ribosomal concentration, the translation efficiencies of the two forms of the mRNA began to diverge, with the CAT mRNA reaching a maximum efficiency between 0.1X and 0.2X and the Ω-CAT mRNA continuing to increase in efficiency up to 2X the physiological ribosomal concentration. As a result, the fold-enhancement observed for Ω also increased with the ribosomal concentration.

Similarly, when rabbit reticulocyte HSW Rib were present in the homologous S-100 at a concentration 0.01X to 0.1X that which is physiological, the translational efficiencies of the two forms of the mRNA were equivalent; Ω-containing mRNA was progressively more efficiently

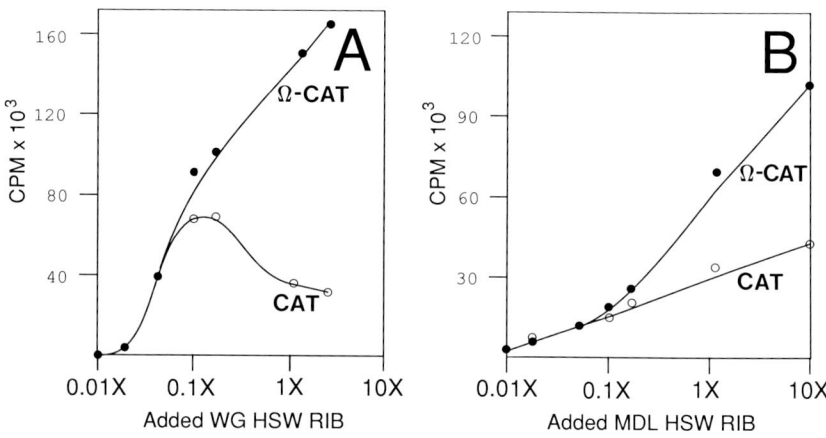

Fig. 4. The effect of ribosomal concentration on Ω-associated enhancement.

translated above 0.1X the normal ribosome concentration (Fig. 4B). Neither the CAT nor the Ω-CAT mRNA reached a maximal efficiency but continued to increase in expression up to 10X the physiological ribosomal concentration. Here as in the WG system, the fold-enhancement afforded by Ω continued to increase with the ribosomal concentration.

As the Ω-associated enhancement of an in vitro translation system seemed dependent on the ribosome concentration, the concentration and perhaps the source of the ribosome may be the key to a lysate's characteristic response to Ω. The following series of experiments identified the Ω responsive component through the fractionation of the MDL and WG lysates, followed by homologous or heterologous reconstitution. The WG lysate was fractionated into S-100, the ribosome/initiation factor fraction (P-100), HSW Rib, and the A and B fractions, each of which contain a partially purified subset of the wheat germ initiation factors (eIFs). The MDL was fractionated into the MDL S-100, P-100, HSW Rib, and HSW (containing the MDL eIFs). Because the ribosomes of one lysate operate at a much reduced efficiency in the S-100 of the other, it was necessary to use a high mRNA input (2.5 pmole or 1 µg). At these high input levels, the Ω-associated enhancement is artificially low in each reconstituted homologous system as the available ^{35}S-met becomes exhausted. As a result, the level of enhancement of the reconstituted WG system, i.e., experiments 2-4 (Table 8), ranged from 2.3 to 3.9 fold, although the overall level of translation was quite high. When the MDL P-100 was substituted for the WG P-100, the level of translation dropped by over an order of magnitude, and the enhancement afforded by the presence of Ω on the CAT mRNA was only marginally detected or lost (1.03 fold). Once the MDL ribosomes had been subjected to a high salt wash and thereby further depleted of their MDL eIFs, translational efficiency in the WG S-100 dropped another order of magnitude. Nevertheless, the enhancement by Ω was only 1.3 fold, again similar to that seen in the MDL system. If the WG eIFs, in the form of the A and B fractions were added to the MDL HSW Rib, they could not functionally replace the MDL eIFs (and as shown above, the B fraction is, in fact, inhibitory) as there was no stimulation of translational activity observed (compare experiment 7 to 6). The fold-enhancement remained close to 1.0. The addition of MDL HSW to MDL HSW Rib substantially restored translational activity but no enhancement was observed for Ω (0.89 fold). Similar results were obtained when MDL HSW was added to MDL P-100 (experiment 9).

When the MDL lysate was fractionated and reconstituted in various fashions (experiments 11-13), no Ω-associated enhancement was observed, i.e., the ratio remained close to 1. If the MDL P-100 was replaced by WG P-100, the enhancement increased to 4.8 fold. If WG HSW Rib were used instead, the level of translation dropped almost an order of magnitude but the level of enhancement remained high at 3.7 fold. Addition of the WG A and B fractions to the components of experiment 15 increased the overall level of translation (experiment 16) and the enhancement observed was 5.1 fold. If MDL HSW was added instead to the components of experiment 15, a reduction in the level of translation was observed but the enhancement was still 3.1 fold. Finally, if the WG A and B fractions were added to the components of experiment 17, there was some restoration of activity, and the enhancement remained high at 4.3 fold. These data suggest that the component which is responsible for a lysate's characteristic response to Ω is contained in the high-salt-washed ribosomal fraction.

TABLE 8
ISOLATION OF THE LYSATE FRACTION WHICH IS REPONSIVE TO Ω

Experi-ment	Components in assay	mRNA	^{35}S-met incorporation into protein (cpm)	Relative fold stimulation
1	WG S-100	CAT	0	-
		Ω-CAT	0	-
2	WG S-100 + WG P-100	CAT	705043	1
		Ω-CAT	1598112	2.3
3	WG S-100 + WG HSW Rib	CAT	252564	1
		Ω-CAT	987422	3.9
4	WG S-100 + WG HSW Rib + A + B	CAT	256342	1
		Ω-CAT	968076	3.8
5	WG S-100 + MDL P-100	CAT	28398	1
		Ω-CAT	29379	1.03
6	WG S-100 + MDL HSW Rib	CAT	1945	1
		Ω-CAT	2525	1.3
7	WG S-100 + MDL HSW Rib + A + B	CAT	1859	1
		Ω-CAT	1795	0.97
8	WG S-100 + MDL HSW Rib + MDL HSW	CAT	11155	1
		Ω-CAT	9982	0.89
9	WG S-100 + MDL P-100 + MDL HSW	CAT	33367	1
		Ω-CAT	38659	1.2
10	MDL S-100	CAT	0	-
		Ω-CAT	0	-
11	MDL S-100 + MDL P-100	CAT	313543	1
		Ω-CAT	307190	0.98
12	MDL S-100 + MDL HSW Rib	CAT	185377	1
		Ω-CAT	193858	1.05
13	MDL S-100 + MDL HSW Rib + MDL HSW	CAT	354490	1
		Ω-CAT	327006	0.92
14	MDL S-100 + WG P-100	CAT	203875	1
		Ω-CAT	978833	4.8
15	MDL S-100 + WG HSW Rib	CAT	25737	1
		Ω-CAT	95227	3.7
16	MDL S-100 + WG HSW Rib + A + B	CAT	44178	1
		Ω-CAT	225308	5.1
17	MDL S-100 + WG HSW Rib +MDL HSW	CAT	14651	1
		Ω-CAT	45432	3.1
18	MDL S-100 + WG HSW Rib + MDL HSW + A + B	CAT	25091	1
		Ω-CAT	107891	4.3
19	MDL S-100 + MDL HSW + A + B	CAT	0	-
		Ω-CAT	0	-

DISCUSSION

The stimulatory effect of various viral leader sequences on mRNA translation has been demonstrated in several in vitro and in vivo systems. One viral leader sequence in particular, Ω from TMV genomic RNA, has been shown to be the most enhancing as well as being the only leader which enhances translation in both prokaryotic and eukaryotic cells and translation assays. Ω is a highly structured sequence containing three copies of an eight base repeat, the 5' copy of which has been shown to act as a ribosome binding site (2). In tobacco mesophyll protoplasts, Ω derivatives with just one copy of the direct repeats or with the CAA repetitive sequence removed were still stimulatory. Although this same series of deletion derivatives suggested that the 5' direct repeat was responsible for the enhancement observed in Xenopus oocytes and the 3' direct repeats were responsible for the enhancement observed in E. coli, the subsequence responsible for the enhancement in plants has yet to be identified. A more thorough mutational analysis will be carried out to pinpoint the nucleotides involved. Interestingly, one derivative actually resulted in additional enhancement over the seen for Ω itself. This raises the possibility that further mutational analysis, in addition to determining the enhancing element of Ω, may optimize its ability to stimulate translation.

What then is the mechanism by which a leader causes mRNA expression to be enhanced? Sequence comparisons between the leaders tested have shown no significant homologous sequences other that a high AU content, a common feature of viral leader sequences. It is tempting to speculate that Ω acts as an enhancing element for ribosome association with the mRNA. Yokoe et al. (5) reported that, at least for the TMV leaders, there is a possible eukaryotic equivalent of a Shine-Dalgarno region.

The lack of sequence homology between the leaders may indicate that no one strategy is followed by all but that there may be several ways to achieve enhancement. Because of the high AU content, these leaders might possess little secondary structure. This lack of secondary structure may allow more rapid ribosome scanning and might be the mechanism by which translational enhancement occurs in eukaryotes. This hypothesis must take into account several predictions. The limited secondary structure of the reporter mRNA leader (in this case the mRNA from the β-glucuronidase gene (GUS)) without Ω present consists primarily of palindromic polylinker sequence. Kozak (1) has shown that eukaryotic ribosomes can melt secondary structures with a ΔG of -30 Kcal/mol. The polylinker leader would, therefore, present little problem to a ribosome. Moreover, there is disagreement on the role of secondary structure in a leader sequence. Secondary structure within a leader of an mRNA has been hypothesized to promote initiation (7, 20), and in other cases to inhibit initiation (21).

In prokaryotes, Ω and its derivative $\Omega_{\Delta 3}$ can substitute for the Shine-Dalgarno region of an mRNA to bring about efficient translation. This might suggest that in prokaryotes, $\Omega_{\Delta 3}$ is functionally equivalent to an S-D region and interacts with 16S rRNA 3' terminal sequence to bring about mRNA-ribosome binding. S-D sequences are G rich and as $\Omega_{\Delta 3}$ is devoid of G residues, $\Omega_{\Delta 3}$ could not interact with the sequence, 5'-CCTCC-3, close to the 16S rRNA 3'-terminus, as the S-D region does. Moreover, there are significant functional differences between an S-D region and Ω-associated enhancement. Whereas an S-D sequence must be positioned 3 to 8 bases upstream from the initiator AUG to function properly, Ω-derived sequences enhance translation whether they are present immediately upstream of the initiator AUG (16), or as far upstream as 110 bases (D. Gallie, personal observations; tests further upstream have not yet been performed). Moreover, in E. coli, the enhancement afforded by one copy of Ω can be more than doubled when two copies are present in a tandem arrangement. It may well be that there are several ways for an mRNA to interact with the translational machinery to promote efficient translation.

ACKNOWLEDGEMENTS

We wish to thank David Sleat, John Watts, and Philip Turner for helpful discussions and contributions to this work, and Alexandra Bloom for the preparation of the manuscript.

REFERENCES

1. Kozak M (1986). Point mutations define a sequence flanking the AUG initiator codon that modulates translation by eukaryotic ribosomes. Cell 44:283.
2. Konarska M, Filipowicz W, Domdey H, Gross HJ (1981). Binding of ribosomes to linear and circular forms of the 5'-terminal leader fragment of tobacco-mosaic-virus RNA. Eur J Biochem 114:221.
3. Tyc K, Konarska M, Gross HJ, Filipowicz W (1984). Multiple ribosome binding to the 5'-terminal leader sequence of tobacco mosaic virus RNA. Eur J Biochem 140:503.
4. Hunter TR, Hunt T, Knowland J, Zimmern D (1976). Messenger RNA for the coat protein of tobacco mosaic virus. Nature 260:759.
5. Yokoe S, Tanaka M, Hibasami H, Nagai J, Nakashima K (1983). Cross-linking of tobacco mosaic virus RNA and capped polyribonucleotides to 18S rRNA in wheat germ ribosome-mRNA complexes. J Biochem 94:1803.
6. Filipowicz W, Haenni A-L (1979). Binding of ribosomes to 5'-terminal leader sequences of eukaryotic messenger RNAs. Proc Natl Acad Sci USA 76:3111.
7. Ahlquist P, Dasgupta R, Shih DS, Zimmern D, Kaesberg P (1979). Two-step binding of eukaryotic ribosomes to brome mosaic virus RNA3. Nature 281:277.
8. Darlix J-L, Spahr P-F, Bromley PA, Jaton JC (1979). In vitro, the major ribosome binding site on Rous sarcoma virus RNA does not contain the nucleotide sequence coding for the N-terminal amino acids of the gag gene product. J Virol 29:597.
9. Pinck M, Fritsch C, Ravelonandro M, Thivent C, Pinck L (1981). Binding of ribosomes to the 5'-leader sequence (N 258) of RNA3 from alfalfa mosaic virus. Nucl Acids Res 9:1087.
10. Wilson TMA (1986). Expression of the large 5'-proximal cistron of tobacco mosaic virus by 70S ribosomes during co-translational disassembly in a prokaryotic cell-free system. Virology 152:277.
11. Shine J, Dalgarno L (1975). Determinant of cistron specificity in bacterial ribosomes. Nature 254:34.
12. Gallie DR, Sleat DE, Watts JW, Turner PC, Wilson, TMA (1987). The 5'-leader sequence of tobacco mosaic virus RNA enhances the expression of foreign gene transcripts in vitro and in vivo. Nucl Acids Res 15:3257.

13. Gallie DR, Sleat DE, Watts JW, Turner PC, Wilson, TMA (1987). In vivo uncoating and efficient expression of foreign mRNAs packaged in TMV-like particles. Science 236:1122.
14. Sleat DE, Gallie DR, Jefferson RA, Bevan MW, Turner PC, Wilson TMV (1987). Characterization of the 5'-leader sequence of tobacco mosaic virus RNA as a general enhancer of translation in vitro. Gene 217:217.
15. Jefferson RA, Burgess SM, Hirsh D (1986). β-Glucuronidase from Escherichia coli as a gene fusion marker. Proc Natl Acad Sci USA 83:8447.
16. Gallie DR, Sleat DE, Watts JW, Turner PC, Wilson, TMA (1987). A comparison of eukaryotic viral 5'-leader sequences as enhancers of mRNA expression in vivo. Nucl Acids Res 15:8693.
17. Kukla BA, Guilley HA, Jonard GX, Richards KE, Mundry KW (1979). Characterization of long guanosine-free RNA sequences from the Dahlemense and U2 strains of tobacco mosaic virus. Eur J Biochem 98:61.
18. Goelet P, Lomonossoff GP, Butler PJG, Akam EE, Gait MJ, Karn J (1982). Nucleotide sequence of tobacco mosaic virus RNA. Proc Natl Acad Sci USA 79:5818.
19. Gallie DR, Sleat DE, Watts JW, Turner PC, Wilson TMA (1988). Mutational analysis of the tobacco mosaic virus 5'-leader for altered ability to enhance translation. Nucl Acids Res 16:883.
20. Darlix J-L, Zucker M, Spahr P-F (1982). Structure-function relationship of Rous sarcoma virus leader RNA. Nucl Acids Res 10:5183.
21. Kozak M (1986). Influences of mRNA secondary structure on initiation by eukaryotic ribosomes. Proc Natl Acad Sci USA 83:2850.

INVOLVEMENT OF 3'-POLY(A) TRACTS IN PROTEIN SYNTHESIS: A ROLE IN mRNP ASSEMBLY[1]

David Munroe and Allan Jacobson

Department of Molecular Genetics and Microbiology
University of Massachusetts Medical School
Worcester, Massachusetts 01655

ABSTRACT We have utilized synthetic mRNAs and a rabbit reticulocyte cell-free translation system to evaluate the hypothesis that the 3'-poly(A) tract of mRNA plays a role in translational initiation. We have constructed derivatives of pSP65 which direct the synthesis of synthetic mRNAs with different poly(A) tail lengths and compared the relative efficiencies with which such mRNAs are recruited into polysomes and translated *in vitro*. Using this assay, poly(A)-deficient mRNAs (including poly(A)$^-$ mRNAs) are found to have a significantly reduced translational capacity when compared to mRNAs with poly(A) tail lengths of at least 30 adenylate residues. This difference in translational capacity is not due to differences in mRNA decay rates, but is attributable to a marked reduction in the relative efficiency with which poly(A)$^-$ mRNAs are recruited into polysomes. The defect in poly(A)$^-$ mRNAs reduces the rate of translational initiation, is distinct from the phenotype associated with CAP-deficient mRNAs, and appears to have a profound effect on the formation of mRNPs.

[1]This work was supported by a grant (GM27757) to AJ from the National Institutes of Health.

INTRODUCTION

Although it has been almost two decades since the discovery of poly(A) tracts in eukaryotic mRNAs (1-5), the function of these mRNA "tails" has yet to be elucidated. We have proposed a translational model for poly(A) function (6,7) which postulates that: a) an interaction between the poly(A) tract of mRNA and a cytoplasmic poly(A)-binding protein enhances the efficiency of translational initiation; b) mRNAs with relatively long poly(A) tails have a translational advantage over mRNAs with shorter poly(A) tails; and c) the regulatory mechanisms which ensure the efficient translation of poly(A)$^+$ and poly(A)$^-$ mRNAs may be quite different. The results of several different experimental approaches have provided indirect evidence in support of this model. For example, it has been shown that: a) there is a correlation between the adenylation status of mRNA and its translatability *in vivo* and *in vitro* (8-12); b) exogenously added poly(A) is a potent inhibitor of the initiation of translation of poly(A)$^+$, but not poly(A)$^-$ mRNAs *in vitro* (6,13,14); and c) there is a correlation between the stability of the *Dictyostelium* poly(A)-binding proteins and the rate of translational initiation (15,16). In this paper we describe experiments which provide direct evidence in support of our model. We have constructed plasmids which direct the *in vitro* synthesis of a set of synthetic mRNAs differing only in their respective poly(A) tail lengths and compared the efficiencies with which such synthetic mRNAs are assembled into mRNPs and recruited into polysomes in a cell-free translation system.

METHODS

In vitro transcription. *In vitro* transcription and co-transcriptional capping, directed by SP6 RNA polymerase (Boehringer), were as described previously (17,18). After transcription, DNA templates were removed by digestion with RQ1

DNase (1 U/ug DNA; Promega). Transcripts were further purified by two phenol:CHCl$_3$ (1:1) extractions, CHCl$_3$ extraction, Sephedex G-50 spin-chromatography, and ethanol precipitation. The extent of capping was monitored by 2D TLC analysis (18) and the integrity of all RNA samples was verified by gel electrophoresis.

In vitro protein synthesis. mRNA-dependent translation extracts were prepared from commercial rabbit reticulocyte lysate (Promega) as described previously (19).

Polysome gradients. Samples were diluted 1:10 in ice cold buffer A (25mM Hepes, pH 7.0, 50mM KCl, and 2mM MgOAc) containing 50ug/ml cycloheximide and fractionated on gradients of 15-50% sucrose in buffer A. Gradients were centrifuged at 175,000g for 110 min at 4°C in an SW41 rotor. Fractions were collected from the bottom, transferred directly into ice cold 5% TCA, and filtered.

Identification of mRNP proteins by label transfer. Rabbit reticulocyte lysates were preincubated for 1 min at 37°C with 20mM EDTA, then supplemented with 0.1 ug of ^{32}P-labeled, synthetic poly(A)$^+$ or poly(A)$^-$ mRNAs and incubated for an additional 5 min at 37°C. To crosslink mRNP proteins to mRNA, samples were UV-irradiated at 22°C (flux=2000 uW/cm^2; dose=3.6 x10^5 ergs/mm^2). Following irradiation, label was transferred from mRNA to mRNP proteins by digestion with either RNases A, T$_1$, and T$_2$ or with RNase A alone (15).

RESULTS

To test the hypothesis that poly(A) may be involved in protein synthesis we have constructed derivatives of plasmid pSP65 which direct the synthesis of synthetic mRNAs differing only in their respective poly(A) tail lengths. This was accomplished by insertion of (dA:dT) tracts of different length between the Pst I and Hind III sites of the polylinker (DM and AJ, ms. in prep.). In vitro transcription of RNA from these vectors (after linearization with Hind III) leads to the synthesis of polyadenylated RNAs with precisely defined lengths of poly(A) (as measured by gel electrophoretic analysis of the products resulting from digestion with RNases A and T_1).

The experiments shown in Figure 1 assess the effects of differences in 3'-poly(A) tail length and/or 5'-cap structure on the translatability of otherwise identical synthetic mRNAs transcribed from the vesicular stomatitus virus N gene (VSV N). Figure 1 shows that, in rabbit reticulocyte lysates, capped, poly(A)$^+$ VSV N mRNA has 1.4-2.5 times the translational activity of capped, poly(A)$^-$ VSV N mRNA, 1.4-1.9 times the translational activity of the uncapped, poly(A)$^+$

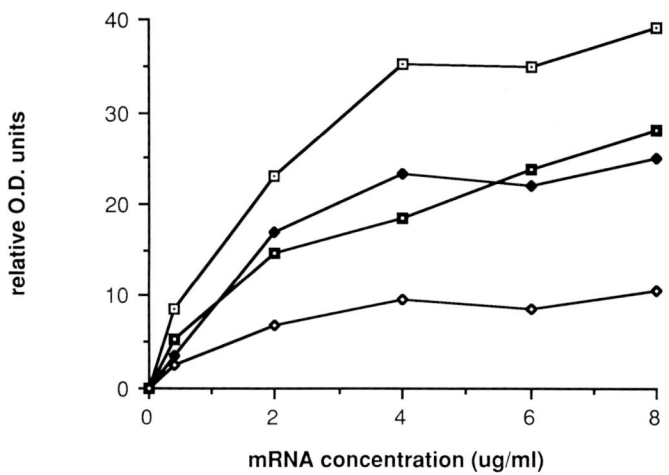

Figure 1. Relative translational efficiency of mRNAs differing in poly(A) tail lengths and/or cap structure. Synthetic VSV N mRNAs, with or without 5'-caps or 3'-poly(A) tails (61 As), were translated for 30 minutes at 37°C in mRNA-dependent rabbit reticulocyte lysates supplemented with ^{35}S-methionine. Equal volumes of each sample were digested with RNases A, T1, and T2 and translation products were analyzed by SDS polyacrylamide gel electrophoresis and fluorography (20,21). The resulting autoradiographs were quantitated by densitometry. (□):capped, A$^+$; (◆): capped, A$^-$; (■):uncapped, A$^+$; (◇):uncapped, A$^-$.

mRNA, and 3.3-4.0 times the translational activity of the uncapped, unadenylated transcript. These data suggest that poly(A)⁻ mRNAs have a significantly reduced translational capacity that is comparable in magnitude to the defect associated with cap-deficient mRNAs. To ensure that the observed differential formation of protein product by poly(A)⁺ and poly(A)⁻ mRNAs was due solely to differences in their respective efficiencies of translation, their decay rates were compared. Degradation of synthetic poly(A)⁺ and poly(A)⁻ VSV N (61As vs 0As) and rabbit B-globin (RBG) (68As vs 0As) mRNAs was monitored in reticulocyte lysates. In the course of a 60 min incubation no differences were observed in the decay rates of the individual mRNAs (data not shown).

A comparison of the polysomes formed by poly(A)⁺ and poly(A)⁻ B-globin mRNAs demonstrates that the adenylated mRNA is recruited into polysomes to a greater extent than its poly(A)⁻ counterpart (Figures 2A,B). Moreover, a comparison of the ratio of A⁺:A⁻ mRNA in the polysomal region indicates that those poly(A)⁻ mRNAs which do form polysomes are associated with a smaller average number of ribosomes than poly(A)⁺ mRNAs (Figures 2A,B). Identical results have been obtained with poly(A)⁺ and poly(A)⁻ VSV N mRNAs (data not shown). Figure 2 also shows that the polysomal distribution of poly(A)⁻ B-globin mRNA is comparable to that observed with cap-deficient B-globin mRNA (see 2A and 2B). Collectively, these data suggest that poly(A)⁻ mRNAs have a reduced efficiency of translational initiation. This conclusion has been confirmed by experiments which indicate that there are no differences in translational elongation rates between poly(A)⁺ and poly(A)⁻ VSV N mRNAs (data not shown). The additivity of the initiation-defective phenotypes of cap-deficient and poly(A)-deficient mRNAs (Figure 2B) suggests that the 5'-cap and the 3'-poly(A) tail affect different steps in the initiation process.

In an effort to characterize further the initiation defect associated with poly(A)-deficient mRNAs we examined the ability of poly(A)⁺

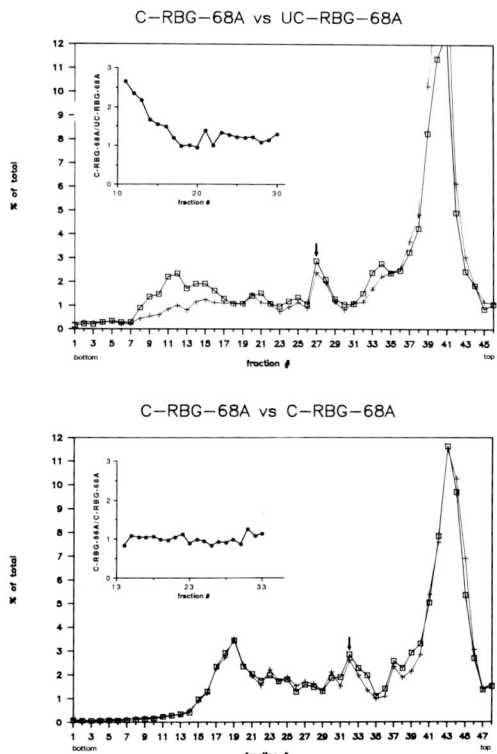

Figure 2A. Polysomal distribution of synthetic mRNAs differing in poly(A) tail lengths and/or cap structures. Equal amounts of a ^{32}P-labeled, capped, rabbit B-globin synthetic mRNA containing 68 3'-adenylate residues (□) and ^{3}H-labeled synthetic B-globin mRNAs of indicated structure (+) were mixed and then incubated for 15 minutes at 37°C in mRNA-dependent rabbit reticulocyte lysates. The polysomal distribution of each combination of mRNAs was analyzed by fractionation on 15-50% sucrose gradients.

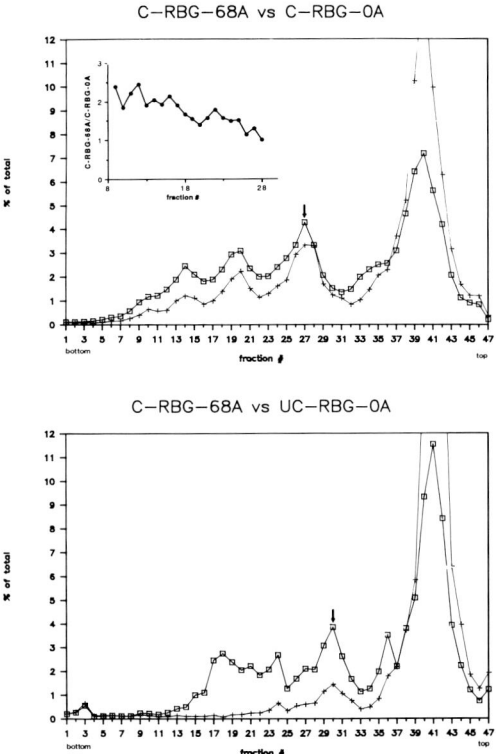

Figure 2B. Polysomal distribution of synthetic mRNAs differing in poly(A) tail lengths and/or cap structures. See legend to Figure 2A for details.

and poly(A)⁻ B-globin mRNAs to form mRNPs, 48S pre-initiation complexes, and 80S monosomes in reticulocyte lysates. We found that, in sucrose gradients, the poly(A)⁻ mRNA forms anomalously sedimenting mRNPs and 48S complexes (data not shown). To investigate the RNA and protein composition of the respective mRNP structures, they were fixed with formaldehyde and banded in Cs_2SO_4 density gradients. mRNPs formed by capped, adenylated B-globin mRNA band at the expected density of 1.31gm/cm³. In contrast, mRNPs formed by capped, unadenylated B-globin mRNA show a diffuse distribution with densities ranging from that of naked mRNA (1.66gm/cm³) to 1.4gm/cm³ (data not shown). These results indicate that the poly(A)⁻ mRNAs are associated with signifigantly less protein than the poly(A)⁺ mRNAs. This conclusion was confirmed by using a label transfer procedure to identify the individual mRNP proteins bound to poly(A)⁺ or poly(A)⁻ B-globin mRNAs (Figure 3). The results of this experiment indicate that poly(A)⁺ mRNPs contain a significantly larger amount of total protein and a larger number of different polypeptides than poly(A)⁻ mRNPs. Since previous studies (22) have indicated that there is only one polypeptide (p78) associated with the poly(A) tracts of rabbit reticulocyte mRNAs, the observed differences between poly(A)⁺ and poly(A)⁻ mRNPs cannot be due simply to the loss of poly(A)-binding proteins.

DISCUSSION

By using synthetic mRNAs which differ only in the lengths of their poly(A) tails we have demonstrated directly that the polyadenylation status of an mRNA is an important determinant of its efficiency of translational initiation. Poly(A)⁺ mRNAs are recruited into polysomes to a greater extent, and are associated with a larger average number of ribosomes, than poly(A)⁻ mRNAs. These differences are attributable to bona fide differences in the respective rates of translational initiation since the two types of mRNA do not differ in their rates of decay or their rates of

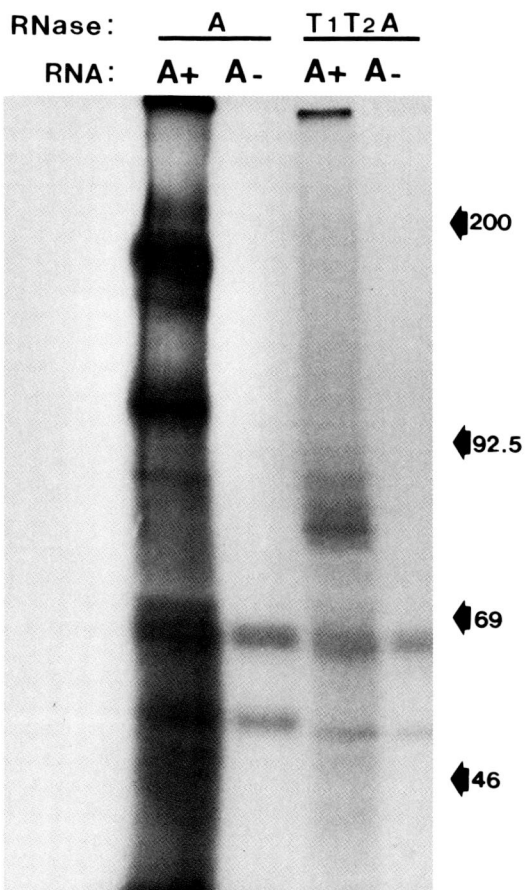

Figure 3. Protein content of mRNPs differing in poly(A) tail length. mRNPs were assembled in vitro by incubating synthetic, ^{32}P-labeled, poly(A)$^+$ or poly(A)$^-$ B-globin mRNAs with EDTA-treated reticulocyte lysates. Label was transferred from mRNA to mRNP proteins by UV-irradiation and RNase digestion and the labeled proteins were resolved by SDS polyacrylamide gel electrophoresis and autoradiography (15). A$^+$, A$^-$: proteins associated with mRNAs having poly(A) tails of either 68 or 0 adenylate residues.

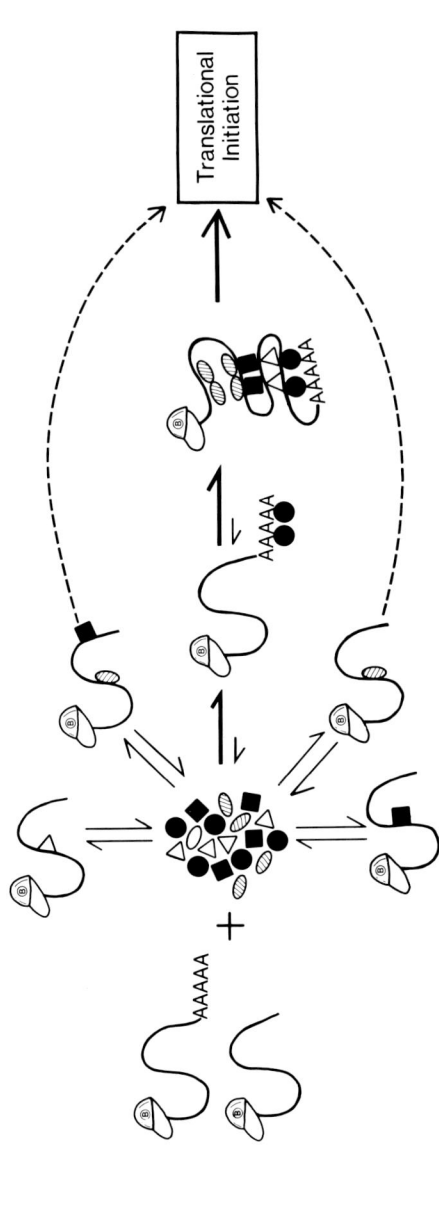

Figure 4. A model for the role of poly(A) in mRNP assembly. The cartoon depicts a pool of mRNP proteins interacting with capped mRNAs that are either poly(A)+ or poly(A)−. Poly(A)+ mRNAs are postulated to assemble mRNPs efficiently and to rapidly enter the translational initiation pathway whereas poly(A)− mRNAs are postulated to form incomplete mRNPs that are less suitable substrates for translational initiation.

translational elongation. Experiments not shown here indicate that the initiation enhancement contributed by poly(A) requires poly(A) tracts in excess of 30 adenylate residues (DM and AJ, in prep.). Below this length the enhancement effect diminishes significantly.

What, then, is the actual contribution of the poly(A) tract? Our data suggest that the presence of a poly(A) tract enables an mRNA to form, or to stabilize, a "complete" mRNP structure and that such a structure constitutes the optimal substrate for translational initiation. In the absence of a poly(A) tail the assembly, or maintainence, of such structures may be compromised (Figure 4). The poly(A) tail could participate in this process by providing a binding site for the poly(A)-binding protein. The requirement for poly(A) tracts of a minimal length would thus reflect the minimal binding site for a stable interaction with this protein. Interaction of the poly(A) tract with the poly(A)-binding protein may be a requisite first step in mRNP formation or, alternatively, the presence of the poly(A)-binding protein may facilitate the binding, or otherwise influence the activity, of other proteins with well-documented roles in translational initiation, e.g., cap-binding proteins.

REFERENCES

1. Kates J (1970). Transcription of the vaccinia virus genome and the occurrence of polyriboadenylic acid sequences in messenger RNA. Cold Spg Hrbr Symp Quant Biol 35:743.
2. Lim L, Canellakis ES (1970). Adenine-rich polymer associated with rabbit reticulocyte messenger RNA. Nature 227:710.
3. Darnell JE, Wall R, Tushinski RJ (1971). An adenylic acid-rich sequence in messenger RNA of HeLa cells and its possible relationship to reiterated sites in DNA. Proc Natl Acad Sci USA 68:1321.
4. Edmonds M, Vaughn MH Jr, Nakazoto H (1971). Polyadenylic acid sequences in the heterogeneous nuclear RNA and rapidly-labelled polyribosomal RNA of HeLa cells: possible

evidence for a precursor-product relationship. Proc Natl Acad Sci USA 68:1336.
5. Lee SY, Mendecki J, Brawerman G (1971). A polynucleotide segment rich in adenylic acid in the rapidly-labelled polysomal RNA component of mouse sarcoma 180 ascites cells. Proc Natl Acad Sci USA 68:1331.
6. Jacobson A, Favreau M (1983). Possible involvement of poly(A) in protein synthesis. Nuc Acids Res 11:6353.
7. Palatnik CM, Wilkins C, Jacobson A (1984). Translational control during early Dictyostelium discoideum development: possible involvement of poly(A) sequences. Cell 36:1017.
8. Doel MT, Carey NH (1976). The translational capacity of deadenylated ovalbumin messenger RNA. Cell 8:51.
9. Deshpande AK, Chatterjee B, Roy AK (1979). Translation and stability of rat liver messenger RNA for α_{2u}-globulin in Xenopus oocyte: the role of terminal poly(A). J Biol Chem 254:8937.
10. Rosenthal ET, Tansey TR, Ruderman JV (1983). Sequence specific adenylations and deadenylations accompany changes in the translation of maternal messenger RNA after fertilization of Spisula oocytes. J Mol Biol 166:309.
11. Steel LF, Jacobson A (1987). Translational control of ribosomal protein synthesis during early Dictyostelium discoideum development. Mol Cell Biol 7:965.
12. Shapiro RA, Herrick D, Manrow RE, Blinder D, Jacobson A (1988). Determinants of mRNA stability in Dictyostelium discoideum amoebae: poly(A) tail length, ribosome loading, and mRNA size cannot account for the heterogeneity of mRNA decay rates. Mol Cell Biol 8:1957.
13. Bablanian R, Banerjee AK (1986). Poly(riboadenylic acid) preferentially inhibits in vitro translation of cellular mRNAs compared with vaccinia mRNAs: possible role in vaccinia virus cytopathology. Proc Natl Acad Sci USA 83:1290.

14. Lemay G, Millward S (1986). Inhibition of translation in L-cell lysates by free polyadenylic acid: differences in sensitivity among different mRNAs and possible involvement of an initiation factor. Arch Biochem Biophys 249:191
15. Manrow RE, Jacobson A (1986). Identification and characterization of developmentally regulated mRNP proteins of Dictyostelium discoideum. Devel Biol 116:213.
16. Manrow RE, Jacobson A (1987). Increased rates of decay and reduced levels of accumulation of the major poly(A)-associated proteins of Dictyostelium during heat shock and development. Proc Natl Acad Sci USA 84:1858.
17. Melton DA, Krieg PA, Rebagliati MR, Maniatis T, Zinn K, Green MR (1984). Efficient in vitro synthesis of biologically active RNA and RNA hybridization probes from plasmids containing a bacteriophage SP6 promoter. Nucl Acids Res 12:7035.
18. Konarska MM, Padgett RA, Sharp PA (1984). Recognition of cap structure in splicing in vitro of mRNA precursors. Cell 38:731.
19. Palatnik CM, Storti RV, Jacobson A (1979). Fractionation and functional analysis of newly synthesized and decaying messenger RNAs from vegetative cells of Dictyostelium discoideum. J Mol Biol 128:371.
20. Laemmli UK (1970). Cleavage of structural proteins during the assembly of the head of bacteriophage T4. Nature 227:680.
21. Laskey RA, Mills AD (1975). Quantitative film detection of ^{3}H and ^{14}C in polyacrylamide gels by fluorography. Eur J Biochem 56:335.
22. Greenberg JR, Carroll E (1985). Reconstitution of functional mRNA-protein complexes in a rabbit reticulocyte cell-free translation system. Mol Cell Biol 5:342.

GLUTAMYL-tRNAS FROM CHLOROPLASTS AS COFACTORS IN NON-RIBOSOMAL ENZYMATIC REACTIONS

Astrid Schön, Gary O'Neill, David Peterson, and Dieter Söll

Department of Molecular Biophysics & Biochemistry
Yale University
New Haven, CT 06511, USA

ABSTRACT In chloroplasts as well as in several groups of prokaryotes the universal precursor of porphyrins, δ-aminolevulinic acid (ALA), is synthesized from glutamate in a multi-step pathway subject to a complex pattern of regulation. The conversion of glutamate to ALA requires chloroplast tRNA^{Glu} and ATP and involves an NADPH-dependent reduction of the glutamyl moiety to glutamate-1-semi-aldehyde (GSA) and a subsequent transamination. In the reduction step the Glu-tRNA reductase requires a specific tRNA^{Glu} as cofactor. The priming reaction of the overall pathway is the glutamylation of tRNA^{Glu} by chloroplast glutamyl-tRNA synthetase. The same enzyme also glutamylates the organelle-specific tRNA^{Gln}, which is then converted to Gln-tRNA^{Gln} by a tRNA-dependent amidotransferase. As this pathway is the only route to Gln-tRNA^{Gln} used in protein biosynthesis in chloroplasts, a misacylation of tRNA is involved in normal protein biosynthesis in these organelles. Since the plastid-specific glutamyl-tRNA synthetase charges a number of tRNAs that are not active in the biosynthesis of ALA, the Glu-tRNA^{Glu} reductase must be able to discriminate between the different substrates.

INTRODUCTION

Tetrapyrroles, active components in many biochemical reactions, are composed of eight molecules of δ-aminolevulinic acid (ALA). In chloroplasts and several groups of bacteria ALA is synthesized by a pathway different from the one used in non-photosynthetic eukaryotes and purple bacteria (for a review, see 1). Instead of being synthesized by condensation of succinyl-CoA and glycine, the C_5 skeleton of ALA is directly derived from glutamate in these cases (2-7).

The conversion of glutamate to ALA involves a complex series of reactions that requires ATP and includes a NADPH-dependent reduction

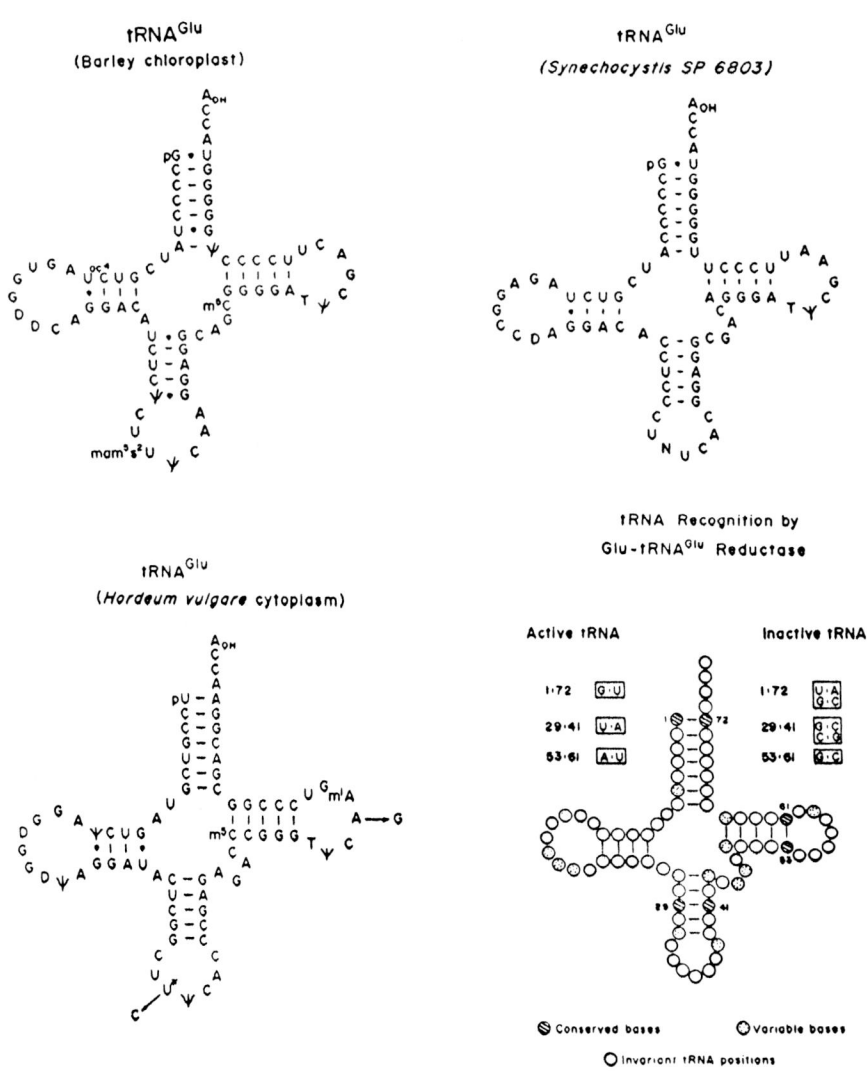

Figure 1. Nucleotide sequences of Glu-tRNAs from barley and *Synechocystis*. **Top left:** Sequence of tRNA^{Glu} (ALA-RNA); the primary sequence is identical to the chloroplast tRNA^{Glu} gene from wheat (33). **Top right:** Sequence of tRNA^{Glu} from *Synechocystis* SP 6803. **Lower left:** Sequence of the two major tRNA^{Glu} species from barley cytoplasm; the

and a transamination. The postulated intermediate in this pathway is GSA (8-11). The synthesis of ALA is the rate limiting step of chlorophyll biosynthesis and is light regulated in the chloroplasts of angiosperm plants (for a summary, see 12). Dissection of the pathway of ALA biosynthesis showed that the components required for activity could be separated into three fractions. Unexpectedly, one of the three "enzyme" fractions contained RNA as the active component, as demonstrated by RNase inactivation experiments using a reconstituted assay system for the overall reaction (13-15). The two other fractions contained the postulated enzyme activities, i.e. glutamyl-tRNA reductase (synonymous with GSA dehydrogenase), GSA aminotransferase, and an ATP-dependent enzyme required for activation of glutamate. Fractionation of chloroplast tRNA showed that only one of the three chloroplast Glu-tRNA species was active in ALA formation. Therefore, we set out to investigate the tRNA requirements of ALA-formation by determining the sequence of the active and inactive glutamate accepting RNA species present in barley and the cyanobacterium *Synechocystis 6803*.

RESULTS

The RNA required for Biosynthesis of δ-Aminolevulinic Acid is a Chloroplast tRNA^{Glu}.

Purification from barley chloroplasts and sequence analysis of ALA-RNA, the RNA active in ALA synthesis, revealed that it is a tRNA^{Glu} (16,17). Its sequence (Fig. 1, top left) is highly homologous to the known chloroplast tRNA^{Glu} genes of tobacco, pea, broad bean, liverwort and *Euglena* (18,19). The tRNA is extensively modified; the modified base, mam^{5}s^{2}U, in the first position of the anticodon is identical to the one present in the *E. coli* tRNA. As the cyanobacterium *Synechocystis 6803* forms ALA also in a tRNA-dependent reaction (20) we purified and sequenced the tRNA^{Glu} responsible (Fig. 1, top right).

RNA Specificity of Glu-tRNA^{Glu} Reductase.

As described above the Glu-tRNA^{Glu} reductase from barley chloroplasts displays a pronounced substrate specificity. Besides *E. coli* tRNA^{Glu} and the glutamate-accepting tRNA^{Gln} from barley chloroplasts, glutamate specific tRNAs of yeast or of plant cytoplasmic origin are inactive in ALA

Legend to Fig. 1 continued:
differences between them are indicated by arrows. **Lower right:** Comparisons of glutamate-specific tRNAs that are active or inactive in ALA biosynthesis. Nucleotide positions conserved in active species are hatched; nucleotides variable among the active tRNAs are stippled.

biosynthesis (13). To rule out the possibility that inefficient charging is responsible for the inability of these tRNAs to reconstitute ALA synthesis, we performed *in vitro* ALA biosynthesis experiments with cytoplasmic tRNAs that had been glutamylated by their homologous (i.e. cytoplasmic) synthetases (21). These experiments confirmed that the reductase can discriminate among different Glu-tRNA species. We determined the sequence of two cytoplasmic glutamate tRNAs isolated from barley embryos (21); their sequences are shown (Fig. 1, lower left). It is intriguing to point out differences between the sets of tRNAs active and inactive in ALA formation in barley. The nucleotide positions stippled (Fig. 1, lower right) are obviously not important for recognition of various tRNA^{Glu} molecules by Glu-tRNA reductase as different bases are found in these positions in the tRNAs from *Synechocystis 6803*, barley and other plant chloroplasts. However, the positions hatched (Fig. 1, lower right) may be important as they represent the nucleotides different in the set of inactive tRNAs, the tRNA^{Glu} species from barley cytoplasm and *E. coli* (18), and the glutamate-accepting tRNA^{Gln} species from barley chloroplasts (22) and from *Bacilli* (18). In addition to some changes in the acceptor stem and anticodon stem region, the most striking feature of the active tRNA^{Glu} species is an A:U pair at the position of the otherwise highly conserved G_{53} : C_{61} base pair. Clearly, studies with *in vitro* created mutant tRNAs will be needed to define unambiguously tRNA features that are important for recognition by the Glu-tRNA^{Glu} reductase.

Some Glutamate Isoaccepting tRNAs from Barley Chloroplasts are tRNA^{Gln} Species.

Purification and sequencing of the two glutamate acceptors inactive in ALA synthesis revealed that they are in fact very different from tRNA^{Glu} (22); they are both chloroplast tRNA^{Gln} species, differing from each other only by one ribose methylation. Both show a high degree of homology to chloroplast tRNA^{Gln} genes as well as to prokaryotic tRNA^{Gln} (18). Thus, these tRNAs are misacylated Glu-tRNA^{Gln} species (22). Further analysis revealed that there is no glutaminyl-tRNA synthetase detectable in barley chloroplasts. This is strongly reminiscent of the mischarging phenomenon that has been described for a number of gram-positive bacteria (23-25) where no glutaminyl-tRNA synthetase is present, and glutamine for protein biosynthesis is provided by mischarging of tRNA^{Gln} followed by transamidation. In these cases, a single enzyme is responsible for the glutamylation of both tRNA^{Glu} and tRNA^{Gln} (26).

Mischarging of a tRNA is Required for Protein Synthesis in Organelles and many Prokaryotes.

A systematic search for the occurrence of this pathway of Gln-tRNA^{Gln} formation revealed that chloroplasts, plant and animal mitochondria, and cyanobacteria utilize this mischarging phenomenon; they do not contain glutaminyl-tRNA synthetase, but a Glu-tRNA^{Gln} dependent amidotransferase (22). There is also evidence for the presence of this mischarging pathway in yeast mitochondria (27) and in archaebacteria (28).

Chloroplast GluRS is a Mischarging Aminoacyl-tRNA Synthetase with unique Substrate Specificity.

The demonstration that two of the three glutamate isoacceptors in barley chloroplasts are tRNA^{Gln} species prompted us to determine the efficiency of aminoacylation of these three tRNAs with glutamate and glutamine by homologous and heterologous aminoacyl-tRNA synthetases. As Fig. 2 shows, barley chloroplast aminoacyl-tRNA synthetases charge chloro-

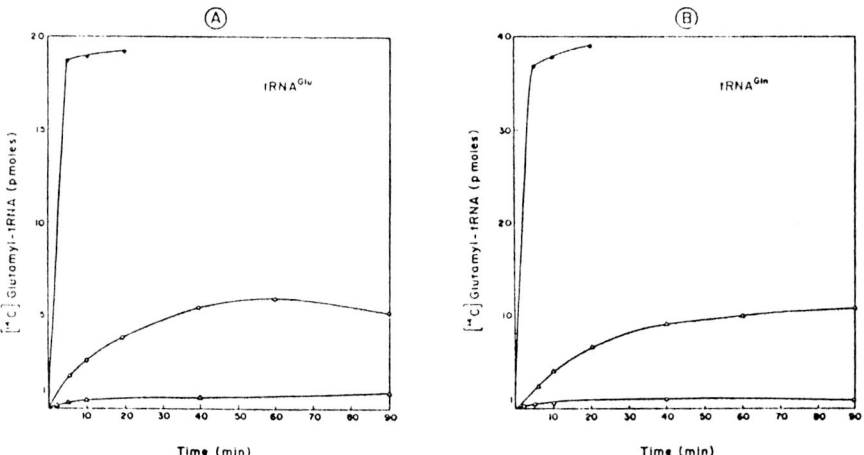

Figure 2. Time course of glutamylation of barley chloroplast tRNA^{Glu} (A) and tRNA^{Gln} (B) with homologous and heterologous aminoacyl-tRNA synthetases. The purity of the tRNA^{Glu} and tRNA^{Gln} preparations was 12 and 25%, respectively. Closed circles, barley chloroplast enzyme; open circles, *E. coli* S-100; triangles, *B. subtilis* S-100.

plast tRNA^{Glu} and tRNA^{Gln} equally well with glutamate. In contrast, the tRNA^{Gln} species cannot be glutamylated by a synthetase preparation from *E. coli* that charges chloroplast tRNA^{Glu}. On the other hand, a *B. subtilis* extract charges both tRNA^{Gln} species, but not tRNA^{Glu}, with glutamate. Barley chloroplast GluRS efficiently charges the tRNA^{Glu} from the cyanobacterium *Synechocystis* (29). Thus, this enzyme is a naturally occurring mischarging enzyme.

DISCUSSION

δ-Aminolevulinic Acid Synthesis Requires three Enzymes and a tRNA^{Glu}. The identification of the nature of the tRNA involved in ALA synthesis and of the glutamylation of this RNA allows to define a plausible description of the enzymes and tRNAs involved in ALA biosynthesis (16,17) in chloroplasts. This pathway is outlined in Fig. 3. In the first, ATP-dependent step, glutamate is esterified to tRNA^{Glu} by a ligase in a normal

THE FIRST STEPS OF CHLOROPHYLL BIOSYNTHESIS

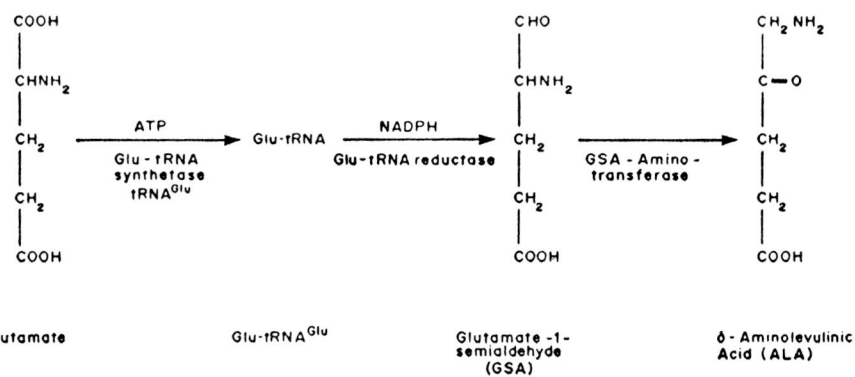

Figure 3. The pathway of ALA synthesis in chloroplasts. For explanations, see text.

aminoacyl-tRNA bond (16). Subsequently, this glutamyl-tRNA serves as a substrate for a Glu-tRNA^{Glu} specific, NADPH dependent reductase. Chloroplast tRNA^{Glu} is a required cofactor in this reaction; the two other glutamate accepting tRNA species present in barley chloroplasts are not recognized as substrate by the reductase (13,16).

Chloroplast Glutamyl-tRNA Synthetase is a Mischarging Enzyme.

Glutamyl-tRNA synthetase from barley chloroplasts shows an unusual pattern of substrate specificity (22,29); it efficiently charges both tRNA^{Glu} and tRNA^{Gln} from chloroplasts, but only tRNA^{Glu} and not tRNA^{Gln} from *E. coli* and *B. subtilis*. This behavior may be helpful in defining features of the tRNA that are involved in recognition by glutamyl-tRNA synthetase. Comparison of the sequences of the tRNAs that are glutamylated (i.e. barley chloroplast tRNA^{Glu} and tRNA^{Gln}, *E. coli* tRNA^{Glu}, and *B. subtilis* tRNA^{Glu}) to those that are not recognized by the enzyme (i.e. *B. subtilis* tRNA^{Gln} as well as both tRNA^{Gln} species from *E. coli*) allows conclusions about nucleotides that might be involved in enzyme:substrate interaction. For the tRNAs compared, only one base pair in the acceptor stem and one base pair in the anticodon stem meet these criteria (29): all tRNAs that are efficiently glutamylated have a U_6:G_{67}, C_6:G_{67} or G_6:C_{67} and a U_{31}:G_{39}, G_{31}:C_{39} or C_{31}:G_{39}, whereas the tRNAs that are not recognized have an A:U and an A:U or U:A base pair in these respective positions. The primary sequence of *Synechocystis* tRNA^{Glu} (20) and the published chloroplast tRNA^{Glu} and tRNA^{Gln} gene sequences also meet these criteria. Thus, it can be concluded that these two base pairs may play an important role in substrate recognition by barley chloroplast glutamyl-tRNA synthetase.

Aminoacylation in Chloroplasts needs subsequent Error Correction.

The pathway of Gln-tRNA^{Gln} formation in chloroplasts is very similar to the one described for *Bacilli*: tRNA^{Gln} is misacylated with glutamate by glutamyl-tRNA synthetase, presumably the same enzyme that charges tRNA^{Glu} (30). In the presence of a suitable amide donor, Glu-tRNA^{Gln} is then converted to Gln-tRNA^{Gln} by a tRNA-dependent amidotransferase. Thus, two enzymes are required for the synthesis of Gln-tRNA^{Gln}; a mischarging GluRS and a Glu-tRNA^{Gln} specific amidotransferase. The amidotransferase is formally similar to GlnRS: both enzymes possess specific binding sites for tRNA^{Gln}, glutamine and ATP. A functional similarity seems to exist between the amidotransferase and glutamine synthetase from *B. subtilis*: this enzyme is able to perform the tRNA-dependent amidotransferase reaction, albeit at a much lower rate than the amidotransferase proper (31).

It is tempting to speculate how the cells cope with this mischarging phenomenon while maintaining the required accuracy in protein biosynthesis in the organelle. Clearly, the mischarged Glu-tRNA^{Gln} must not be present in the free pool of aminoacyl-tRNAs in the cell. One possibility is that the ribosome has the ability to discriminate against the mischarged RNA, or the affinity for the ribosomal elongation factor EF-Tu is lower for the mischarged tRNA than for Gln-tRNA^{Gln}. It is known that the affinity for *E. coli* EF-Tu is strongest for Gln-tRNA^{Gln} and much weaker for Glu-tRNA^{Glu} (32); but the individual contributions of the amino acid and the tRNA to the binding strength are not known.

The occurrence of the pathway of Gln-tRNA^{Gln} formation via mischarging and transamidation is widespread. Evidence for its presence exists in gram-positive eubacteria (23,25), archaebacteria (28), yeast mitochondria (27), *Synechocystis*, and plant and animal organelles (22). In contrast, the direct pathway of tRNA glutaminylation occurs in the cytoplasm of eukaryotic cells and in gram-negative eubacteria. It is tempting to speculate that the mischarging/transamidation pathway was already present before the presumed common ancestors of modern prokaryotes and organelles diverged, and that the GlnRS of eukaryotic cytoplasm and gram-negative bacteria evolved independently, possibly to assure a greater degree of fidelity in protein biosynthesis.

Outlook.

It was very gratifying to see that the study of a biochemical pathway of intermediary metabolism has led to the finding of the involvement of tRNA and aminoacyl-tRNA synthetases in this process. The role of tRNA in a reductase reaction has no precedent in biochemical reactions. It underscores the variety of reaction mechanisms cells have at their disposal to carry out the complex reactions of intermediary metabolism. The study of the enzymatic mechanism of this reduction process and as well as the detailed investigation of the conversion of Glu→Gln when attached to tRNA may uncover novel principles which may also operate in other pathways yet to be discovered.

ACKNOWLEDGEMENTS

This work was supported by a grant from the Department of Energy and from the National Science Foundation.

REFERENCES

1. Beale SI (1978). Ann Rev Plant Physiol 29: 95-120.
2. Beale SI, Gough SP, Granick S (1975). Proc Natl Acad Sci USA 72: 2719-2723.
3. Friedmann HC, Thauer RK (1986). FEBS Lett 207: 84-88.
4. McKie J, Lucas C, Smith A (1981). Phytochem 20: 1547-1549.
5. Oh-Hama T, Seto H, Miyachi S (1986). Eur J Biochem 159: 189-194.
6. Oh-Hama T, Seto H, Miyachi S (1986). Arch Biochem Biophys 246: 192-198.
7. Oh-Hama T, Stolowich NJ, Scott AJ (1988). FEBS Lett 228: 89-93.
8. Huang D-D, Wang W-Y, Gough SP, Kannangara CG (1984). Science 225: 1482-1484.
9. Kannangara CG, Schouboe A (1985). Carlsberg Res Comm 50: 179-191.
10. Mau Y-H, Wang W-Y, Tamura RN, Chang T-E (1987). Arch Biochem Biophys 255: 75-79.
11. Kannangara CG, Gough SP (1978). Carlsberg Res Commun 43: 185-194.
12. Castelfranco PA, Beale SI (1983). Annu Rev Plant Physiol 34: 241-278.
13. Kannangara CG, Gough SP, Oliver RP, Rasmussen SK (1984). Carlsberg Res Commun 49: 417-437.
14. Wang W-Y, Huang D-D, Stachon D, Gough SP, Kannangara CG (1984).

Plant Physiol 74: 569-575.
15. Huang DD, Wang W-Y (1986). J Biol Chem 261: 13451-13455.
16. Schön A, Krupp G, Gough SP, Berry-Lowe S, Kannangara CG, Söll D (1986). Nature (London) 322: 281-284.
17. Schön A, Krupp G, Gough SP, Kannangara CG, Söll D (1987). In Inouye M, Dudock B (eds): " The Molecular Biology of RNA: New Perspectives," New York: Academic Press, pp 295-303.
18. Sprinzl M, Hartmann T, Meissner F, Moll J, Vorderwülbecke T (1987). Nucl Acids Res 15: r53-r188.
19. Ohyama K, Fukuzawa H, Kohchi T, Shirai H, Sano T, Sano S, Umesono K, Shiki Y, Takeuchi M, Chang Z, Aota S, Inokuchi H, Ozeki H (1986). Nature (London) 322: 572-574.
20. O'Neill GP, Peterson DM, Schön A, Chen M-W, Söll D (1988). J Bacteriol, in press.
21. Peterson DM, Schön A, Söll D (1988), manuscript submitted.
22. Schön A, Kannangara CG, Gough SP, Söll D (1988). Nature (London) 331: 187-190.
23. Wilcox M, Nirenberg M (1968). Proc Natl Acad Sci USA 61: 229-236.
24. Wilcox M (1969). Eur J Biochem 11: 405-412.
25. Schön A, Hottinger H, Söll D (1988). Biochimie, in press.
26. Lapointe J, Duplain L, Proulx M (1986). J Bacteriol 165: 88-93.
27. Martin NC, Rabinowitz M, Fukuhara H (1977). Biochemistry 16: 4672-4677.
28. Gupta R (1984). J Biol Chem 259: 9461-9471.
29. Schön A, Söll D (1988). FEBS Lett 228: 241-244.
30. Bruyant P, Kannangara CG (1987). Carlsberg Res Comm 52: 99-109.
31. Strauch MA, Zalkin H, Aronson AI (1988). J Bacteriol 170: 916-920.
32. Louie A, Ribeiro S, Reid BR, Jurnak F (1984). J Biol Chem 259: 5010-5016.
33. Quigley F, Weil JH (1985). Curr Genetics 9: 495-503.

HOST tRNA REPROCESSING: A PHAGE T4-SYSTEM OF RNA-SPLICING MEDIATED BY PROTEIN ENZYMES[1]

G. Kaufmann[2], M. Amitsur[2], D. Chapman[2],
M. J. Gait[3], R. Levitz[2], L. Jorissen[4],
I. Morad[2] and L. Snyder[4],

Tel-Aviv University[2], Israel 69978,
Medical Research Council[3], Cambridge, CB2 2QH, U.K.
Michigan State University[4], MI 48824.

ABSTRACT During phage T4-infection of E. coli cells that harbor the prr gene (prr^r), the host tRNA^{Lys} is cleaved by anticodon nuclease (ACNase) and is then repaired by polynucleotide kinase and RNA ligase. Genes implicated with ACNase are: (i) prr, which restricts T4 mutants lacking polynucleotide kinase or RNA ligase; and (ii) stp, a T4 gene whose mutant alleles suppress prr-restriction. ACNase was reconstituted in an extract of T4-infected E. coli-prr^r and a synthetic polypeptide deduced from the stp nucleotide sequence.

INTRODUCTION

Phage T4 induces cleavage-ligation (**reprocessing**) of at least two host tRNA species. The enzymes participating in the reprocessing pathway are anticodon nuclease (ACNase), polynucleotide kinase and RNA ligase (1,2). The more abundant of the reprocessing substrates, a tRNA^{Lys} species, is cleaved by ACNase 5' to the wobble base, yielding 2':3'-P> and 5'-OH termini. These termini are converted by polynucleotide kinase into 3'-OH & 5'-P pair that is joined by RNA ligase to restore the original form of tRNA^{Lys} (ref. 3, Fig. 1).

[1]Supported by NIH grants GM28001 to LS and GM34124 to GK

Host tRNA reprocessing is manifested specifically in
E. coli CTr5X and in derived strains which restrict phage
T4 mutants that are deficient in polynucleotide kinase
(pnk⁻) or RNA ligase (rli⁻) (1,4-6). This restriction is
specified by the host prr locus (5) and is suppressed by
mutations in the T4 stp gene (6-8). Both prr and stp are
required for in vivo manifestation of ACNase (4). In
vitro complementation studies suggest that prr and stp
encode components needed for ACNase activity (Amitsur,
Morad and Kaufmann, submitted; and below).

FIGURE 1
THE T4-INDUCED HOST tRNA REPROCESSING PATHWAY

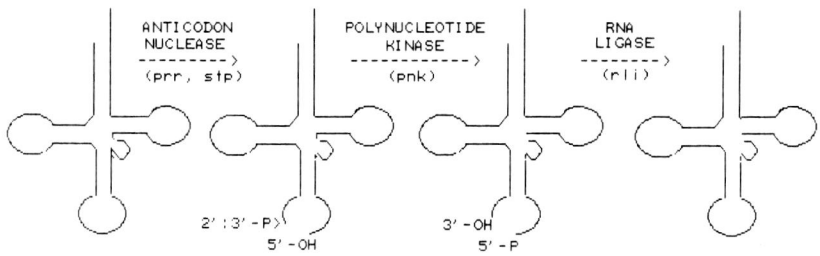

stp was cloned and the nucleotide sequence of its
wild-type and mutant alleles determined. Their
comparison defined an stp open reading frame of 29 codons
at 162.8-9 Kb of T4 DNA (9). The host prr locus was
mapped to 29 min of the E. coli CTr5X chromosome and was
transduced into permissive (prr^p) E. coli B and K 12
strains (5). prr is associated with a unique DNA
region found only in the restrictive (prr^r) hosts,
presumably a cryptic prophage or large transposon.
Linked to prr is an unusual restriction modification
system that prevents genetic expression of phage lambda
without degrading its DNA (5). Recently prr was marked
by inactivating transposon insertions and cloned in a
plasmid vector. The insertions also inactivated the
linked lambda restriction system, suggesting that both
traits may be encoded by the same gene (Levitz, Kaufmann,
Green and Snyder; unpublished results).

RESULTS

In vitro Reconstitution of Anticodon Nuclease

The _in vitro_ ACNase assay (Amitsur, Morad and Kaufmann, submitted) is based on known _in vivo_ properties of this activity (3) as well as on the identification of the stp gene (9). The ACNase reaction mixture contained a crude S-30 extract from T4 pnk$^-$ infected E. coli CTr5X (prr^r). The pnk mutation served to prevent further processing of the ACNase reaction products. The preparation of the S-30 extract and the composition of the reaction mixture were adapted from _in vitro_ protein translation protocols (cf. 10) because previous _in vivo_ studies indicated that ACNase is sensitive to chloramphenicol (Morad and Kaufmann, unpublished results). The ACNase reaction mixture was supplemented with a synthetic polypeptide of the deduced amino acid sequence of stp (pstp, Fig. 3). pstp stimulated the reaction even with extracts derived from T4 stp$^+$-infected cells. As a substrate for ACNase cleavage served ^{32}P-tRNA^{Lys}, labeled in the cleavage-ligation junction.

As shown, incubation of the substrate under these conditions resulted in the appearance of a labeled product migrating with tRNA^{Lys} fragment 1-33 (Fig. 2d). Subsequent treatment with nuclease P1 indicated that this tRNA fragment contained the 2':3'-P> end-group (not shown), as in the _in vivo_ ACNase reaction product (3). ATP and GTP were both required for ACNase activity. The S-150 supernatant and ribosome subfractions were devoid of ACNase activity. However, activity could be restored by combining these fractions (data not shown). These results are consistent with the dependence of ACNase activity on ongoing protein synthesis. Presumably, the anticodon cleavage is coupled to a translation-step in which the vulnerable tRNA participates. Alternatively, a short-lived ACNase protein has to be replenished.

By itself, pstp had no ACNase activity (Fig. 2a) but it markedly stimulated ACNase when added to the standard extract (Fig. 2d). An stp mutation rendered the extract ACNase-deficient but activity could be restored by adding synthetic pstp to the stp$^-$ extract (Fig 2c).

FIGURE 2.
ACNase COMPLEMENTATION IN VITRO

ACNase reaction mixtures containing synthetic pstp as indicated were incubated for 1 hour at 25°. The mixtures were then extracted with aqueous phenol, the RNA precipitated with ethanol, dissolved in 10 M urea and separated by electrophoresis on a 15% polyacrylamide-7M urea gel. (a) no extract, (b) extract of uninfected E. coli CTr5x, (c) extract from E. coli CTr5X infected with T4 pseT2Rb (pnk⁻, stp⁻, ref. 8), (d) standard extract of E. coli CTr5X infected with T4 pseTΔ1 (pnk⁻, ref.13), (e) extract from E. coli K10 (prr^p) infected with T4 pseTΔ1, (f) a 1:1 combination of extracts (b) and (e). 1-33 designates the position of the in vivo derived tRNA^{Lys} fragment 1-33.

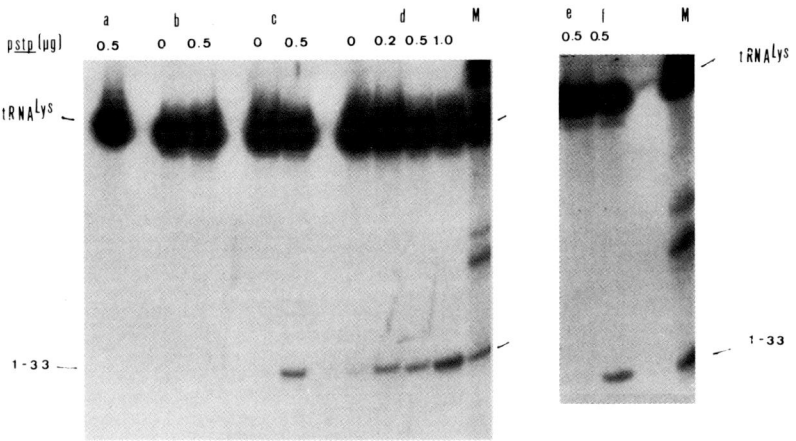

Extracts from uninfected prr^r cells or from T4-infected E. coli K10 (prr^p) were each inactive, even when supplemented with pstp (Fig. 2b, 2e; respectively). However, mixing these inactive extracts at 1:1 ratio restored ACNase activity to a level seen with the standard extract from T4-infected E. coli CTr5X (Fig. 2f). Based on these data it was concluded that ACNase is multicomponent, comprising at least physiological pstp, additional T4-induced product(s) and component(s) encoded by prr.

DISCUSSION

Is pstp a Catalytic Component of Anticodon Nuclease?

ACNase transesterifies its phosphodiester substrate, similar to the first step in an RNase A reaction (3). Inspection of the deduced amino acid sequence of stp revealed a subset of functional residues (shown in boldface in Fig. 3) matching those forming the catalytic site of RNase A. The coincident presence of this subset in a basic, 29 residue polypeptide needed for ACNase activity hints at a possible catalytic role of pstp. This idea is reinforced by a similarity between the stp-subset and a motif in the RNA-binding domain of satellite tobacco necrosis virus (STNV) coat protein which, we also suspect, harbors an RNase activity (9).

Regardless of its precise role, pstp is unlikely to fold by itself into an active structure and assume all the functions of ACNase, including tRNA substrate recognition. Additional ACNase functions may be provided by other subunits, e.g., the prr product(s), as indicated by the in vitro ACNase complementation assay described above. This complementation assay may serve now to test the proposed role of pstp, using engineered derivatives; and to isolate other ACNase components.

FIGURE 3.
DEDUCED AMINO ACID SEQUENCE OF stp

NH_2-Met-Ser-Asn-Phe-**His**-Asn-**Glu**-**His**-Val-Met-
Asn-Phe-Tyr-Arg-Asn-Asn-Leu-**Lys**-**Thr**-**Lys**-Arg-
Leu-Arg-Thr-Pro-Thr-Val-Arg-**Lys**-Ile-COOH

Significance of Host tRNA Reprocessing

Depletion of needed host tRNAs in the absence of pnk and rli-mediated repair (3) can account for defects in DNA replication and true-late protein synthesis during pnk^- and rli^- infections of E. coli prr^r strains (6-8). This interpretation portrays stp as a cofactor of prr restriction and pnk and rli merely as means to overcome it. Yet, it is conceivable that stp, pnk and rli evolved to benefit T4 directly. One possible role of host tRNA reprocessing might be to adjust the cellular tRNA ensemble to T4-codon usage, by analogy to roles proposed for the T4-encoded tRNAs and the T4-induced removal of the host tRNA$^{Leu}_1$ (cf. 11). Specifically, reprocessing might adjust the lysine decoding potential to post-infection codon usage. This assumption is based on a transient reduction in the level of the reprocessed tRNA^{Lys} during T4 infection of E. coli prr^r strains and the removal of another tRNA^{Lys} variant by ACNase (3 and unpublished results). However, in the prr^p context, other roles have to be invoked for stp, pnk and rli, roles that are perhaps also related to RNA cleavage-ligation.

Host tRNA reprocessing is a prokaryotic system of RNA-splicing that is mediated by protein-enzymes. It resembles in many respects the splicing of nuclear pre-tRNAs (12). However, unlike all known RNA splicing reactions, reprocessing does not entail the removal of an intron. Furthermore, the reprocessing substrates are mature and functional RNAs whose cleavage-ligation is triggered by a cellular change.

REFERENCES

1. David M, Borasio GD, Kaufmann G (1982). Bacteriophage T4-induced anticodon-loop nuclease detected in a host strain restrictive to RNA ligase mutants. Proc Natl Acad Sci USA 79: 7097.
2. David M, Borasio GD, Kaufmann G (1982). T4 bacteriophage-coded polynucleotide kinase and RNA ligase are involved in host tRNA alteration or repair. Virology 123: 480.
3. Amitsur M, Levitz R, Kaufmann G(1987). Bacteriophage T4 anticodon nuclease, polynucleotide kinase and RNA ligase reprocess the host lysine tRNA. EMBO J 6:2499.
4. Kaufmann G, David M, Borasio GD, Teichmann A, Paz A, Green R, Snyder L (1986). Phage and host determinants of the specific anticodon loop cleavages in bacteriophage T4 infected Escherichia coli CTr5X. J Mol Biol 188: 15.
5. Abdul-Jabbar M, Snyder L (1984). Genetic and physiological studies of an Escherichia coli locus that restricts polynucleotide kinase and RNA ligase-deficient mutants of bacteriophage T4. J Virol 51:522.

6. Depew RE and Cozzarelli NR (1974). Genetics and physiology of bacteriophage T4 3'-phosphatase: evidence for the involvement of the enzyme in T4 DNA metabolism. J. Virol. 13:888.
7. Sirotkin K, Cooley W, Runnels J and Snyder L (1978). A role in true-late gene expression for the T4 bacteriophage 5'polynucleotide kinase 3'phosphatase. J. Mol. Biol. 123:221.
8. Runnels J, Soltis D, Hey T and Snyder L (1982). Genetic and physiological studies of the role of RNA ligase of bacteriophage T4. J. Mol. Biol. 154:273.
9. Chapman D, Morad I, Kaufmann G, Gait MJ, Jorissen L, Snyder L (1988). Nucleotide and deduced amino acid sequence of stp: the bacteriophage T4 anticodon nuclease gene. J Mol Biol 199: 373.
10. Chambliss GH, Henkin T, Levinthal J (1983). Bacterial in vitro protein synthesizing systems. Meth. Enzymol. 101: 598.
11. Schmidt FJ, and Apirion D (1983). T4 transfer RNAs: Paradigamtic system for the study of RNA processing in T4 bacteriophage. In (EK Mathewes et al., eds.) Bacteriophage T4: p. 208. ASM Publications, Washington DC.
12. Greer CL, Peebles C, Geggenheimer P, Abelson J (1983). Mechanism of action of a yeast RNA ligase in tRNA splicing. Cell 32: 536.
13. Snyder L, Gold L, Kutter E, (1976). A gene of bacteriophage T4 whose product prevents true-late transcription on cytosine-containing T4 DNA. Proc. Natl. Acad. Sci. U.S.A. 73: 3908.

REGULATION OF GENE EXPRESSION BY A SMALL ANTISENSE RNA IN BACTERIOPHAGE P22[1]

Sha-Mei Liao and William R. McClure

Department of Biological Sciences
Carnegie Mellon University, Pittsburgh, Pa. 15213

ABSTRACT The small antisense regulatory RNA (sarRNA) of bacteriophage P22 has been investigated with experiments performed in vitro. We propose a secondary structural model for the sarRNA based on limited digestion by RNases of the 5' ^{32}P-labelled sarRNA prepared in vitro. The secondary structural model is consistent with most possibilities for complementary base pairing possible in this small RNA. A principal feature of the model is the location in a loop region of the complement to the ant mRNA Shine-Dalgarno sequence. Additional experiments performed in vitro suggest that this region is important in the initial pairing of sarRNA and, ant mRNA and may be the site of interaction between these RNA's that leads to negative control of antirepressor (Ant) synthesis during bacteriophage P22 development.

INTRODUCTION

A variety of small RNA's found in prokaryotes have been shown to regulate an increasing number of biological processes including plasmid copy number control, DNA transposition, and protein synthesis of key proteins in DNA replication and bacteriophage development. (See ref. 1 for a recent review.)

The regulation of gene expression by sarRNA in bacteriophage P22 development is a part of the regulatory circuitry

[1]Research on this project is supported by NIH (GM30375).

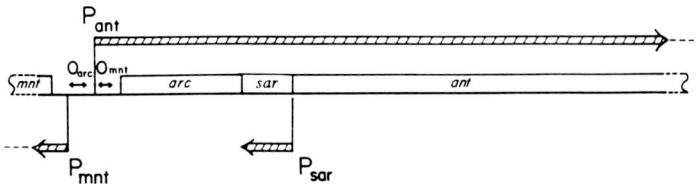

FIGURE 1. Patterns of transcription in the imml region of bacteriophage P22. The arc and mnt genes encode protein repressors that turn off transcription from the major operon promoter, Pant. The sar gene encodes a small antisense regulatory RNA that is required for the inhibition of ant gene expression from the P$_{late}$ transcript (5). The mechanism of sarRNA inhibition of ant expression involves pairing of the two RNA's in the region of complementarity. (See text for details.)

of the immunityI (imml) region of this bacteriophage. Shown in Figure 1 is a schematic diagram of the transcriptional patterns found within the imml region. Bacteriophage P22 is analogous in most respects to bacteriophage λ (2). For example, the lysis-lysogeny decision is controlled by analogous gene functions and the organization of essential genes is similar in both of these temperate phages. Unique to P22 is the presence of the imml region. (See ref. 3 for review.) The principal function of this portion of the P22 genome is to produce a controlled and temporally regulated amount of antirepressor (the product of the ant gene). The control of Ant synthesis is effected by three negative regulators. The arc gene product is synthesized early in infection and has as its effect the turndown of ant transcription from the major operon promoter, Pant. In the lysogenic state the product of the mnt gene, the Mnt repressor, binds to its operator immediately downstream from Pant and blocks all rightward transcription. In so doing the Mnt repressor activates the Pmnt promoter. In addition to these two negative effectors of Pant transcription, the synthesis of a small antisense RNA from Psar is also essential for the complete turnoff of Ant synthesis late in infection during which time the entire imml region is transcribed from the left by RNA polymerases that initiate from the late promoter (4, 5). The sar gene encodes an RNA product that is either 68 or 69 nucleotides long. This gene encompasses the entire intercistronic region between the arc and ant genes. Indeed the 5' end of sarRNA begins immediately upstream from the translation initiation start point of ant mRNA and proceeds leftward, terminating within the translation termination codon of the arc gene.

The location and orientation of the Psar promoter and its RNA product suggested a negative control function for Ant synthesis in P22. The published results to date suggest that sarRNA acts principally in trans to effect this negative control. In addition, the requirement for sarRNA late in infection has been established by Susskind and coworkers, who selected for and obtained mutations that allowed late Ant synthesis (5). These mutations were sequenced and found to lie within the -10 region of the Psar promoter, which is embedded within the 5' end of the ant gene. It is difficult to exclude any cis component of the negative role played by the Psar promoter. However, when we looked for evidence to support a cis function of sar transcription, we found to our surprise that the Pant promoter in fact had a cis negative effect on full length transcription from Psar (4). Thus, any cis effect that we have been able to observe in the in vitro studies has been shown to work in the direction opposite of the negative role for Psar in the regulatory circuitry. Therefore we have concentrated our efforts on examining the components of the sarRNA structure that are responsible for its trans negative regulatory effects.

RESULTS AND DISCUSSION

Secondary Structure of sarRNA.

The secondary structure of sarRNA was examined in vitro using sarRNA that was labelled at its 5' end with ^{32}P. The purified RNA was then subjected to limited RNAse digestions and examined on polyacrylamide gels under denaturing conditions. The pattern of cleavages revealed a set of accessible sites for each of the RNases employed in this series of experiments. A summary of the results is shown in Figure 2 as a proposed secondary structure model for sarRNA. Based on the substrate specificities for the single stranded specific ribonucleases (RNase A, T1, and T2) we find two regions that are very likely to exist as loops in the solution structure of the RNA. These are shown at the top and bottom of Figure 2. In addition, both the 5' end and the 3' end were found to be accessible to the single strand specific ribonucleases, even though the 3' end is complementary to the 5' end. Ribonuclease V1, which is a double strand specific ribonuclease, was found to cleave the

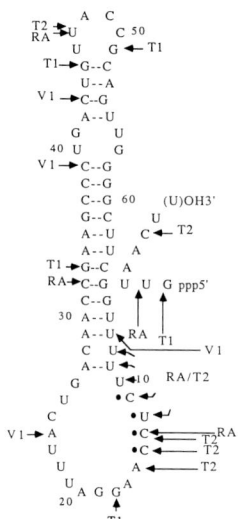

FIGURE 2. Proposed sarRNA secondary structure. The secondary structure of sarRNA is based on sequence complementarity and accessibility to various RNases. The sites of cleavage of the RNases used are shown with arrows; preferred cleavage positions are indicated with longer arrows. See text for details and a discussion of certain discrepancies in the simple model shown here schematically.

stem structure as proposed near the 5' end of the sarRNA molecule. The above results as summarized in Figure 2 are all consistent with the proposed structure shown. It is important as well to emphasize that several results are not consistent with the simplest structure we can draw at this time. In particular, the large loop from position 10 to position 26 was found also to be sensitive to RNase V1 at position 24. Moreover although half of the region in this loop contains many pyrimidines, there was no detectible RNAse A cleavage in this region. Thus we suggest that this portion of the molecule may be paired in some nonstandard way into a secondary or tertiary structure that cannot easily be rationalized based on simple Watson-Crick pairing rules. Another caveat is that the structure shown in Figure 2 is displayed with the two stem-loop structures comprising a coaxial helix over the entire length of the molecule. This representation is neither unique nor necessarily preferred as a tertiary structure. Indeed the representation is merely a convenience for showing where the sites of cleavage occur within the molecule.

The secondary structural model is preliminary at this point and requires additional testing with chemical reagents and perhaps even additional enzymatic digestion studies. The definitive method for establishing RNA secondary structure, namely comparative sequence analysis, is not available to us because sarRNA is unique to P22 and not found in other temperate bacteriophages. Thus the importance of using sar mutations for establishing the solution structure becomes of particular interest. Nevertheless the information gained from the secondary structural studies allowed us to investigate further the pairing properties of sarRNA with its complement, the ant mRNA.

Pairing of sarRNA with ant mRNA.

The ant mRNA was prepared in vitro and its kinetics of pairing with sarRNA was followed under pseudo first-order conditions with unlabelled ant mRNA in excess. The 5' end-labelled sarRNA was mixed and samples were applied to a polyacrylamide gel under nondenaturing conditions. We found that the radioactively labelled sarRNA formed a complex with ant mRNA and that the rates of complex formation increased with increasing concentrations of ant mRNA. When analyzed quantitatively, these data yielded an association rate constant for the reaction of 3.5×10^5 M^{-1} sec^{-1}. The overall kinetics are comparable to those obtained by Tomizawa (6) and by J. Kittle and N. Kleckner (personal communication). We were also interested in knowing whether any intermediates occurred along the RNA pairing pathway.

The pairing kinetics determined as described above in combination with the secondary structural analysis as determined by RNase accessibility enabled us to follow in greater detail the pairing process between ant mRNA and the 5' end-labelled sarRNA. In these experiments the two complementary partners were mixed; at short times thereafter RNase A was added such that a brief (15 sec) reaction resulted in partial cleavage at one or the other of two preferred sites at the 5' end of the sarRNA. These two sites, as indicated in Figure 2, were at positions 2 and 13. We found that immediately after mixing, the cleavage site at position 13 of sarRNA was rapidly protected from RNase A digestion, whereas the cleavage at position 2 was actually enhanced. At longer times the cleavage at position 13 was

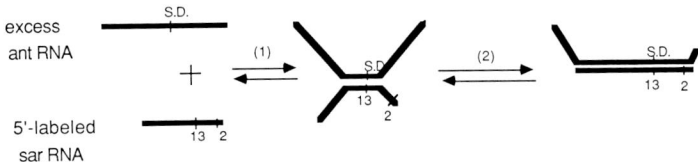

FIGURE 3. Pathway for sarRNA pairing to ant mRNA. The two steps shown schematically correspond to the protection and enhancement patterns of RNase A cleavage at positions 13 and 2 of sarRNA. The two RNA's first interact near position 13 of sarRNA and the Shine-Dalgarno sequence of ant mRNA. This initial paired complex is slowly converted to a fully formed duplex following propagation of the helix to the 5' and 3' ends of sarRNA. Not shown is the cleavage by RNase III of the sarRNA-ant mRNA duplex near positions 13-15 of sarRNA.

completely protected, as was the cleavage at position 2. The difference in the kinetics of accessibility to these two sites suggests an intermediate on the pathway of sarRNA-ant mRNA pairing.

Our current view of the pairing pathway based on these results is shown in Figure 3. In the first step we suggest that initial pairing occurs in the region near position 13 of sarRNA and its complement, the Shine-Dalgarno sequence of ant mRNA. A slower step then follows during which position 2 becomes completely protected from RNase A attack; ultimately the entire sarRNA is found to be completely resistant to RNAse A digestion. The end product of the pairing pathway then is a fully formed duplex between the two RNA molecules. It is necessary as well to qualify the simplicity of the conclusions drawn based on these experiments. In particular the reagent used to judge protection and accessibility, RNase A, is sensitive only to sites at positions 2 and 13 of the sarRNA molecule. Thus, we are unable to draw any strong conclusions at this time about the extent and/or rate of pairing from positions downstream of nucleotide 13. A second point is that the pairing from position 13 to the 5' end of the sarRNA molecule, which we propose to be a slow step in the reaction, is unusual in the sense that duplex propagation is ordinarily not a slow process once a double helix has been nucleated. A slow rate of helix propagation may be involved in the regulatory properties, depending on which complex species is responsible for the negative regulatory role. The slow step may also be caused by the requirement to dissolve a secondary or unknown tertiary structure in the sarRNA or indeed even in the ant mRNA (Note that we have no information about the possible secondary structural folding of the ant mRNA at this time.).

A third step in the pairing process between sarRNA and ant mRNA could be added to the pathway of Figure 3, namely the cleavage of the duplex by E. coli RNase III. This step has been shown to be the end result in vitro using partially purified RNase III. A surprising finding was that the preferred site of cleavage on the sarRNA was about 13 to 15 nucleotides downstream from the start of sarRNA. This was clearly the preferred cleavage over any of the other duplex regions within the sarRNA-ant mRNA duplex. To our knowledge the cleavage near the 5' end of one partner in a RNA-RNA duplex has not been reported to be a preferential site of RNase III hydrolysis. Since we find this cleavage in vitro we suggest that it may also play a role in the in vivo negative control. However, until the experiment is performed in a RNase III⁻ host, we cannot be certain that this step is required for negative regulation or whether it merely corresponds to the ultimate processing and salvage pathway for the sarRNA-ant mRNA duplex.

Results from Other Pairing Studies and Limitations of the Current Model.

Models for the RNA-RNA pairing process have been proposed in studies of ColE1 plasmid replication and Tn10 transposition. Three major steps are involved in the pairing process between ColE1 RNAI and RNAII. The two RNAs initially interact via complementary loop-loop contacts (nucleation). The initial contact, however, does not appear to initiate via propagation from loop interactions; rather, it serves to bring the 5' tail of RNA I into position to initiate hybrid formation. The second step involves breaking loop-loop contacts and dissolving the stem-loop structures. In the last step, base pairing starting from the 5' end of RNA I propagates to its 3' end to produce a complete duplex. The times required to complete helix propagation from 5' to 3' of RNA I are generally not short probably due to the times required for breaking loop to loop contact and for dissolving or unfolding the stem-loop structures (6). The proposed pairing model in Tn10 RNA-OUT and RNA-IN is much simpler; the pairing begins with a rate-limiting interaction between the 5' end of RNA-IN and the loop domain of RNA-OUT (cited in

Green et al. ref. 1). After this initial pairing, the 5'-strand of the RNA-OUT stem is displaced as hybrid formation proceeds.

The pathway proposed for the pairing process between sarRNA and ant mRNA shown in Figure 3 is a useful starting point for discussion of the role of sarRNA in negative control of ant gene expression. The individual steps indicated also expose our current ignorance about how the detailed mechanism of negative control is actually effected. For example, it could be that step 1 alone is sufficient to provide negative control. In other words the simple and reversible pairing of the intermediate complex between position 13 and the ant mRNA Shine-Dalgarno could compete with ribosomes, thereby reducing ant gene expression. This model would be similar to other more detailed examples worked out by L. Gold and his coworkers (see chapter in this volume) and predicts that the sarRNA is a competitive inhibitor of ribosome function. If negative control were localized solely to the second step shown in Figure 3, then the formation of a complete duplex would be required; in this model one would imagine that ribosomes would have no chance of attaching to the mRNA, thereby blocking ant gene expression completely. If this model obtains, then the time required for complete duplex formation may be relevant in vivo since the second step is slow in the pairing pathway. Finally, if negative control requires the cleavage of the message, then RNase III would be involved in the pathway of negative regulatory control, as has been found by Krinke and Wulff for the OOP RNA in the control of λ cII gene expression (7). It is also not impossible to imagine that varying extents of negative control could be exerted at each of the steps shown in the pathway. The final answer to these questions will only be obtained by examining mutations in the sar gene or in the bacterial host that result in alterations or aberrations of the negative control process.

ACKNOWLEDGEMENTS

We are grateful to Kathryn Galligan for her assistance in the preparation of this manuscript.

REFERENCES

1. Green, PJ, Pines, O, Inouye, M (1986). The role of antisense RNA in gene regulation. Ann Rev Biochem 55:569.
2. Susskind, MM, Botstein, D (1978). Molecular genetics of bacteriophage P22. Microbiol Rev 42:385.
3. Susskind, MM and Youderian, P (1983). Bacteriophage P22 antirepressor and its control. In Hendrix, RW, Roberts, JW, Stahl, FW and Weisberg, RA (eds.) "Lambda II" Cold Spring Harbor Laboratory, Cold Spring Harbor, New York, p 347.
4. Liao, S-M, Wu, T, Chiang, CH, Susskind, MM, McClure, WR (1987). Control of gene expression in bacteriophage P22 by a small antisense RNA. I. Characterization in vitro of the Psar promoter and sar RNA transcript. Genes & Development 1:197.
5. Wu, T, Liao, S-M, McClure, WR, Susskind, MM (1987). Control of gene expression in bacteriophage P22 by a small antisense RNA. II. Characterization of mutants defective in repression. Genes & Development 1:204.
6. Tomizawa, J (1984). Control of ColE1 plasmid replication: the process of binding of RNA I to the primer transcript. Cell 38:861.
7. Krinke, L, Wulff, DL (1987). OOP RNA, produced from multicopy plasmids, inhibits λ cII gene expression through an RNase III-dependent mechanism. Genes & Development 1:1005.

ANTISENSE RNA-MEDIATED INHIBITION OF VIRAL INFECTION IN TISSUE CULTURE AND TRANSGENIC MICE[1]

Kathy M. Takayama, Shigeki Kuriyama, Susan Weiss[2], Kiran Chada, Sumiko Inouye, and Masayori Inouye

Department of Biochemistry, UMDNJ-Robert Wood Johnson Medical School
Piscataway, New Jersey 08854

ABSTRACT We have previously demonstrated, using a prokaryotic model, that the production of phage SP is effectively inhibited in E. coli by the inducible production of antisense RNA targeted against the phage genome. We have hence attempted to examine the feasibility of an antisense RNA immune system in tissue culture cells and transgenic mice against mouse hepatitis virus. Portions of the viral genome were linked to the inducible mouse metallothionein promoter in the antisense orientation. The constructs were transfected into mouse L2 cells and also introduced into the mouse germline. In preliminary experiments virus was grown in either transfectant cells containing antisense sequences to viral genes 5 and 6 or in control L2 cells, and cellular extracts were subsequently titered on uninfected L2 cells.

[1]This work was supported by grants from the NJ Commission for Cancer Research and from Enzo Biochem.
[2]Present address: Department of Microbiology, University of Pennsylvania, Philadelphia, Pennsylvania 19104

A decrease in plaque-forming titer was observed for virus grown in the transfectant line as compared to the titer for virus grown in control cells. The study of the tissue culture and transgenic mouse systems in parallel enables us to examine the antisense RNA effect both on a cellular level and in the whole organism.

INTRODUCTION

Antisense RNA (or micRNA: mRNA-interfering complementary RNA) has been utilized in both prokaryotes and eukaryotes to regulate the expression of specific genes as well as to elucidate the functions of uncharacterized genes (1,2,3,4). A particularly interesting application of antisense RNA regulation is the complementary mRNA-mediated inhibition of virus production. Such an antisense RNA-immune system (or micRNA-immune system) is characterized by the capability of the cell to produce antisense RNAs against viral RNAs, thus interfering with translation of the viral message. We have successfully demonstrated in a prokaryotic model that the production of phage SP (positive, single-stranded RNA coliphage) is effectively inhibited by antisense RNA produced in the presence of inducer. We were able to show that the micRNA-immune system can be engineered in such a way that E. coli cells were rendered immune not only to a specific phage (SP) but also to related phages as well (5). Further studies have led us to identify the most effective region for targeting antisense RNA to SP phage, which includes the Shine-Dalgarno sequence and a 13-base sequence immediately upstream.

Based on the above studies, we have attempted to apply the micRNA immune system to eukaryotes. We have chosen to examine the effects of antisense RNAs against mouse hepatitis virus (MHV) in tissue culture cells and transgenic mice. Mouse hepatitis virus, which replicates exclusively in the cytoplasm

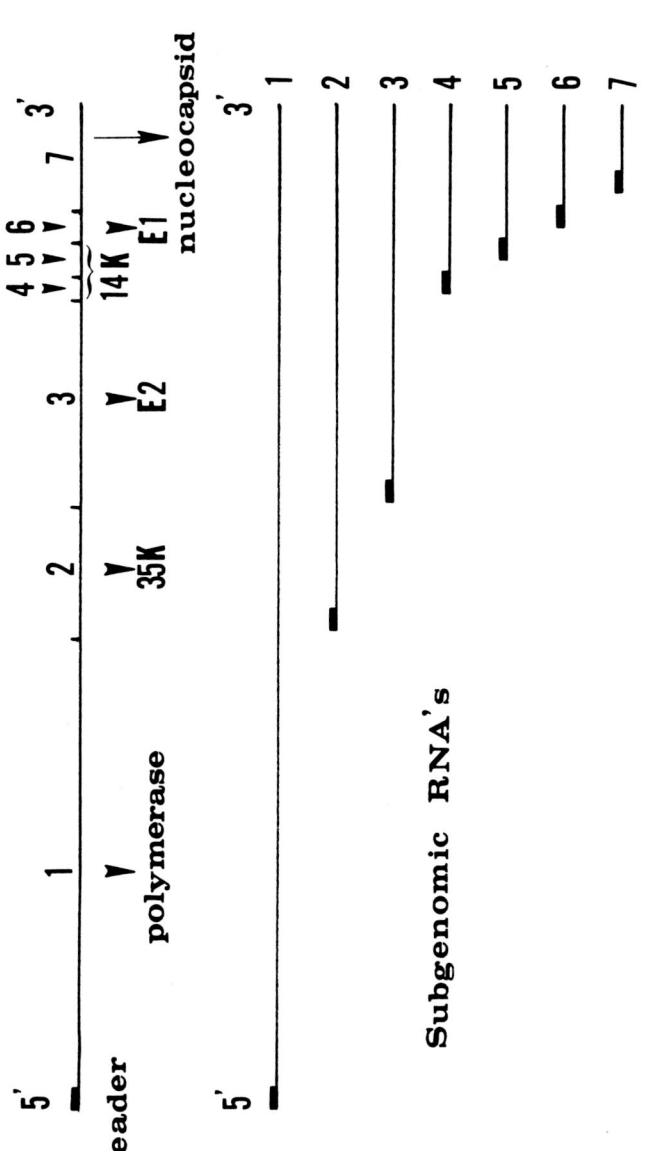

FIGURE 1. Schematic representation of Mouse Hepatitis Virus RNA and subgenomic mRNAs. The 72-base leader sequence is indicated by a box at the 5' end (Adapted from 22).

(6,7), contains an 18 kb, positive single-stranded RNA genome (8) that is first transcribed into a genomic-sized negative-stranded RNA upon infection (9). This vegetative RNA species serves as the template for the synthesis of positive-stranded genomic RNA and six subgenomic mRNA species having a nested-set structure extending for various distances from the 3' end toward the 5' end (10; see Figure 1). A leader sequence of 72 bases is present at the 5' ends of all viral mRNAs and is thought to be synthesized independently as a free leader that serves as a primer for mRNA synthesis (11, 12). The production of a micRNA targeted to this leader may therefore exert an effective immune function against MHV. In addition we have attempted to investigate the mic immune effects of various antisense constructs of the MHV genome in order to determine those sequences which confer the greatest resistance.

RESULTS

micRNA Gene Constructs

cDNA fragments derived from the genome of mouse hepatitis virus were cloned in the antisense orientation into the micRNA cloning vector, pMTSV (Fig 2). The vector includes a fragment of ~600-bp containing the inducible metallothionein promoter, followed by a removable 210-bp fragment derived from an SV40 intron and a 237-bp fragment containing an SV40 polyadenylation signal.

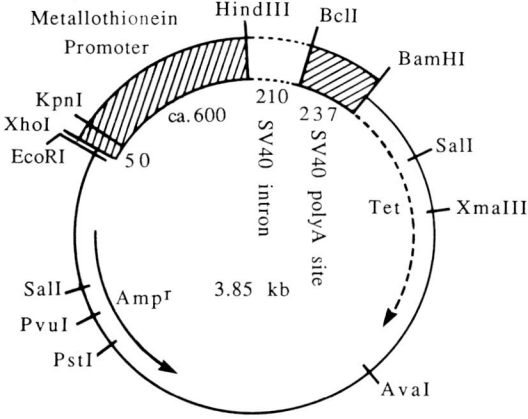

FIGURE 2. Map of pMTSV

Our initial micRNA gene constructions have been targeted to gene 5 of MHV. The levels of mRNA for gene 5 are lower than those of mRNAs 6 and 7 (13), and the gene 5 protein exists in a very small amount (14). pMISmhvα1 (Fig 3), derived from pMTSV, contains a 1.1-kb gene fragment in the antisense orientation encompassing: (a) a 523-bp fragment encoding all but the 19 amino-terminal amino acid residues of gene 5, (b) the 13-bp intergenic region between gene 5 and gene 6, and (c) the 563-bp fragment encoding the first 188 amino acid residues of the gene 6 product, El glycoprotein. pMISmhvβ1 contains the same 1.1-kb fragment in the sense orientation. Plasmids pMISmhvα2 and pMISmhvα3 contain the 72-base leader in the antisense orientation in single copy and four tandem repeats, respectively. pMISmhvα4 contains the 1.1-kb fragment from pMISmhvα1 at the 5' end and a single copy of the leader sequence at its 3' end, all in the antisense orientation. Similarly, pMISmhvα5 is designed to produce an antisense RNA against

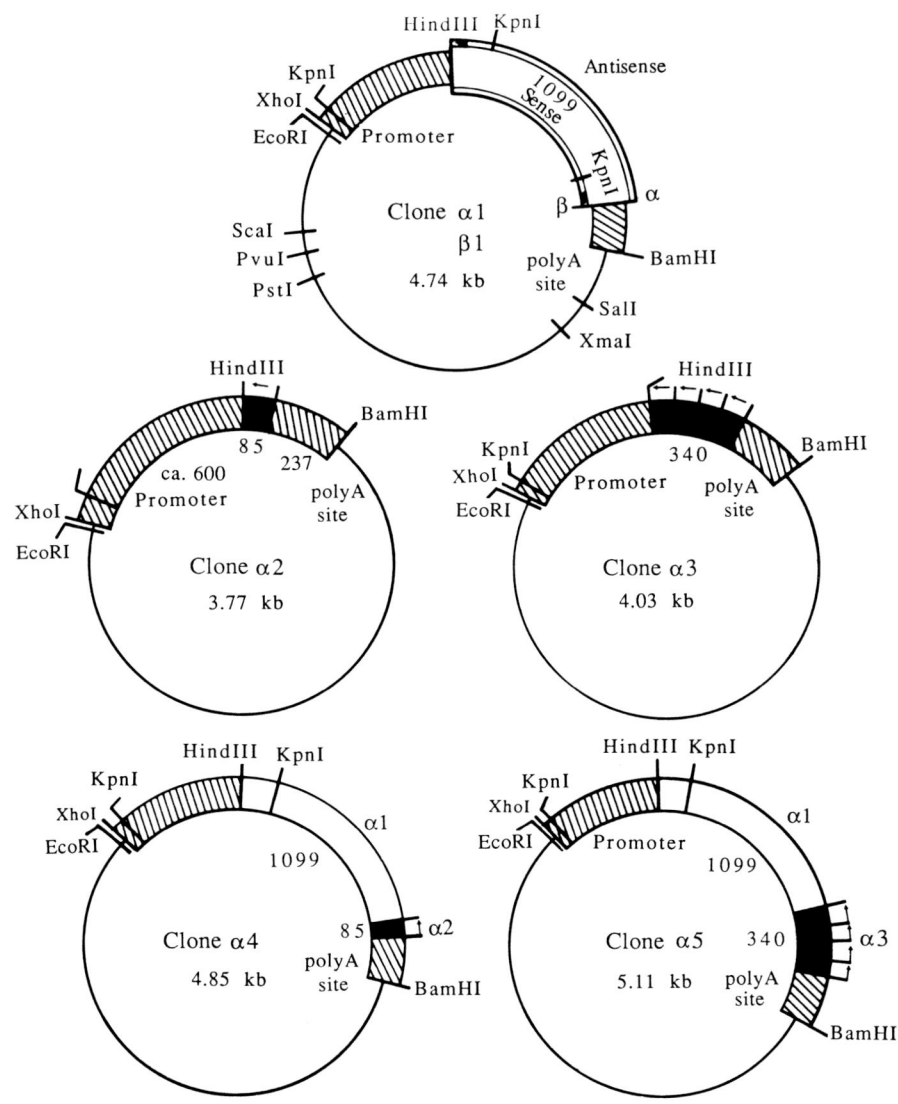

FIGURE 3. Various constructs carrying micRNA-immune sequences against MHV.

the 1.1-kb fragment at the 5' end and the tetramer of the leader sequence at the 3' end. By repeating the antisense sequence against the leader sequence, one can examine whether gene dosage would result in a stronger micRNA immune effect.

Transfections with mic Gene Constructs

pMISmhvα3 and pMISmhvα5 were cotransfected with the selective marker plasmid pSV2neo (15) into mouse L2 cells by the standard calcium phosphate coprecipitation method. Southern blot analysis of genomic DNA isolated from G418-resistant clonal colonies indicated several transfectant lines containing integrated antisense sequences. 3α2-7, 3α2-11, and 5α1-10 contained approximately one, three, and ten copies of their designated antisense sequences, respectively (data not shown). However, the expected 0.6-kb message was detectable on Northern gels for 3α2-11 but not for 3α2-7 (Fig 4). Degrees of immunity to MHV infection will be measured by comparing plaque-forming titers for virus grown in transfectant lines with the titer for virus grown in control L2 cells.

Production of Transgenic Mice

EcoRI-BamHI restriction fragments of pMISmhvα3 and pMISmhvα5 containing the mic gene sequences were injected into one of the pronuclei of fertilized one-cell mouse eggs derived from a C567BL/6J x CBA/J mating. Several founder mice have been identified by

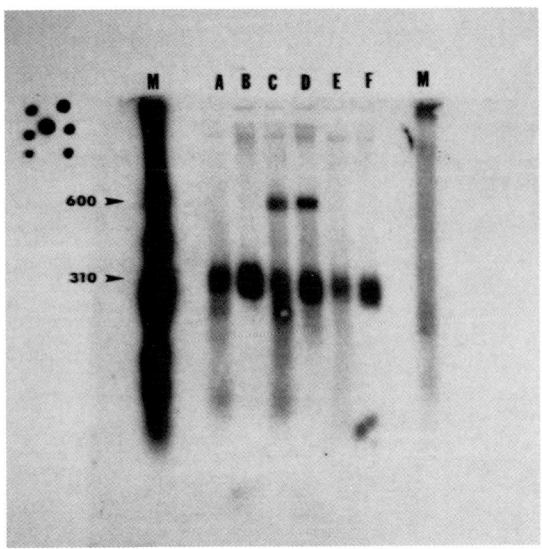

FIGURE 4. Northern analysis of total RNA prepared from positive transfectant cells. A: 3α 2-7 uninduced; B: 3α2-7 induced; C: 3α2-11 uninduced; D: 3α2-11 induced; E: control L2 cells uninduced; F: control L2 cells induced; M: size markers.

Southern analysis of genomic DNA. Two founders contained 1 to 2 copies of head-to-tail integrated α3 sequences and two founders contained ~100 copies of α5 sequences (data not shown). Litters produced from the above founder mice also consistently harbored the micRNA sequences. These F2 progeny are currently being tested for micRNA expression via Northern analysis of RNA prepared from various tissues, and will subsequently be challenged with virus to check for immunity.

DISCUSSION

In our studies of the SP phage micRNA-immune system it was determined that antisense RNA targeted against the maturation protein, which is present at low levels (~1 molecule/phage particle), was most effective in comparison to micRNAs against other regions of the genome such as those coding for coat protein and replicase (5). Our initial efforts with MHV have therefore been focused on gene 5, due to the scarcity of both the message and protein. Comparisons of degrees of immunity to MHV for L2 cells transfected with the various clones mentioned above may give insight into those genetic elements required for most effective inhibition of viral replication by antisense RNA. Contingent upon the results from these experiments are further studies testing additional mic gene constructs currently being made, including those against the 5' end of the genome. It is also of interest to examine the effects of overproduction of the 72-base sense leader sequence. The idea of "pathogen-derived resistance"; i.e., disruption of a pathogen's life cycle by expression of an essential gene or viral sequence at the wrong time, in aberrant amounts, or in altered form (16) has been effectively demonstrated in prokaryotes (17). It is possible that excessive amounts of the leader sequence (or other mouse hepatitis viral sequences) may titrate out viral (or cellular) components required for virus production. This may result in a new type of immune system against mouse hepatitis virus.

It has been proposed that the inhibition of gene expression in eukaryotic systems by antisense RNA may occur by hybridization of antisense message to sense message in the cytoplasm, preventing the translation of a specific mRNA species, or by hybridization in the nucleus, preventing the processing and/or transport of the targeted message (18, 19, 20, 21). In our experimental system, the replication cycle of mouse hepatitis virus does not include a nuclear phase (6,7), leading us to predict the inhibition by $\alpha 3$ and $\alpha 5$ RNAs to be localized in the cytoplasm. Analysis of the molecular basis of antisense RNA inhibition in tissue culture, together with the analysis of micRNA activity in the whole animal where tissue specificity and physiology become factors, should provide insightful information on the applicability and mechanistics of the micRNA immune system.

ACKNOWLEDGMENTS

We thank Dr. Paul March for critically reading the manuscript, and Carol Plisco for her secretarial assistance.

REFERENCES

1. Weintraub H, Izant JG, Harland RM (1985). Antisense RNA as a molecular tool for genetic analysis. Trends Genet 1:22.
2. Green PJ, Pines O, Inouye M (1986). The role of antisense RNA in gene regulation. Annu Rev Biochem 55:569.
3. Pines O, Inouye M (1986). Antisense RNA regulation in prokaryotes. Trends Genet 2:284.
4. Houba-Herin N, Inouye M (1987). Antisense RNA. Nucleic Acids and Mol Biol 1:210.
5. Hirashima A, Sawaki S, Inokuchi Y, Inouye M (1986). Engineering of the mRNA-interfering complementary RNA immune system against viral infection. Proc Natl Acad Sci 83:7726.
6. Brayton PR, Gauges RG, Stohlman SA (1981). Host cell nuclear function and murine hepatitis virus replication. J Gen Virol 56:457.
7. Wilhelmsen KC, Leibowitz JL, Bond CW, Robb JA (1981). The replication of murine coronaviruses in enucleated cells. Virology 110:225.
8. Lai MMC, Stohlman SA (1978). RNA of mouse hepatitis virus. J Virol 26:236
9. Lai MMC, Patton CD, Stohlman SA (1982). Replication of mouse hepatitis virus:

negative-stranded RNA and replicative form RNA are of genome length. J Virol 44:487.
10. Lai MMC, Brayton PR, Armen RC, Patton CC, Pugh C, Stohlman SA (1981). Mouse hepatitis virus A59: messenger RNA structure and general localization of the sequence divergence from a hepatotropic strain MHV-3. J Virol 39:823.
11. Lai MMC, Patton CD, Baric RS, Stohlman SA (1983). Presence of leader sequences in the mRNA of mouse hepatitis virus. J Virol 46:1027.
12. Lai MMC, Baric RS, Brayton PR, Stohlman SA (1984). Characterization of leader RNA sequences on the virion and mRNAs of mouse hepatitis virus, a cytoplasmic RNA virus. Proc Natl Acad Sci 81:3626.
13. Leibowitz JL, Wilhelmson KC, Bond CW (1981). The virus specific intracellular species of two murine coronaviruses: MHV-A59 and MHV-JHM. Virology 114:39.
14. Budzilowicz CJ, Weiss SR (1987). In vitro synthesis of two polypeptides from a nonstructural gene 5 of coronavirus mouse hepatitis virus strain A59. Virology 157:509.
15. Southern P, Berg P (1982). Transformation of mammalian cells to antibiotic resistance with a bacterial gene under control of an SV40 early region promoter. J Mol Appl Genet 1:327.
16. Sanford JC, Johnston SA (1985). The concept of parasite-derived resistance - deriving resistance genes from the parasite's own genome. J Theor Biol 113:395.
17. Grumet R, Johnston SA, Sanford JC (1986), personal communication.
18. Crowley TE, Nellen W, Gomer RH, Firtel RA (1985). Phenocopy of discoidin I-minus mutants by antisense transformation in *Dictyostelium*. Cell 43:633.
19. Izant JG, Weintraub H (1985). Constitutive and conditional suppression of exogenous and endogenous genes by anti-sense RNA. Science 229:345.
20. Kim SK, Wold BJ (1985). Stable reduction of

thymidine kinase activity in cells expressing high amounts of anti-sense RNA. Cell 42:129.
21. Melton DA (1985). Injected anti-sense RNAs specifically block messenger RNA translation in vivo. Proc Natl Acad Sci 82:144.
22. Budzilowicz CJ, Wilczynski SP, Weiss SR (1985). Three intergenic regions of coronavirus mouse hepatitis virus strain A59 genome RNA contain a common nucleotide sequence that is homologous to the 3' end of the viral mRNA leader sequence. J Virol 53:834.

SITE-SPECIFIC ENDONUCLEOLYTIC CLEAVAGES IN THE 5' REGION OF THE E. coli ompA AND bla mRNA SEEM TO INITIATE DEGRADATION[1]

A. von Gabain[2]*, U. Lundberg, Ö. Melefors, G Nilsson,[3]

Department of Bacteriology, Karolinska Institute, S-104 01 Stockholm, Sweden.

ABSTRACT Degradation of the monocistronic bla and ompA gene transcripts was comparatively analysed to identify mechanisms that control mRNA stability in E. coli. The stability of the ompA mRNA is depending on the growth rate, while the stability of the bla messenger is constant at different bacterial doubling times. In cells growing at fast growth rate the ompA mRNA is about five times more stable than the bla transcript. The 5' noncoding region seems to act as a determinant of mRNA stability. Analysis of the efficiency of translation and the effect of premature stop codons revealed that changes and differences in stability are not explained by alterations of the ribosomal protection. However, a minimal number of ribosomes (i.e. one) at the 5' part of the bla transcript is sufficient for its normal stability. We have identified cleavages in the 5' region of the bla and ompA mRNA that may well explain the initial step of degradation. The stability of the ompA mRNA follows the rate of these cleavages. The same cleavages were identified in an in vitro system of mRNA decay: the cleavages are catalysed by enzymes which are not identical to the endoribonucleases RNaseIII, RNaseE or RNaseP.

[1]This work was supported by RMC, STU and SBF.
[2]The authors are in alphabetical order.
[3]Present address: Department of Cell Biology, University of Basel, CH-4056 Basel, Switzerland.

INTRODUCTION

The mechanism of mRNA degradation has been relatively ignored in studying elements controlling gene expression. mRNA stability controls the level of gene expression by influencing the concentration of mRNA in the cytoplasm that in turn determines the rate of protein synthesis. There is a great deal of variability in the stability of transcripts which can vary in E. coli from seconds to 20 minutes (1,2). Regulation of degradation of mRNA is also important because it allows cells to adjust rapidly in response to physiological alterations. In both prokaryotes and eukaryotes, the half-life of certain mRNA species has been shown to be growth-dependent. An example of this is the ompA message of E. coli (2).

The stability of a mRNA is controlled by the interaction between regions of the molecule (such as 5' and 3' noncoding regions) and outside factors such as ribosomes and by specific enzymes (e.g.RNases). (For review see,3). An example is the 3' stem loop structure in many E. coli transcripts that has been found to impede 3'- 5' exonucleolytic degradation. However, this structure does not in general control the initial step of degradation. 3'- 5' exnucleases have been shown to be involved in mRNA degradation while it is not believed that they control the rate-limiting step of messenger decay (3). A number of site-specific endonucleases have been characterized in E. coli such as RNase III, RNase E and RNase P. They seem normally to be involved in stable RNA processing and only exceptionally in the degradation of certain mRNA species (3).

We have studied the decay of two monocistronic mRNA species in E. coli: the ompA mRNA which encodes a major outer membrane protein (OmpA) and is transcribed from a gene located on the chromosome and the bla mRNA of plasmid pBR322 which encodes the periplasmic enzyme β-lactamase. These messages are approximately equal in length while their rates of decay differ considerably in rapidly growing cells. The present communication summarizes the results obtained by analysing and comparing the decay of these two mRNA species.

MATERIAL AND METHODS

Material and methods used for the experiments as well as description of strains, constructions of genes and plasmids have been published elsewhere (2,4,6,7,15,16,18).

Results and Discussion

Stability of the bla and ompA Gene Transcripts at Different Bacterial Growth Rates.

E. coli strain C600 harboring the plasmid pBR322 was cultured in one of the two bacterial growth media L-broth or MOPS-Acetate. After blocking transcription with rifampicin the decay of the ompA and bla gene transcripts was analysed by S1 nuclease protection or Northern blotting as described previously (4). Table 1 shows that the half-life of the ompA transcript falls from 15 min. in cells growing with a doubling time of 40 min. to about 4 min. in cells growing with a doubling time of 200 min. The half-life of the bla mRNA did not differ at the two different growth rates. The amount of OmpA protein (per cell mass) in cells grown under different growth conditions is constant (5), while the amount of Bla protein increases by roughly a factor of three when the two growth rates are compared (2). Consistent with this finding is that the rate of OmpA protein synthesis dropped fivefold when the doubling time increased by a factor of five. The decreased rate of OmpA protein synthesis is accompanied by a reduction of the amount of ompA mRNA (2,6). Thus, most of the reduction of OmpA protein synthesis can be attributed to the decreased stability of its encoding transcript. An other E. coli mRNA species has been found that follows the pattern of the ompA mRNA: the cat transcript of the plasmid pACYC184. Its half-life dropped from 2 to less than 0.5 minutes, when the two growth conditions were compared (2). The stability of various E. coli transcripts can differ by at least a factor of 50 and their rates of decay are differently affected by changes in the rate of cell growth (1,2).

TABLE 1

Half-lives of the transcripts

at different doubling times

	40 min. (in L-broth)	200 min. (in MOPS-Acetate)
ompA	15 min.	4 min.
bla	3 min.	3 min.

The 5' Noncoding Region of the ompA Message is Capable of Conferring Stability to Downstream Fused bla Transcript.

With the goal of identifying the structural determinants responsible for the disparate decay characteristics of the ompA and bla transcripts, we constructed gene fusions where specific segments of one transcript were replaced by the respective segment of the other (7). All hybrid ompA-bla genes shown in Figure 1 preserved the correct translational reading frame of the β-lactamase protein. The untranslated RNA segments are indicated as straight or wavy lines. The bars (open or filled) indicate the coding regions of the bla, ompA and bla-ompA hybrid transcripts. The half-lives obtained for the respective wild type and hybrid transcripts are depicted in Figure 1. The analysis of the ob2 hybrid mRNA disclosed a half-life of 17 min (the degradation profile was biphasic, half of the population of molecules decayed more rapidly)(7).

The result presented in Figure 1 demonstrates that the 3' hairpin structure of the stable ompA messenger does not confer stability to the bla transcript, while fusion of the 5' terminal (147 nucleotide) segment of the ompA messenger to the 5' truncated bla transcript increased the stability of the bla segment about 5 fold. Thus, a model explaining the stability of a transcript by the time it takes for 3'- 5' exonucleases to overcome the 3' terminal hairpin structure is not compatible with this finding. Such a view is also opposed by the result that the three bla transcripts, differing only by the number of the 3' terminal hairpin structures (one, two or three, as a result of leaky termination), decay with identical rate (4).

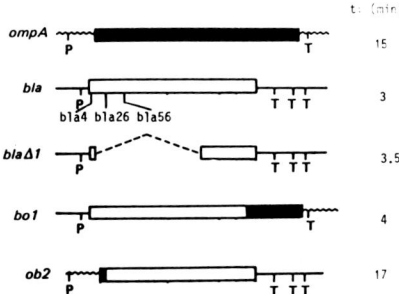

FIGURE 1. Physical maps of ompA and bla genes

However, it should be mentioned that endonucleolytic cleavage of an RNA stem-loop structure by RNase III has been shown to affect the level of the upstream mRNA segments for the phage-Lambda encoded int transcript (8). Other examples have shown that 3' structures are important for stability of the upstream mRNA segment (9,10,11). In contrast, the 5' noncoding region act more as a determinant of mRNA stability. Several prokaryotic examples are known by now where the 5' noncoding region of a stable transcript is capable of conferring stability to downstream fused labile messenger (12,7,13,14).

The Influence of Translational Efficiency and Premature Stop Codons on mRNA Stability

Translating ribosomes have been considered to be involved in the control of mRNA stability. To analyse the effect of premature termination of translation on mRNA stability, translational stop codons were introduced at various locations in the coding region of the bla gene transcript (15). The positions of the stop codons in the different mutant transcripts are depicted in Figure 1.

Premature termination of translation at codon position 26 reduced the stability of the entire transcript (half-life 0.9 minutes), whereas release of ribosomes just 30 codons downstream (codon 56) resulted in a almost normal stability of the transcript (half-life 2.0 minutes). Thus, depriving an mRNA segment of translating ribosomes does not necessarily invoke endonucleolytic degradation.

Analysis of a bla gene lacking the critical codons 26 to 56 (bla 1 in Figure 1), indicated that this part of the bla transcript seems not to act as a destabilizer. Thus, we conclude that the short half-life observed for bla4 (half-life 0.8 minutes) and bla26 mRNA results from the complete absence of ribosomes on a substantial fraction of these mutant transcripts. This conclusion is based on the consideration that the rate of translation initiation by ribosomes is slow compared to the rate of peptide elongation. Hence, the average number of bound ribosomes will be less that one per transcript for the bla4 and bla26 messages. (For details see Discussion in ref. 15).

The different stabilities of the bla and ompA gene transcripts were also related to the efficiency of translation per transcript (6). The relative ratio of the bla and ompA gene transcripts was determined in cells growing at

TABLE 2

Relative rate of protein synthesis related to transcript per cell mass at different growth rates

	40 min.	200 min.
ompA	0.44	1.0
bla	0.19	0.1

either of the two growth rates described in the first paragraph of Results and Discussion.

Efficiency of translation was quantified by short-term radio-labelling of the proteins in cells grown in either of the two growth media. Proteins were precipitated by antibodies and analysed by gel-electrophoresis. The relative rate of OmpA and Bla protein synthesis was related to the relative amount of the transcripts. Table 2 shows the relative rate of protein synthesis per transcript for the two transcripts at the two different growth rates. The efficiency of translation of the ompA transcript increases while it decreases for the bla messenger when cells grown at fast growth rate are compared versus cells grown at slow growth rate. It is apparent that the decrease in stability of the ompA mRNA at slow growth-rate cannot be attributed to a decrease in efficiency of translation.

In summary, the result on ribosomes indicate that translating ribosomes at the coding region are not necessarily important to protect the transcript against degradation. However, a minimal number of ribosomes needs to be assembled at the 5' part of the transcript in order to maintain the normal stability. Such a view on ribosomes is in accordance with the finding obtained for the ermC mRNA in B. subtilis where ribosomal assembly at the 5' region seems to stabilize the transcript even in the absence of ribosomes in the coding region (14).

Site-Specific Endonucleolytic Cleavages and the Control of Stability of the ompA and bla Transcripts

The detection of mRNA cleavage products should be facilitated by transient over-production of the messages. Such an over-production has been achieved by inserting the ompA and bla gene downstream of an isopropyl-β-D-thiogaloactoside (IPTG)-inducible promotor (16,18) and analysing the mRNA after induction in the cell (the induced ompA mRNA was analysed in an E. coli strain which does not express the chromosomal ompA gene). Northern blot analysis showed that, as expected, the ompA-specific transcript increased after ITPG induction. Interestingly, in addition to the full-length ompA mRNA an 100-150 nucleotides smaller transcript became visible only after the IPTG-induction. Thus, the smaller mRNA species is the apparent product of a processing of the full-length transcript and not a result of multiple promoters (16).

To elucidate the origin of the smaller mRNA species, we used S1 nuclease mapping to identify the 5' endpoints (Figure 2: 5' probe). The full length transcript was identified (*) and in addition five shorter transcripts (B-E), reflecting multiple 5' endpoints (Figure 3A). If the different 5' ends of ompA mRNA are a result of endonucleolytic cleavages, the complementary cleavage products toward the 5' end of the transcript should be identifiable. A DNA probe labelled at the 3' end (Figure 2) was used to search for such molecules. S1 analysis revealed such upstream cleavage products (not shown, see 16) to have 3' ends complementary to the 5' ends of the above identified mRNA species. Analysis of the bla transcript after ITPG-induction revealed site-specific cleavages in the 5' noncoding region of the bla messenger (18).

The biological importance of such cleavages was assayed for the wild type ompA mRNA where the same cleavages could be identified that were found for the IPTG-induced ompA messenger. The ratio of cleavage products to full-length mRNA was analysed at different bacterial growth rates when the wild type mRNA shows different stabilities (Table 1). Figure 3A and Table 1 show that the rate of degradation of the ompA mRNA parallels the

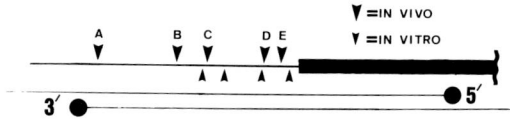

FIGURE 2. Physical map of 5' cleavage in the ompA transcript.

ratio of cleavage products to full-length transcript which in turn seems to reflect the rate of endonucleolytic cleavages. Thus, the 5' noncoding region appears to be the target of endonucleolytic cleavages that may well regulate the expression of the ompA gene.

There are several mechanisms by which 5' endonucleolytic cleavages could initiate degradation of the ompA transcript. Most probably the cleavages disturb the loading of ribosomes onto the transcript, since some cleavages disturb the ribosome binding site. Complete deprivation of a transcript of its translating ribosomes leads to a rapid degradation (15). Another way of explaining the decay of the ompA mRNA after the initial cleavage is that the cleavage opens the transcript to a wave of 5' to 3' degradation as it has been proposed for β-galactosidase mRNA (17).

The cleavages observed in the ompA and bla gene transcripts are likely to be the results of an endonucleolytic activity. In order to identify such an activity both messages were transcribed in vitro using radiolabelled nucleotides. The radiolabelled transcripts were submitted to crude, ribosome free, cellular E. coli extracts and the degradational products were analysed by gel-electrophoresis (18). A number of specific degradational products were identified, some of which even accumulated during the decay period. The accumulation of the degradational products was dependent on the presence of tRNA which is likely to compete out the 3'- 5' exonucleases.

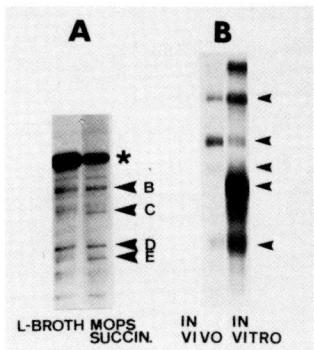

FIGURE 3. in vivo and in vitro analysis of the cleavages in the ompA mRNA.

Treating the E. coli extracts with proteinase K or with micrococcal nuclease indicated that the cleavages were catalyzed by a protein and not by a catalytic RNA. Analysis of extracts from E. coli mutants that are deficient for certain characterized riboendonucleases disclosed that the in vitro activity is not identical to RNase III, RNase E or RNase P (18). The in vitro degradational products of the ompA mRNA were compared to their in vivo counter-products using the 3' end labelled probe described above (Figure 2) and it was found that the cleavage products extending from the cleavage site towards the 5' end (upstream cleavage products) are mostly the same as in vivo (Figure 3B). The in vivo cleavage product extending from the cleavage site towards the 3' end (downstream cleavage products) were also compared to their in vivo counter-products and the same downstream cleavage products were found in vitro and in vivo (not shown, see 18). However, the in vitro departs from the in vivo situation in that the downstream cleavage products were more diversified and that the upstream cleavage products were more abundant in vitro than in vivo. An explanation for this discrepancy could be the inhibition of the 3'- 5' exonucleases in vitro and the absence of ribosomes that may facilitate and hence diversify the downstream cleavage products.

REFERENCES

1. Pedersen S, Reeh S (1978). Functional mRNA half-lives in E. coli. Molec gen Genet 166:329.
2. Nilsson G, Belasco JG, Cohen SN, von Gabain A (1984). Growth-rate dependent regulation of mRNA stability in Escerichia coli. Nature 312:75.
3. Belasco JG, Higgins CF (1988). Mechanisms of mRNA decay in Bacteria: a perspective. Gene (in press).
4. von Gabain A, Belasco JG, Schottel JL, Chang ACY, Cohen SN (1983). Decay of mRNA in E. coli: investigation of the fate of specific segments of transcripts. Proc Natl Acad Sci USA 80:653.
5. Lugtenberg B, Peters R, Bernheimer H, Berendsen W (1976). Influence of cultural conditions and mutations on the composition of the outer membrane proteins of Escherichia coli. Molec gen Genet 147:251.

6. Lundberg U, Nilsson G, von Gabain A (1988). The differential stability of the E. coli ompA and bla mRNA at various growth rates is not correlated to the efficiency of translation. Gene (in press).
7. Belasco JG, Nilsson G, von Gabain A, Cohen SN (1986). The stability of E. coli gene transcripts is dependent on determinents of localized to specific segments. Cell 46:245.
8. Guarneros G, Montanez C, Hernandez T, Court D (1982). Posttranscriptional control of bacteriophages lambda int gene expression from a site distal to the gene. Proc Natl Acad Sci USA 79:238.
9. Belasco JG, Beatty JT, Adams CW, von Gabain A, Cohen SN (1985). Differential expression of photosynthesis genes in R. capsulata results from segmental difference in stability within the polycistronic rxcA. Cell 40:171.
10. Wong HC, Chang S (1986). Identification of a positive retroregulator that stabilizes mRNA in bacteria. Proc Natl Acad Sci USA 83:3233.
11. Newbury SF, Smith NH, Robinson EC, Hiles ID, Higgins CF (1987). Stabilization of translational active mRNA by procaryotic REP sequences. Cell 48:297.
12. Yamamoto T, Imamoto F (1975). Differential stability of the trp promoter and pl promoter of lambda trp phage. J Mol Biol 92:289
13. Gorski K, Roch J-M, Prentki P, Krisch HM (1985). The stability of bacterio phage T4 gene32 mRNA: a 5' leader sequence that can stabilize mRNA transcript. Cell 43:461.
14. Bechhofer DH, Dubnau D (1987). Induced mRNA Stability in Bacillus subtilis. Proc Natl Acad Sci USA 84:498.
15. Nilsson G, Belasco JG, Cohen SN, von Gabain A (1987). Effect of premature termination of translation on mRNA stability depends on the site of ribosome release. Proc Natl Acad Sci USA 87:4890.
16. Melefors Ö, von Gabain A (1988). Site-specific endonucleolytic cleavages and the regulation of E. coli ompA mRNA. Cell 52:893.
17. Cannistraro VJ, Kennel D (1985). Evidence that the 5' end of lac mRNA starts to decay as soon as it is synthesized. J Bact 161:820.
18. Nilsson G, Lundberg U, von Gabain A (1988). In vivo and in vitro identity of site-specific cleavages in the 5' non-coding region of ompA and bla mRNA in Escherichia Coli. EMBO J 7:(in press).

DISCUSSION SUMMARY: TRANSCRIPTION TERMINATION AND RNA 3' END FORMATION

Claire L. Moore

Department of Molecular Biology and Microbiology
Tufts University School of Medicine
136 Harrison Avenue
Boston, Massachusetts 02111

The focus of the first half of this session was transcription termination in both prokaryotic and eukaryotic organisms. A unifying theme in these presentations was the concept that events affecting the processivity of RNA polymerase once it reaches a termination region happen before this point is encountered. As a consequence, factors influencing termination most likely interact with the RNA transcript and the RNA polymerase rather than with the DNA at the termination site. A brief summary of the talks presented at the workshop is given below. For more details on each, please refer to the authors' abstracts published in the Journal of Cellular Biochemistry, Supplement 12D, 1988.

Factors can act to cause termination or to prevent it. Examples of the first situation were described for rho-dependent termination of bacterial polymerase (T. Platt) and for termination of transcription by RNA polymerase II (N. Hernandez and L. Ryner). In a cumulative model described by T. Platt, rho-mediated termination depends on rho binding to nascent RNA strands at a specific site. Cytosine bases are important for this interaction, but the exact sequence required for binding has been hard to define. Rho then migrates down the RNA strand and releases RNA from the transcriptional complex by unwinding the RNA/DNA helix. Release is dependent on ATP hydrolysis and requires the rho binding site (1). One puzzle not yet explained by the model is the existence of sites which are good rho-dependent terminators but poor stimulators of rho's helicase activity, and vice versa. Further experiments defined separate binding domains for RNA and ATP (2). An N-terminal fragment of rho is sufficient for specific RNA binding. The C terminal portion has strong homology to consensus elements of other nucleotide binding proteins, but this fragment alone or in combination with the N-terminal fragment exhibits no ATP hydrolysis.
Suprisingly, the unique cysteine, thought to be essential for rho function, can be changed to serine or glycine with no detectable effects on the protein's hexameric structure, RNA binding ability, or ATPase, helicase and transcription termination activities.

Two examples of termination by eukaryotic RNA polymerase II were described. Results from J. Manley's laboratory (3) and a recent joint paper from J. Darnell's and T. Shenk's groups (4) indicate that proper termination requires the presence of a functional polyadenylation site on the RNA precursor. Mutation of the AAUAAA upstream of the poly(A) site or deletion of critical signals downstream of this prevent processing of the RNA 3' end as well as termination of transcription. The mechanism of this effect on termination is not understood. It could require only assembly of a polyadenylation processing complex on the precursor RNA. Alternatively, in a model proposed by Manley's group, it may require cleavage at the poly(A) site to expose an uncapped 5' terminus susceptible to exonuclease degradation. Regions defined as termination sites might simply be places where RNA polymerase stalls long enough for the exonuclease to catch up and destabilize the transcriptional complex. In any event, this is likely a mechanism to insure that coding regions get transcribed before termination occurs. N. Hernandez described how in the transcription of snRNAs, RNA polymerase II terminates appropriately only if transcription was initiated at an snRNA promoter and not if a promoter for an mRNA transcriptional unit was used. Again, the mechanism is not clear, and it has not been possible to separate promoter elements necessary for termination from those required for initiation.

Two presentations discussed situations in which the prevention of termination allowed expression of downstream coding regions. J. Greenblatt described yet another cellular factor (NusG) required for invitro antitermination by the N protein of phage λ. Further research will obviously be directed at understanding RNA-protein and protein-protein interactions in this large elongation complex. It is intriguing that while NusG stimulates elongation in the early operons of the phage, it will facilitates rho-dependent termination. B. Peterlin described how the HIV-1 encoded trans-activator TAT functions as an antiterminator of RNA polymerase II transcription. In the absence of TAT, transcripts initiated at the HIV-1 cap site terminate about 60 bases downstream. In the presence of TAT, full length transcripts were found, and viral gene expression greatly increased. It will be interesting to see if this viral factor interacts directly with the RNA transcript.

The discussion on RNA 3' end processing dealt only with the formation of the polyadenylated ends of eukaryotic mRNA. L. Ryner described recent fraction of crude nuclear extracts (5). Two activities were separated - one for cleavage and one containing nonspecific poly(A) polymerase. Mixing of both fractions restored AAUAAA-dependent poly(A) addition.

Hopefully, further research will define how these activities interact with RNA precursor and with each other, and perhaps identify the still elusive snRNP implicated in this processing. The last two talks described how ultraviolet crosslinking can identify specific proteins which interact with polyadenylation precursor. J. Wilusz from T. Shenk's group described a 64 kd protein which requires AAUAAA for binding (6). He also reported on how the binding of the C hnRNP protein required sequence downstream of the SV40 late poly(A) site. Experiments described by C. Moore detected the 64 kd protein as well as a 155 kd protein whose binding depended on AAUAAA. Both proteins were found in polyadenylation-specific complexes and disassociated from the RNA once processing occured. The 155 kd protein was crosslinked to a 15 n. oligomer containing the AAUAAA hexanucleotide and did not require extended RNA sequence for binding. In addition, it was recognized by antibodies against Sm proteins but was not a constituent of a U snRNP. It will be interesting to determine the role of these proteins in the polyadenylation process.

REFERENCES

1. Brennan CA, Dombroski AJ, Platt T (1987). Transcription termination factor rho is an RNA-DNA helicase. Cell 48:945.
2. Dombroski AJ, Platt T (1988). Structure of rho factor: an RNA-binding domain and a separate region strongly homologous to proven ATP-binding domains. Proc. Natl. Acad. Sci. USA 85:2538.
3. Conway S, Manley J (1988). A functional mRNA polyadenylation signal is required for transcription termination by RNA polymerase II. Genes and Dev., April.
4. Logan J, Falck-Pedersen E, Darnes JE, Shenk T (1987) A poly(A) addition site and a downstream termination region are required for efficient cessation of transcription by RNA polymerase II in the mouse β^{maj}-globin gene. Proc. Natl. Acad. Sci. USA 84:8306.
5. Takagaki Y, Ryner LC, Manley JL (1988). Separation and characterization of a poly(A) polymerase and a cleavage/specificity factor required for pre-mRNA polyadenylation. Cell 52:731.
6. Wilusz J, Shenk T (1988) A 64 kd nuclear protein binds to RNA segments that include the AAUAAA polyadenylation motif. Cell 52:221.

MUTATIONAL AND CHEMICAL MODIFICATIONS OF RHO FACTOR AFFECT ITS DOMAIN CONFORMATIONS AND INTERACTIONS

T. Platt, C. A. Brennan, A. J. Dombroski, and P. Spear[1]

Department of Biochemistry, University of Rochester Medical Center, 601 Elmwood Avenue, Rochester, NY 14642

ABSTRACT The results of chemical modification by NEM and pHMB, site directed mutagenesis of numerous amino acid residues, and active site labeling with an ATP analog support a model for the relationship between rho factor structure and function with several important characteristics. (1) RNA binding specificity resides within the first 151 amino acids of the polypeptide. (2) A domain spanning residues 150-350 strongly resembles the structure of adenylate kinase and related ATPases, and contains the ATP binding site. (3) The C-terminal 70 amino acids are important for solubility and possibly for the hexameric quaternary structure of rho. (4) ATP hydrolysis requires interactions between the RNA and ATP binding domains of the protein. (5) Native rho exists in a "closed" conformation, that becomes "open" upon RNA binding, and changes to an "active" conformation when both RNA and ATP are bound.

INTRODUCTION

In E. coli, the termination factor rho is an essential cellular protein that helps catalyze transcription termination at particular sites (see 1 for review). The mechanism of its action is not fully understood but requires interaction of rho with the nascent RNA chain upstream of the termination site(s), within regions that are relatively unstructured, rich in cytosine residues, and untranslated.

[1]This work was supported in part by NIH Grant GM35658 to T.P., postdoctoral award GM11203 to C.A.B, and Genetics Training Grant GM07102 to A.J.D.

The binding of rho to RNA activates an RNA-dependent nucleoside triphosphatase activity of rho (2), which is essential for rho-dependent termination (3, 4), though its role in the termination event has only partly been determined (5).
In vitro, this RNA-dependent ATPase activity can be observed "uncoupled" from transcription termination (2). Though ATP hydrolysis is best activated by polyribocytosine, any single strand RNA containing cytosine residues and lacking guanosine will work at some level (6). How the energy from ATP hydrolysis is utilized to terminate transcription and release the nascent RNA from the ternary complex has not been understood. The ATPase activity has been suggested to provide energy to translocate rho along the RNA chain toward an RNA polymerase molecule paused at the termination site (7). Rho undergoes a conformational change upon binding RNA and ATP as demonstrated by stabilization of the hexameric form of the protein (8) and a decrease in its susceptibility to cleavage by trypsin (9, 10) in the presence of RNA and ATP.
When examined under transcription conditions in vitro, rho can release the RNA from artificially paused ternary complexes, dependent on ATP hydrolysis (4, 11, 12). In a purified system lacking RNA polymerase, rho can act as a helicase to catalyze the release of RNA from an RNA·DNA duplex (5). This activity is dependent upon ATP hydrolysis and requires a 5' single strand region of RNA that contains the signal(s) needed for rho-dependent termination.

CHEMICAL MODIFICATION AT CYSTEINE-202

Lowery and Richardson (13) reported that the ATPase activity of rho factor was inactivated by the sulfhydryl-modifying reagents N-ethylmaleimide (NEM) and p-hydroxymercuribenzoate (pHMB). This led to our assumption that cys-202 (the sole cysteine in rho) was important for ATPase activity and/or RNA binding, and to a plausible hypothesis that cys-202 might form a transient covalent intermediate with cytosine residues in the RNA (1). Surprisingly, a fluorescent modification at cys-202 has no inhibitory effect on the ATPase activity (14) and conversion of cys-202 to serine or glycine fails to reveal any functional defects in rho activity (15). To resolve this apparent paradox, and in hopes of gaining further understanding of the relationship between RNA binding and ATPase activation, we performed a more detailed study of these chemical inactivations.

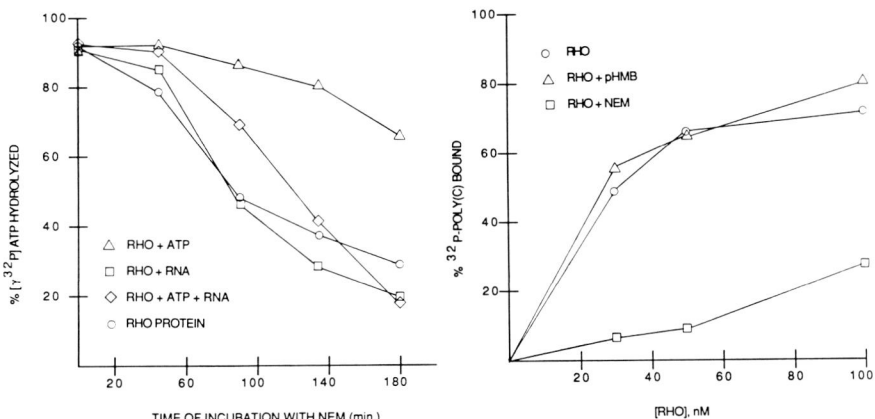

FIG. 1. (a) Inactivation of ATPase activity by NEM. 70 nM rho was incubated with 100 mM NEM in the presence and absence of 20 μM ATP and/or 5 μM poly(C). ATPase assays were performed as described (16) with 10 nM rho for 10 min. (b) Binding of chemically-modified rho to RNA. Nitrocellulose filter binding assays were performed as described (15) using 5' end-labeled poly(C) and rho that had been inactivated for ATPase activity by NEM or pHMB.

Rho protein was incubated with NEM both with and without ATP and RNA, and ATPase activity was monitored as a function of incubation time. Fig. 1a shows that rho alone and rho plus poly(C) are both modified by NEM similarly, indicating that bound RNA offers no protection from NEM. In the presence of ATP however, rho is substantially protected from NEM inactivation shown by a reduced rate of ATPase inactivation. If both RNA and ATP are present, such that rho is actively hydrolyzing ATP, the inactivation rate is again similar to that of rho alone or rho plus RNA. These results support the notion that rho undergoes specific conformational changes upon binding ATP and RNA, and upon participation in ATP hydrolysis. Similar experiments using pHMB as the modifying reagent showed that neither ATP nor RNA, separately or in combination, could serve to protect rho from pHMB (data not shown).

The fact that NEM, but not pHMB, inactivation is sensitive to the presence of ATP, suggests that the sites of chemical modification differ for the two reagents and could potentially affect properties other than ATPase activity. We therefore compared the abilities of rho inactivated by each reagent to bind to RNA. Rho that had been treated with NEM or pHMB was incubated with ^{32}P-poly(C) and filtered through nitrocellulose to trap rho-RNA complexes. NEM modification of rho prevents the protein from binding to ^{32}P-poly(C) and suggests that NEM is reacting non-specifically at sites other than cysteine-202 to cause interference with RNA binding as well as ATP hydrolysis. Rho modified by pHMB retained full RNA binding ability as shown in Figure 1b, suggesting that its inhibitory effect is at a subsequent stage of activation, as discussed below.

CHARACTERIZATION OF CYSTEINE-202 MUTANTS

We used oligonucleotide site-directed mutagenesis to change cys-202 to either glycine (CG202) or serine (CS202) (15). Neither altered rho protein was affected with regard to any of its activities, and inactivation by NEM in each case was indistinguishable from wild-type. The RNA-dependent ATPase activities of the mutant proteins CG202 and CS202 were analyzed kinetically. Initial velocity plots were obtained by varying the concentration of ATP from 5-60 µM with the concentration of poly(C) kept constant at 5 µM (expressed in terms of nucleotides). The reactions were stopped at various times up to 5 minutes and the reaction products separated by thin layer chromatography. Double reciprocal plots of v versus [ATP] were used to calculate the kinetic constants for ATP hydrolysis, and the K_m, V_{max}, and V_{max}/K_m values for CG202 and CS202 differ by less than 30% from those of wild-type protein. Such small deviations are within the error inherent in the measurements, and we conclude that the behavior of CG202 and CS202 is not significantly different from that of wild-type rho protein.

The ability of CG202 and CS202 to function in helicase assays (5) was also determined (Fig. 2a). Here again the CG202 and CS202 proteins do not differ in behavior from unmodified rho. The termination efficiency of each mutant was also tested in vitro (results not shown) and was found to be identical to wild type rho. Cys-202 is therefore not required for the ATP hydrolysis, helicase, or termination functions of rho. A similar result has been seen for an aminoacyl-tRNA synthetase where a sulfhydryl presumed essential from chemical modification studies was shown to be dispensable by site-directed mutagenesis (17).

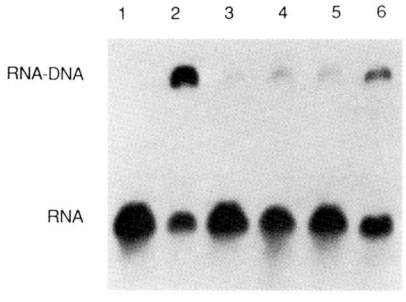 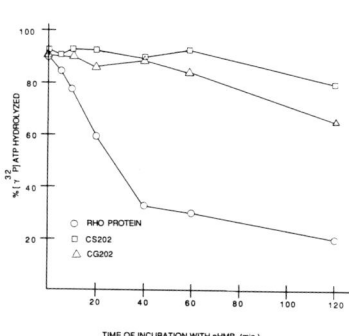

FIG. 2. (a) Helicase activity of cys-202 mutants. Assays were as described by Dombroski et al. (16), modified from Brennan et al. (5). Lane 1 shows the mobility of RNA alone, while lane 2 shows the much slower migrating RNA-DNA hybrid used as the helicase substrate. The activity in releasing RNA from the annealed duplex is shown for pure rho (lane 3), and samples partially purified on sucrose gradients for wild-type rho (lane 4), CG202 (lane 5), and CS202 (lane 6) proteins respectively. (b) Effect of pHMB on ATPase activity of CG202 and CS202. 70 nM rho was incubated with 7.5 mM pHMB, and ATP assays at various times were as described (16) with 20 nM rho for 5 min.

Unlike NEM modification, pHMB inactivation is specific for the cysteine, since proteins altered at this residue (CG202 and CS202) were unaffected by the reagent (Fig. 2b). Rho was not protected from inactivation by pHMB by the presence of ATP, poly(C) or ATP plus poly(C). Cys-202 therefore appears to remain very accessible to pHMB despite the conformational changes induced by ATP and RNA. Rho that has been inactivated for ATPase activity by pHMB nonetheless retains its ability to bind to RNA. pHMB is thus reacting specifically at cys-202 to inactivate ATPase activity but the modification does not interfere with the RNA binding function of the protein.

RHO FACTOR HAS SEPARATE BINDING DOMAINS FOR RNA AND ATP

An RNA-binding domain was previously defined for rho factor to reside within the first 283 amino acids of the protein (10). We have recently shown that the RNA binding function is actually contained within a 151 amino acid N-terminal fragment (N1), generated by hydroxylamine cleavage (15). It and its complementary C-terminal fragment of 268 amino acids (N2) were extracted from SDS gels and renatured. N1 binds to poly(C) nearly as well as the full length protein does, and appears to comprise a rho domain separate from that required for ATP binding. Though uncleaved rho renatures readily to regain its RNA-dependent ATPase activity, neither the N1 nor N2 fragments, renatured separately or together, exhibit any detectable ATP hydrolysis. However, strong homology to consensus sequences shared by a large number of nucleotide binding proteins suggests that a structural domain for rho's ATP binding begins after amino acid 164 (15). This is supported by our demonstration of crosslinking by an ATP analog, pyridoxal 5'-diphospho-5'-adenosine (PLP-AMP) efficiently and uniquely to lysine-181 (18). Further tests of this model utilizing site-directed mutagenesis to change lys-181 and/or lys-184 to glutamine, and asp-265 to asparagine (Fig. 3) reveal reductions in activity consistent with the postulated involvement of these residues in ATPase activity (16).

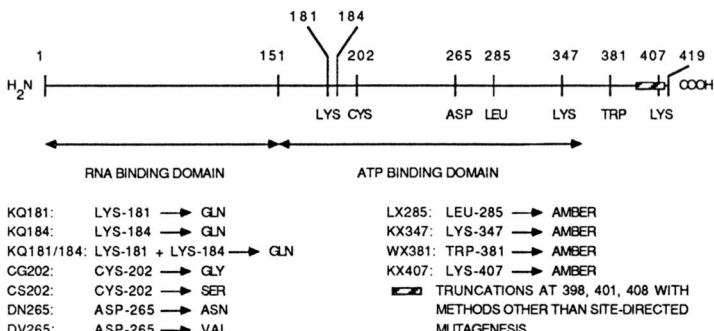

FIG. 3. Map of amino acid substitutions in rho protein. The locations of altered residues are shown on the linear protein diagram. The RNA and ATP-binding domains are labeled. Mutant nomenclature (using the one letter amino acid code, nonsense codon - X) and respective amino acid substitutions are listed; arrows indicate the mutational changes.

S P S P S P S P

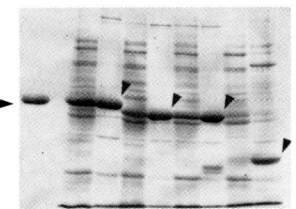

Figure 4. Solubility of rho proteins altered at the carboxy-terminus. Overproduced rho factors in supernatant (S) and pellet (P, resuspended to the same volume) fractions of cell extracts were compared on SDS polyacrylamide gels. Pure rho is on the left, arrows indicate the insoluble rho in each pair: from left to right are wild-type (WT, 419 aa), KX407 (amber mutation, 406 aa), XM397 (fusion at residue 397, 401 aa) and LX285 (nearly equivalent to the F1 fragment of Bear et al. [10], 284 aa); see Fig. 3.

We have in addition created both amber mutations and fusions (replacing rho sequences with other amino acids) in the carboxy-terminal region of rho (Fig. 3). Upon analyzing these proteins after overproduction in vivo, we encountered severe solublity problems, finding that most of the proteins remained with the insoluble membrane and cell debris fraction (Fig. 4), and could not be extracted efficiently enough for further analysis, despite the use of various non-ionic detergents and salt or buffer variations. When the small amounts of protein that could be extracted were run on sucrose gradients, there was no trace of rho at the normal hexameric position. It may be significant that on an α-helix wheel, the last 10-15 residues of rho display predominantly hydrophobic side chains on one side (including two met and two phe), and highly charged residues on the other side. We are trying to test whether this region is involved in inter-subunit interactions, and that its removal exposes another hydrophobic region, which is responsible for the insolubility of the proteins with C-terminal alterations.

From the evidence summarized above, we propose a domain structure for rho where amino acids 1-150 have the RNA binding specificity, 150-350 comprise the ATP binding domain, and 350-419 may be involved in oligomeric association.

CONCLUSIONS

We have shown that the RNA-binding domain of rho resides within the first 151 amino acids of the protein and that cys-202 does not participate in binding. Chemical modification and oligonucleotide site-directed mutagenesis shows that cysteine-202 is also not required for catalytic activity and that inactivation by NEM is different from that of pHMB. Seifried et al. (14) have found that attachment of sulfhydryl-specific fluorescent labels at cys-202 does not significantly change the properties of the protein. Thus pHMB may be inactivating by introduction of a negatively charged species at cys-202 rather than by introducing a bulky group that sterically hinders function. In our model for conformational change (Fig. 5), we imagine that cys-202 is poised at the interface between N1 and N2, where a charge change could have a major effect. This is supported by our finding that pHMB-modified rho still binds both RNA and PLP-AMP with normal affinity (data not shown). NEM eliminated the ATPase activity of CG202 and CS202 as rapidly as it did wild-type rho, indicating that alkylation at sites other than cysteine must be involved. Inactivation by NEM is slowed considerably if ATP is present while poly(C) offers no protection. This observation agrees with conformational studies that showed ATP binding protects rho from total digestion by trypsin while the presence of poly(C) alone actually affords trypsin greater access and results in complete degradation of rho (10). NEM appears to be inactivating rho at sites that become much less accessible when ATP is bound, and once rho has been inactivated by NEM it is no longer capable of binding to RNA. In summary, both NEM and pHMB inactivate rho but their specificity and modes of interaction are distinct. pHMB is specific for cys-202 and unaffected by the conformational state of the protein while NEM is acting at sites other than cys-202, may be disrupting RNA binding, and is dependent upon the conformation of rho.

As schematically illustrated (Fig. 5), a model for the interaction of the RNA-binding and ATP-binding domains consistent with this data can be envisioned as follows: binding of RNA to rho alters the protein conformation from a "closed" state to an "open" one, where it is very accessible to reagents such as NEM and trypsin. This conformation would also allow greater access of ATP to the ATP-binding domain. Upon binding ATP, the protein assumes a much tighter "active" conformation preventing access by NEM and trypsin. The accessibility of the ATP site may be modulated

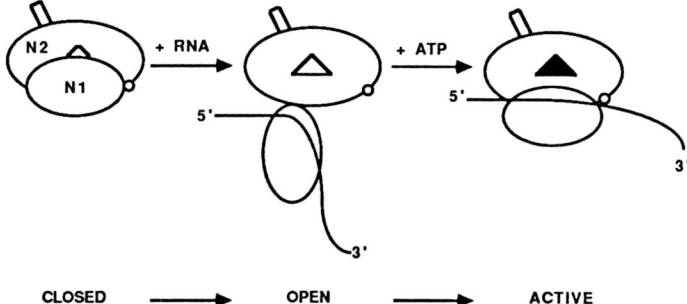

FIG. 5. Cartoon model for conformational changes of a rho subunit. The RNA-binding and ATP-binding domains are N1 and N2 respectively. The open triangle represents an empty ATP-binding site and the closed triangle an occupied one. The small circle is cys-202 and the rectangle represents the α-helical C-terminal region, which may be involved in subunit association. See text for discussion.

by the movement of the flexible loop as seen in adenylate kinase (19). For ATP hydrolysis to take place, the RNA-binding domain must then interact correctly with the ATP-binding domain. pHMB modification shows that cys-202 (shown by the small circle in the model) remains accessible despite the conformation of rho, and suggests that this modification interferes with the interaction between the two domains rather than with the process of substrate binding.

In Richardson's two-site model for RNA activation, affinity for the secondary (oligo) site is highest when the ATP site is occupied (20), i.e. our "active" conformation. This may account for our prior inference from kinetic analysis of the ATPase activity that ATP must bind "before" RNA does (21). It is not known why rho only binds three ATP molecules per hexamer (22), or requires a hexameric quaternary structure for activity (8). Resolution of these problems may lie in higher order structural considerations, since interactions between the domains of adjacent subunits are possible. If only 3 subunits can bind ATP at any one time, a coupled alternation of "open" and "closed" conformational states among subunits could help link ATP hydrolysis to directed translocation along the transcript. In addition, the hydrolysis of ATP provides energy that is presumably used in the process of unwinding the RNA-DNA duplex in the transcription bubble during transcription termination (5). With our increasing knowledge of rho structure and its relationship to function, the answers to some of these questions promise to be forthcoming in the near future.

REFERENCES

1. Platt, T (1986). Ann Rev Biochem 55: 339-372.
2. Lowery-Goldhammer C, Richardson JP (1974). Proc Nat Acad Sci USA 71: 2003-2007.
3. Howard BH, de Crombrugghe B (1976). J Biol Chem 251: 2520-2524.
4. Richardson JP, Conaway R (1980). Biochem 19: 4293-4299.
5. Brennan CA, Dombroski AJ, Platt T (1987). Cell 48: 945-952.
6. Lowery C, Richardson JP (1977). J Biol Chem 252: 1381-1385.
7. Galluppi G, Richardson JP (1980). J Mol Biol 138: 513-539.
8. Finger LR, Richardson JP (1982). J Mol Biol 156: 203-219.
9. Engel D, Richardson JP (1984). Nucl Acids Res 12: 7389-7400.
10. Bear DG, Andrews CL, Singer JD, Morgan WD, Grant RA, von Hippel PH, Platt T (1985). Proc Natl Acad Sci USA 82: 1911-1915.
11. Shigesada K, Wu CW (1980). Nucl Acids Res 8:3355-3369.
12. Morgan WD, Bear DG, von Hippel PH (1983). J Biol Chem 258: 9553-9564.
13. Lowery C., Richardson JP (1977). J Biol Chem 252: 1375-1380.
14. Seifried SE, Wang Y, von Hippel PH (1988). J Biol Chem 263: submitted.
15. Dombroski AJ, Platt T (1988). Proc Natl Acad Sci USA 85: 2538-2542.
16. Dombroski AJ, Brennan CA, Spear P, Platt T (1988). J Biol Chem 263: submitted.
17. Profy AT, Schimmel P (1986). J Biol Chem 261: 15474-15479.
18. Dombroski AJ, LaDine J, Cross RL, Platt T (1988). J Biol Chem 263: submitted.
19. Fry DC, Kuby SA, Mildvan AS (1986). Proc Natl Acad Sci USA 83: 907-911.
20. Richardson JP (1982). J Biol Chem 257: 5760-5766.
21. Sharp JA, Galloway JL, Platt T (1983). J Biol Chem 258: 3482-3486.
22. Stitt B (1988). J Biol Chem 263: in press.

TWO SEPARABLE ACTIVITIES ARE REQUIRED FOR PRE-mRNA 3' END FORMATION: A POLY(A) POLYMERASE AND A CLEAVAGE/SPECIFICITY FACTOR[1]

Lisa C. Ryner, Yoshio Takagaki, Justina Voulgaris and James L. Manley

Department of Biological Sciences, Columbia University
New York, NY 10027

ABSTRACT To study the mechanism and factors required to form the 3' ends of polyadenylated mRNAs, we have fractionated HeLa cell nuclear extracts that carry out the normally coupled cleavage and polyadenylation reactions. Each reaction is catalyzed by a distinct activity that can be completely separated from the other. The partially purified cleavage enzyme retained the specificity displayed in nuclear extracts, as base substitutions in the AAUAAA signal sequence inhibited cleavage. In contrast, the poly(A) polymerase detected after fractionation had lost all specificity. When fractions containing the cleavage and polyadenylation activities were mixed, the efficiency and specificity of the polyadenylation reaction were restored.

INTRODUCTION

The vast majority of mRNAs in higher eukaryotes are polyadenylated at their 3' ends. Although the function(s) of this modification remain(s) unclear, its near universal presence, as well as the fact that mutations in the cis-acting sequences that control 3' end formation block synthesis of mRNA (for review, 1), strongly support the view that this process plays a crucial role in gene expression. In addition, the existence of transcription units that allow the utilization of alter-

---

[1]This work was supported by National Institutes of Health grant GM 28983.

native 3' end sites (reviewed in 2) raises the possibility that selection of polyadenylation signals may play a role in controlling gene expression. It is therefore of considerable importance to obtain a detailed understanding of the mechanism by which a mature mRNA 3' end is formed.

A good deal is now known about the cis-acting sequences that are required for polyadenylation. Probably the most important is the highly conserved hexanucleotide AAUAAA, which is found approximately 10-30 nucleotides upstream of the site of polyadenylation in most mRNAs (3). Analysis of the effects of mutations in this sequence have confirmed the idea that it is a crucial element of the polyadenylation signal (4-8). Also implicated in polyadenylation are sequences that lie just downstream of the site of poly(A) addition. In many cases, deletions that remove these sequences significantly reduce the accumulation of polyadenylated mRNA (9-13), although there are instances in which such deletions appear to influence 3' end formation only slightly (14).

The basic pathway of polyadenylation has been well defined. The 3' ends of mRNAs are formed by endonucleolytic cleavage of longer precursors followed by the addition of poly(A) tracts of approximately 200 nucleotides that are synthesized by addition of adenosine residues onto the 3' end of the RNA at the site of cleavage (15,16). These two reactions (cleavage and polyadenylation) appear to be mechanistically coupled, because cleaved but non-polyadenylated RNA is normally not detected, either in vivo or in vitro. However, this coupling is not obligatory, since RNAs that have only been cleaved can be detected in vitro when polyadenylation is blocked by the addition of an inhibitor, such as EDTA (17). In addition, pre RNAs can be polyadenylated in vitro in the absence of cleavage; ie, at their 3' ends. These include precursors terminated precisely at the in vivo site of polyadenylation (8,15) as well as those that extend beyond this point (eg 18,19). In each case, this "end-addition" polyadenylation requires an intact AAUAAA, suggesting not only that both types of polyadenylation involve the same mechanism, but also that polyadenylation and cleavage both require this signal sequence (8,18,20).

The identity of the factor(s) that carry out the cleavage/polyadenylation reaction is not known. Although poly(A) polymerases have been purified from a number of sources (for reviews, see 21 and 22), their relevance to the in vivo reaction is unclear. These enzymes all polymerize poly(A) onto RNA primers that lack an AAUAAA signal sequence. This could indicate that these purified enzymes have lost a required specificity factor, or alternatively that they are not the same enzyme that carries out sequence specific polyadenylation of pre mRNAs in vivo. The possibility that an snRNP is a required factor in the polyadenylation reaction has received some support (17,23-25). However, it is unlikely that one of the major, abundant snRNPs is involved (26), and it has also been shown that nuclear extracts pretreated with micrococcal nuclease to degrade endogenous nucleic acids remain active in polyadenylation (19). Thus, a number of models for the make-up of the polyadenylation machinery remain viable. These range from, at one end, a single enzyme capable of carrying out both cleavage and polyadenylation reactions to, at the other extreme, a large multicomponent complex containing one or more snRNP particles.

To identify the factors that catalyze pre mRNA polyadenylation, we have fractionated HeLa cell nuclear extracts that contain the required activities. The data presented here show that the polyadenylation and cleavage activities can be separated from each other, and that a factor responsible for the sequence specificity of the reaction cofractionates with the cleavage activity.

RESULTS

Nuclear extracts active in pre mRNA cleavage and polyadenylation were prepared from HeLa cells essentially as described by Dignam et al. (27). Cleavage and polyadenylation of RNA precursors in crude or fractionated extracts were assayed by analyzing on denaturing polyacrylamide gels the products formed following incubation of bacteriophage SP6-generated pre RNAs containing either an adenovirus late (L3) or SV40 late poly(A) signal (see Figure 1). Previous studies have shown that both of these precursors can be processed with high efficiency in vitro (8,17). Because it is difficult to distinguish between polyadenylated but uncleaved RNA and RNA that has been both cleaved and polyadenylated, it was necessary to perform separate reactions to assay each activity. When Mg^{++} was present in reaction mixtures, polyadenylation could be readily detected, while accurately cleaved but not polyadenylated RNAs were observed when Mg^{++} was omitted (15, 16; see Figure 1).

The nuclear extract was first subjected to fractionation by precipitation with different concentrations of $(NH_4)_2SO_4$. Virtually all of both activities remained soluble when $(NH_4)_2SO_4$ was added to nuclear extracts at 20% of saturation, but precipitated when the salt concentration was raised to 40%. Figure 1 shows the results of processing reactions using

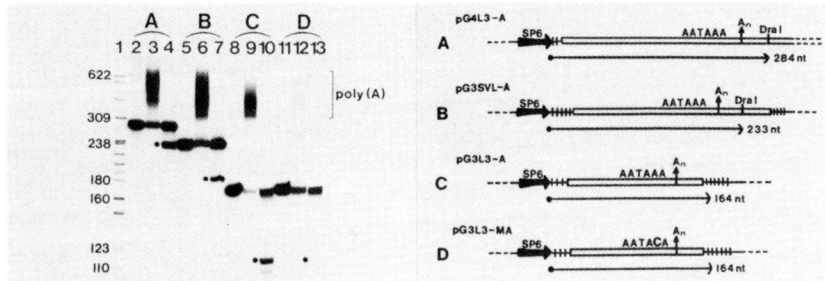

Figure 1. Cleavage and Polyadenylation of Precursor RNAs Processed in an $(NH_4)_2SO_4$-Fractionated Nuclear Extract. Adenovirus L3 and SV40 late precursor RNAs were synthesized in vitro with SP6 RNA polymerase and incubated in a 20-40% $(NH_4)_2SO_4$ fraction of a HeLa cell nuclear extract under standard reaction conditions. The processed RNAs were isolated and subjected to electrophoresis on a 5% polyacrylamide-8.3M urea gel. The structure of each precursor RNA is diagrammed at the right of the figure. Templates for transcription were cleaved with DraI (A and B) or BamHI (C and D). Lane 1, pBR322 HpaII DNA size markers; lanes A, pG4L3-A RNA; lanes B, pG3SVL-A lanes C, pG3L3-A RNA; lanes D, pG3L3-MA RNA. Lanes 2, 5, 8, and 11, precursor RNAs; lanes 3, 6, 9, and 12, precursor RNAs processed in the presence of 1mM Mg^{++}. Lanes 4, 7, 10, and 13, precursor RNAs processed in the presence of 1mM EDTA. The numbers on the left are the sizes of the DNA markers, in nucleotides. The same pBR322 HpaII markers were used in all experiments.

several different precursors and the 20-40% $(NH_4)_2SO_4$ fraction of the nuclear extract (20-40 fraction; see 28 for experimental procedures). The properties of the cleavage/polyadenylation reaction in the 20-40 fraction were identical to those detected in an unfractionated nuclear extract. Thus, Mg^{++} was required in the polyadenylation but not the cleavage reaction (polyadenylated RNAs are indicated by brackets, cleavage products by asterisks); the two reactions were coupled in the presence of Mg^{++} (ie, all of the cleaved RNA was polyadenylated); both the L3 and SV40 late precursors were processed efficiently; and both polyadenylation and cleavage were significantly inhibited by base substitutions in the AAUAAA sequence. The mutant analyzed in the experiment shown in Figure 1 contained a transversion that changed the L3 hexanucleotide to AAUACA, and resulted in an approximately five-fold inhibition of both cleavage and polyadenylation.

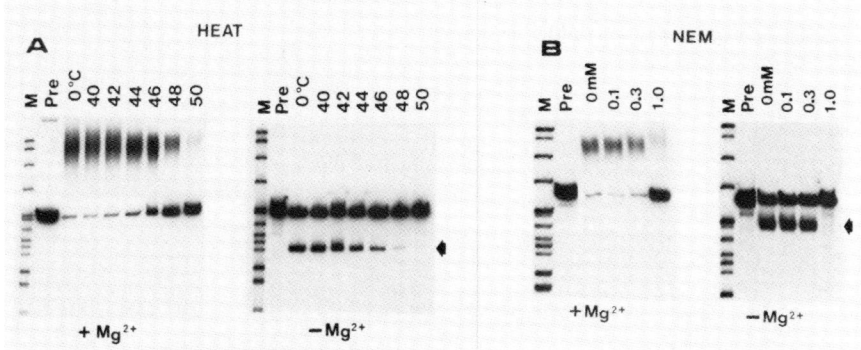

Figure 2. Sensitivities of Cleavage and Polyadenylation to Heat and n-ethylmaleimide. Aliquots of the 20-40 fraction were either pretreated for 5 minutes at various temperatures (panel A), or various concentrations of n-ethylmaleimide were added to processing reactions (panel B), as indicated. Reactions contained either G3SVL-A (panel A) or pG4L3-A (panel B) pre RNAs as substrates and were analyzed as described in Figure 1.

Cleavage and Polyadenylation Have Identical Sensitivities to Inhibitory Treatments.

To explore further the coupled nature of the cleavage and polyadenylation reactions, the sensitivities of each activity to heat and n-ethylmaleimide (NEM) were tested (Figure 2). After treatment, cleavage and polyadenylation were assayed in the presence or absence of Mg^{++}. Both reactions exhibited identical sensitivities to these inhibitory treatments (~50% inactivation occurred after heat treatment at 46°C and ~90% inactivation at 1mM NEM). In addition, various concentrations of the detergents NP40 and sodium deoxycholate inhibited both reactions to the same extent (not shown). These results suggest that the coupled nature of the cleavage and polyadenylation reactions may be due to the existence of a common factor that is required for both reactions.

Cleavage and Polyadenylation Activities Can be Separated From Each Other on the Basis of Size or Ion Exchange Properties.

To characterize further the activities that catalyze pre mRNA 3' end formation, the 20-40 fraction was subjected to gel

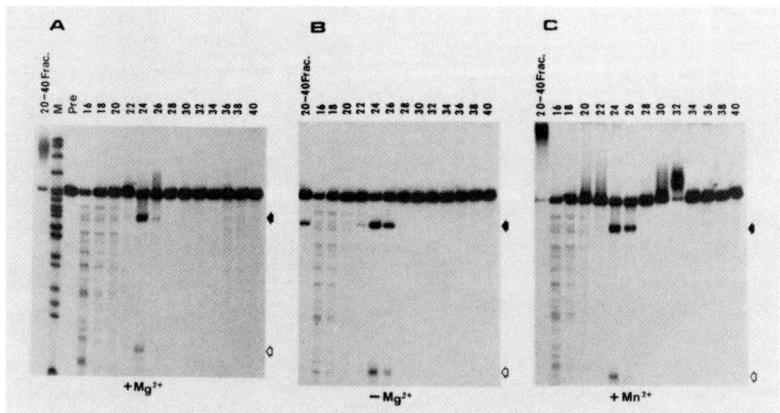

Figure 3. Separation of Cleavage and Poly(A) Polymerase Activities by Gel Filtration. Following Superose 6 gel filtration chromatography, fractions were concentrated by $(NH_4)_2SO_4$ precipitation, resuspended, dialyzed and were incubated with SV40 late (pG3SVL-A) pre RNA under standard reaction conditions in the presence of 1 mM $MgCl_2$ (A), 0.5 mM $MnCl_2$ (C) or no divalent cation (B). Upstream and downstream cleavage products are indicated by closed and open arrows, respectively.

filtration chromatography (see 28 for experimental procedures). Specifically, this fraction was applied to a Superose 6 column equilibrated with buffer containing 0.2M $(NH_4)_2SO_4$. Fractions were collected and assayed using a pre RNA containing the SV40 late poly(A) site. When Mg^{++} was used as a cofactor, poly(A) polymerase activity could not be detected in any of the fractions (Figure 3). However, a single peak of cleavage activity, centered around fraction 24, was observed. Both the upstream and downstream products of the cleavage reaction were present. Similar results were obtained in the absence of a divalent cation, strengthening the view that cleavage does not require a divalent cation.

In a further attempt to detect poly(A) polymerase activity after gel filtration chromatography, the same Superose 6 fractions were assayed in the presence of Mn^{++}. It is well known that a number of template-dependent polymerases become non-specific and are stimulated when Mn^{++} is present as a cofactor instead of Mg^{++} (28,29). Under these conditions quite a different profile was observed. While cleavage was again apparent in the same fractions as above, a single peak of polyadenylation was now also detected (fraction 32, Figure 3C). That the apparent increase in the size of the precursor, a property shown previously to result from polyadenylation when assays were performed with unfractionated extracts (30), in fact reflects the activity of a poly(A) polymerase was verified by oligo (dT) selection and by testing the ability of the enzyme to utilize each of the four ribonucleotides: the size of the precursor increased in the presence of ATP but not in the presence of CTP, UTP or GTP (not shown; see 28).

If the cleavage and polyadenylation activities separated by size fractionation are indeed distinct physical entities, then it may be possible to resolve them by a fractionation method that would distinguish differences other than size. Therefore, the 20-40 fraction was subjected to DEAE Sepharose chromatography. The results (not shown; see 28) indicate that the cleavage and polyadenylation activities could also be effectively separated from each other by anion exchange chromotography and that significant levels of polyadenylation activity were detected only in the presence of Mn^{++}, results virtually identical to those obtained when cleavage and polyadenylation activities were resolved by Superose 6. Polymerase activity was detected almost exclusively in the flow-through fraction (0.05 M $(NH_4)_2SO_4$), while the cleavage activity was retained on the column, with the majority eluted between 0.1 and 0.2 M $(NH_4)_2SO_4$.

Properties of the Separated Cleavage and Poly(A) Polymerase Activity.

Further characterization of the separated cleavage and poly(A) polymerase activities, either by gel filtration or DEAE Sepharose chromatography, indicate that the fraction containing cleavage activity, but not the fraction containing poly(A) polymerase activity, retains the requirement for an intact AAUAAA sequence. A base substitution in the SV40 late poly(A) site that changes the AAUAAA sequence to AAAAAA was not cleaved in the fraction containing cleavage activity, but was efficiently polyadenylated in the fraction containing poly(A) polymerase activity (not shown; see 28). In addition, as was found previously for the cleavage reaction when assayed in crude nuclear extract (eg. 8,17), the partially purified cleavage activity appears to require an energy source, as ATP and/or creatine phosphate were required for activity. A possibility not addressed by the above experiments was that the exclusive Mn^{++} requirement displayed by the fractionated poly(A) polymerase may have reflected the use of a substrate that extended beyond the in vivo poly(A) addition site, and would not be observed with a pre RNA cleaved at the in vivo site. However, this was found not to be the case. When a pre-cleaved precursor was used in assays, again only Mn^{++} dependent poly(A) polymerase was detected (not shown; see 28). Finally, the ability of 3'dATP to inhibit poly(A) synthesis by the non-specific Mn^{++} dependent poly(A) polymerase from Superose 6 chromatography and the specific poly(A) polymerase of the 20-40 fraction were examined. Incorporation of ^{32}P labeled AMP into acid insoluble material was measured at several concentrations of 3'dATP in the presence of Mn^{++}. Specifically, assays were performed in 25 microliter reactions in the presence of 1 mM MnCl 0.2 mg/ml HeLa cell cytoplasmic RNA as substrate, 0.5 mM ATP, 2.5 μCi ATP, various concentrations of 3'dATP and poly(A) polymerase activity capable of generating 17,000 cpm of acid insoluble material in 30 minutes at 30°C in the absence of 3'dATP. Reactions were stopped by adding 50 mM Tris pH 8, 10 mM ETDA, 0.1 M NaCl, 0.2% SDS and 1.5 mg/ml BSA, followed by addition of cold 10% trichloroacetic acid. Acid insoluble material was collected on glass microfibre filters and cerenkov radiation was measured. The Mn^{++} dependent poly(A) polymerase was inhibited 2.2 fold at 100 mM 3'dATP and 5.3 fold at 500 mM 3'dATP. The sensitivity of the poly(A) polymerase in the 20-40 fraction to 3'dATP was very similar, 1.8 and 6.5 fold at 100 and 500 mM 3'dATP respectively, suggesting that the mechanism of inhibition by 3'dATP for each enzyme preparation is the same, supporting the idea that the two enzymes are similar, if not identical.

Efficient and Specific Polyadenylation can be Reconstituted by Mixing the Cleavage and Poly(A) Polymerase Fractions.

The poly(A) polymerase activity obtained by Superose 6 or DEAE Sepharose chromatography was different in two respects from the poly(A) polymerase activity found in the 20-40 fraction. First, only Mn^{++} could function as a cofactor for the former, but either Mg^{++} or Mn^{++} activated the latter. Second, the crude poly(A) polymerase activity required an intact AAUAAA sequence, which was not required by the partially purified enzyme. To determine whether these properties of the unfractionated polymerase could be reconstituted, experiments in which different fractions were mixed with non-specific poly(A) polymerase-containing fractions were performed. Figure 4A shows that the peak Superose 6 fraction containing cleavage activity (fraction 24, see Figure 3) converted the Superose 6

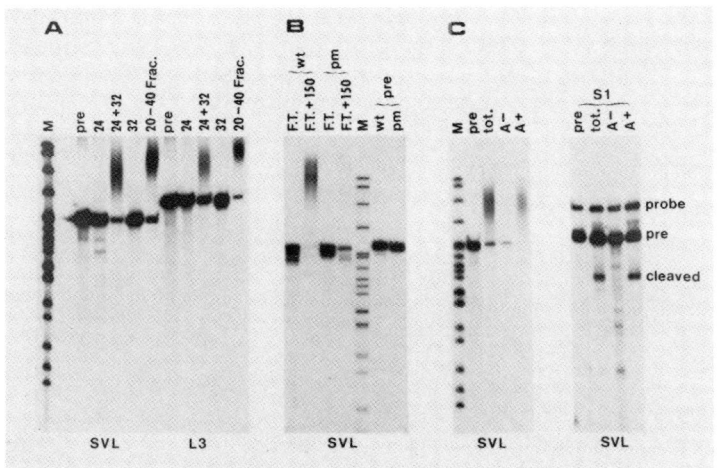

Figure 4. Reconstruction of Specific and Efficient Polyadenylation. (A) 3 μl of fractions containing either cleavage activity (fraction number 24) or Mn^{++}-dependent poly(A) polymerase activity (fraction number 32) obtained by Superose 6 gel filtration chromatography were assayed alone or after mixing (24+32). Polyadenylation reactions were performed under standard conditions using pG3SVL-A (SVL) or pG4L3-A (L3) pre RNAs in the presence of 1mM $MgCl_2$. 3 μl of the 20-40 fraction (20-40 Frac.) was assayed for comparison. (B) Polyadenylation activity was assayed using 3 μl of the DEAE flow-through fraction (F.T.) alone or after mixing with 3 μl of the DEAE- 0.15M $(NH_4)_2SO_4$ fraction (150). pG3SVL-A (SVL) wild-type (wt) and point mutant (pm) precursor RNAs were processed in the presence of 1mM Mg^{++}. (C) SVL pre-RNA processed in a mixture of Superose 6 fractions 24 and 32 (tot.) was selected on oligo (dT)-cellulose into poly (A)- and poly (A)+ fractions. Half of this RNA was analyzed directly (left) and half by S1 nuclease mapping (right).

polymerase (fraction 32) into a Mg^{++} dependent enzyme. Similar results were obtained with both SV40 late and adenovirus L3 precursors. The other Superose 6 fractions did not alter the properties of the poly(A) polymerase, nor did they bring about polyadenylation when mixed with the cleavage activity (results not shown), suggesting that the cleavage activity and non-specific poly(A) polymerase were necessary and sufficient to reconstitute authentic polyadenylation when mixed (see also below).

Figure 4B shows that virtually identical results were obtained when the DEAE Sepharose-poly(A) polymerase (flow-through fraction) and the peak DEAE cleavage fraction (0.15 M $(NH_4)_2SO_4$) were mixed and used to process the SV40 late pre RNA; efficient Mg^{++} dependent polyadenylation was again reconstituted. Additionally, Figure 4B shows that the sequence specificity of the polyadenylation reaction was also restored by mixing poly(A) polymerase and cleavage activities, as the mutant SV40 late pre RNA was polyadenylated very inefficiently. These findings strongly suggest that the Mn^{++} dependent, non-specific poly(A) polymerase detected after fractionation is the same enzyme involved in cleavage and polyadenylation of pre-mRNA.

The above experiments do not prove that the polyadenylation observed in the reconstituted reactions was accurate; ie, that the pre RNA was cleaved prior to polyadenylation. To test this, SV40 late pre RNA was processed using Superose 6 fractions 24 and 32 in the presence of Mg^{++}. RNA was then selected on oligo (dT) cellulose, and the bound and nonbound fractions were analyzed directly or first subjected to S1 nuclease analysis to determine the extent of 3' cleavage. The results (Figure 4C) show that the pre RNA was accurately cleaved, and that 100% of the cleaved RNA was polyadenylated. Virtually identical results (not shown) were obtained when the corresponding DEAE fractions were similarly analyzed. These findings confirm that site-specific cleavage/polyadenylation was effectively reconstituted by mixing two fractions containing cleavage and poly(A) polymerase activities.

DISCUSSION

The data presented here leads to the conclusion that the enzymatic machinery that catalyzes formation of the 3' ends of polyadenylated mRNAs consists of two distinct factors, a poly(A) polymerase and a cleavage/specificity factor (CSF). That these factors could be separated from each other on the basis of different physical characteristics (size and shape and ion exchange properties) suggests that the two complexes are physiologically relevant, and not simply artifacts of the fractionation procedures utilized. Further, based on a variety of mixing experiments, fractions containing these two activities were necessary and sufficient to reconstitute the complete pre mRNA cleavage/polyadenylation reaction.

A number of observations suggest that poly(A) polymerase and CSF interact during the polyadenylation reaction. First, several studies have indicated that the cleavage and polyadenylation reactions are normally coupled. For example, even after very short in vitro incubations, all cleaved pre RNA is polyadenylated (17,31). Second, we have observed identical sensitivities of the two activities in 20-40 fraction to a variety of inhibitory treatments, suggesting that a common factor is crucial for both reactions. Finally, when fractions containing poly(A) polymerase or CSF were mixed, each had a substantial effect on the other. The CSF-containing fraction converted the polymerase into a Mg^{++} dependent, sequence-specific enzyme and also increased the length of the poly(A) tract synthesized, while the polymerase (or a factor copurifying with it) expanded the range of the pre RNAs that could be efficiently cleaved by CSF (not shown; see 28). Taken together, these findings strongly suggest that the two factors physically interact with each other.

We have not yet been able to isolate the poly(A) polymerase - CSF complex as an intact entity. It is conceivable that the two activities associate only in the presence of pre RNA. Skolnik-David et al. (31) have shown that a large complex forms on pre RNAs containing functional poly(A) sites, and have suggested that this complex assembles on the pre RNA.

A number of previous studies have differentiated between Mg^{++} and Mn^{++} dependent poly(A) polymerases using non-specific assays. Perhaps most relevent to the experiments described here, Rose et al. (32) characterized two forms of poly(A) polymerase from rat liver nuclei, a free form and a chromatin-bound form. The free form very strongly preferred Mn^{++} as a cofactor, and could use a variety of small RNAs as primers. In contrast, the chromatin-bound form utilized Mg^{++} and could not polyadenylate exogenously added primers efficiently, using instead endogenous RNA, presumably hnRNA. However, when extracted from chromatin, the properties of the bound form became identical to those of the free form, and purification of the two enzymes suggested that they were identical. These results are strikingly similar to those described here. Thus, it is likely that the non-specific, Mn^{++} dependent polymerase we have described is related to the "free" form of Rose et al. (32), whereas the specific, Mg^{++} dependent polymerase, present in our 20-40 fraction, or reconstituted when the non-specific enzyme was mixed with CSF, corresponds to the "chromatin-bound" form described by these workers. It is possible that the functional association detected previously was not with chromatin, but rather with CSF, and the reason that the "bound" polymerase could not polyadenylate exogenous substrates efficiently is that AAUAAA-containing pre RNAs were not available. One difference, however, between the results of Rose et al. and the experiments presented here is that their chromatin associated or "bound" form of poly(A) polymerase had a much greater sensitivity (10-20x) to inhibition by 3'dATP than the "free" form. In contrast, our results indicate that these two activities have the same sensitivity to this inhibition. This apparent discrepancy may reflect differences in the enzyme preparation or may be due to inherent differences in the enzymes from rat liver and HeLa cell nuclei. In any event, these findings support our belief that the specific and non-specific polyadenylation we have observed are catalyzed by the same enzyme.

The identity of the polypeptides (and possibly nucleic acids) that constitute the poly(A) polymerase and cleavage/specificity factor is unclear. Rose and Jacob (33) purified a protein of monomer molecular weight of 60,000 that may be the only constituent of the polymerase they studied. Whether our preparation of poly(A) polymerase, which appears to be between 170,000 and 300,000 MW, is simply a multimer of this or a related polypeptide, or whether other factors are present, remains to be determined. Recently, Wilusz and Shenk (34) identified a 64,000 MW protein, present in nuclear extracts active in polyadenylation, that could be specifically crosslinked by UV light to pre RNAs containing an intact AAUAAA sequence. These authors showed that this protein elutes from DEAE-Sepharose at a salt concentration similar to that which we found elutes the majority of CSF. These findings suggest that the 64,000 dalton protein may be a constituent of CSF. Zarkower and Wickens (20) have also shown that a nuclear factor binds specifically to the AAUAAA sequences in pre RNAs. Further fractionation of CSF and poly(A) polymerase should reveal the identity of the factors that constitute the cleavage/polyadenylation machinery.

ACKNOWLEDGEMENTS

We thank M. Kapczynski for excellent technical assistance and W. Weast for help in preparing the manuscript.

REFERENCES

1. Birnstiel, M L, Busslinger, M, and Strub, K (1985). Transcription termination and 3' processing: the end is in site! Cell 41:349.
2. Leff, S E, Rosenfeld, M G, and Evans, R M (1986). Complex transcriptional units: diversity in gene expression by alternative RNA processing. Ann Rev Biochem 55:1091.

3. Proudfoot, N J and Brownlee, G G (1976). 3' Non-coding region sequences in eukaryotic messenger RNA. Nature 263:211.
4. Fitzgerald, M and Shenk, T (1981). The sequence 5'-AAUAAA-3' forms part of the recognition site for polyadenylation of late SV40 mRNAs. Cell 24:251.
5. Higgs, D R, Goodbourn, S E Y, Lamb, J, Clegg, J B, Weatherall, D J, and Proudfoot, N J (1983). a-thalassaemia caused by a polyadenylation signal mutation. Nature 306:398.
6. Montell, C, Fisher, E F, Caruthers, M H, and Berk, A J (1983). Inhibition of RNA cleavage but not polyadenylation by a point mutation in mRNA 3' consensus sequence AAUAAA. Nature 305:600.
7. Wickens, M and Stephenson, P (1984). Role of the conserved AAUAAA sequence: four AAUAAA point mutants prevent messenger RNA 3' end formation. Science 226:1045.
8. Zarkower, D, Stephenson, P, Sheets, M, and Wickens, M (1986). The AAUAAA sequence is required both for cleavage and for polyadenylation of simian virus 40 pre-mRNA in vitro. Mol Cell Biol 6:2317.
9. McDevitt, M A, Imperiale, M J, Ali, H, and Nevins, J R (1984). Requirement of a downstream sequence for generation of a poly(A) addition site. Cell 37:993.
10. Conway, L and Wickens, M (1985). A sequence downstream of AAUAAA is required for formation of simian virus 40 late mRNA 3' termini in frog oocytes. Proc Natl Acad Sci USA 82:3949.
11. Sadofsky, M, Connelly, S, Manley, J L, and Alwine, J C (1985). Identification of a sequence element on the 3' sides of AAUAAA which is necessary for simian virus 40 late mRNA 3'-end processing. Mol Cell Biol 5:2713.
12. Gil, A and Proudfoot, N J (1984). A sequence downstream of AAUAAA is required for rabbit b-globin mRNA 3'-end formation. Nature 312:473.
13. Hart, R P, McDevitt, M A, and Nevins, J R (1985). Poly(A) site cleavage in HeLa nuclear extract is dependent on downstream sequence. Cell 43:677.
14. Mason, P J, Elkington, J A, Lloyd, M M, Jones, M B, and Williams, J G (1986). Mutations downstream of the polyadenylation site of a Xenopus b-globin mRNA affect the position, but not the efficiency of 3' processing. Cell 46:263.
15. Moore, C L, Skolnik-David, H, and Sharp, P A (1986). Analysis of RNA cleavage at the adenovirus-2 L3 polyadenylation site. EMBO J 5:1929.

16. Sheets, M D, Stephenson, P, and Wickens, M P (1987). Products of in vitro cleavage and polyadenylation of simian virus 40 late pre-mRNAs. Mol Cell Biol 7:1518.
17. Moore, C L and Sharp, P A (1985). Accurate cleavage and polyadenylation of exogenous RNA substrate. Cell 41:845.
18. Manley, J L, Yu, H, and Ryner, L (1985). RNA sequence containing hexanucleotide AAUAAA directs efficient mRNA polyadenylation in vitro. Mol Cell Biol 5:373.
19. Ryner, L C and Manley J L (1987). Requirements for accurate and efficient mRNA 3' end cleavage and polyadenylation of a simian virus 40 early pre-RNA in vitro. Mol Cell Biol 5:495.
20. Zarkower, D and Wickens, M (1987). Formation of mRNA 3' termini: stability and dissociation of a complex involving the AAUAAA. EMBO J 6:177.
21. Edmonds, M (1982). Poly(A) adding enzymes. In The Enzymes. Vol. XV. Boyer, P.D., (ed.) New York: Academic Press p 217.
22. Jacob, S T and Rose, K M (1983). Poly(A) polymerase from eukaryotes. In Enzymes of Nucleic Acid Synthesis and Modification. Vol. II. Jacob, S T, (ed) Boca Raton, Fla: CRC Press, Inc. p 135.
23. Moore, C L and Sharp, P A (1984). Site-specific polyadenylation in a cell-free reaction. Cell 36:581.
24. Sperry, A O and Berget, S M (1986). In vitro cleavage of the simian virus 40 early polyadenylation site adjacent to a required downstream TG sequence. Mol Cell Biol 6:4734.
25. Hashimoto, C and Steitz, J A (1986). A small nuclear ribonucleoprotein associates with the AAUAAA polyadenylation signal in vitro. Cell 45:581.
26. Berget, S M and Robberson, B L (1986). U1, U2 and U4/U6 small nuclear ribonucleoproteins are required for in vitro splicing but not polyadenylation. Cell 46:691.
27. Dignam, J D, Lebovitz, R M, and Roeder, R G (1983). Accurate transcription initiation by RNA polymerase II in a soluble extract from isolated mammalian nuclei. Nucl Acids Res 11:1475.
28. Takagaki, Y, Ryner, L C and Manley, J L (1988). Separation and characterization of a poly(A) polymerase and a cleavage/specificity factor required for pre-mRNA polyadenylation. Cell 52:731.
29. Kornberg, A (1980). DNA Replication, A.C. Bartlett, (ed) San Francisco: W.H. Freeman and Company.
30. Manley, J L (1983). Accurate and specific polyadenylation of mRNA precursors in a soluble whole-cell lysate. Cell 33:595.

31. Skolnik-David, H, Moore, C L, and Sharp, P A (1987). Electrophoretic separation of polyadenylation-specific complexes. Gene & Develop 1:672.
32. Rose, K M, Roe, F J, and Jacob, S T (1977). Two functional states of poly (adenylic acid) polymerase in isolated nuclei. Biochem Biophys Acta 478:180.
33. Rose, K M and Jacob, S T (1976). Nuclear poly(A) polymerase from rat liver and a hepatoma. Comparison of properties, molecular weights and amino acid compositions. Eur J Biochem 67:11.
34. Wilusz, J, and Shenk, T (1988). A 64kd nuclear protein binds to RNA segments that include the AAUAAA polyadenylation motif in active polyadenylation 4extracts. Cell 52:221.

PROTEINS THAT INTERACT WITH THE SV40 LATE POLYADENYLATION SIGNAL[1]

Jeffrey Wilusz and Thomas Shenk

Department of Molecular Biology, Princeton University
Princeton, New Jersey 08544

ABSTRACT A series of polypeptides were efficiently crosslinked to the SV40 late polyadenylation signal using ultraviolet light. An intact AAUAAA signal was required for the interaction of a 64kd protein with RNA. Downstream element sequences were required for the crosslinking of the hnRNP C proteins, as well as two other proteins. The significance of these results to polyadenylation is discussed.

INTRODUCTION

The maturation of the 3' end of most messenger RNAs involves two concerted events: a site specific endonucleolytic cleavage reaction followed by the addition of 100-200 adenylate residues to the newly formed terminus (1). Two signals have been identified which play a role in the polyadenylation/cleavage event. The hexanucleotide AAUAAA (or a related sequence), located 10-30 bases upstream of the cleavage site, is absolutely required for polyadenylation (2,3). A second signal, located downstream of the cleavage site, controls the efficiency of polyadenylation (4,5). The downstream element appears to be of limited sequence complexity, comprised primarily of GU or U rich sequences, and may in

---

[1] This work was supported by a grant from the National Institutes of Health (CA 38965). J. Wilusz is an American Cancer Society postdoctoral fellow and T. Shenk is an American Cancer Society Research Professor.

some cases be reiterated (6,7,8). Together, these signals may play a role in gene expression by controlling alternative 3' end selection during mRNA processing (9,10).

Recent evidence has indicated a large 40-50S complex may be involved in polyadenylation signal recognition (11,12,13). Immunological evidence suggests snRNPs may form part of this complex (14,15,16). In order to identify components of this polyadenylation signal recognition complex, we have employed the method of UV crosslinking/label transfer analysis. We have identified a 64kd protein which requires an intact AAUAAA motif for interaction with polyadenylation signal containing RNAs (17). In addition, we have recently observed that the hnRNP C proteins bind to polyadenylation signal containing RNAs *in vitro* and that efficient crosslinking of these proteins required downstream element sequences (18). In this report, we describe the set of proteins we have detected which crosslink to the SV40 late polyadenylation signal during incubation of the RNA in the *in vitro* polyadenylation system.

RESULTS

In order to identify proteins associated with the SV40 late polyadenylation signal, a UV crosslinking/label transfer assay was employed. Briefly, RNAs were labeled *in vitro* by incorporation of [α-^{32}P] NTPs using SP6 polymerase (19). After incubation in the *in vitro* polyadenylation system (20) for 10 minutes, samples were irradiated with a germicidal light. Following treatment with ribonuclease A and denaturation, samples were analyzed on SDS/acrylamide gels. Proteins labeled by covalent transfer of [^{32}P]ribonucleotides were observed by autoradiography. Figure 1 shows the results obtained using a 122 base RNA (SVL3) containing the SV40 late polyadenylation signal (17), and a 61 base RNA (SVL8) containing only the SV40 late downstream element (from +8 to +55 relative to the cleavage site) (18). A 64kd protein which requires an intact AAUAAA signal for efficient interaction with the SV40 late polyadenylation signal (17) was found associated with SVL3 but not with SVL8. In addition, when the radiolabel was associated

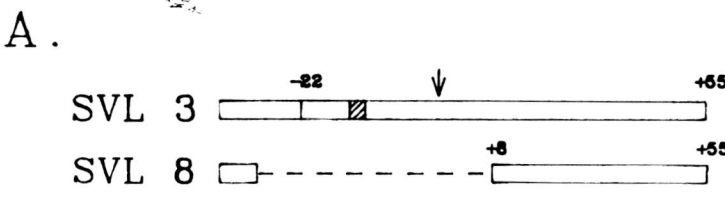

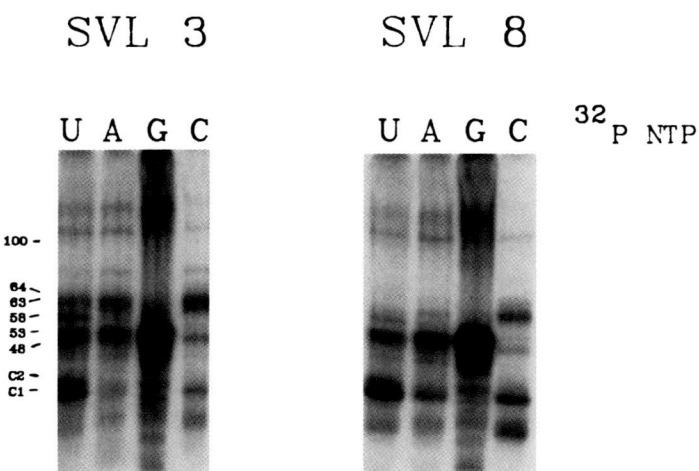

FIGURE 1. Proteins crosslinked to the SV40 late polyadenylation signal. A. Derivatives of the SV40 poly(A) signal analyzed. The arrow indicates the cleavage site. Sequences downstream of this site are designated (+), while sequences upstream are (-). The AAUAAA sequences is shaded. B. RNAs labeled *in vitro* with the indicated [^{32}P] nucleotide were incubated in the *in vitro* polyadenylation system (20) and subjected to UV crosslinking as described in the text (17). Crosslinked proteins were analyzed on 10% acrylamide gels containing SDS.

A. SVL DOWNSTREAM ELEMENT

```
8                             38        46        55
GCAUUCAUUUUAUGUUUCAGGUUCAGGGGGAGGUGUGGGAGGUUUUUU
```

B.

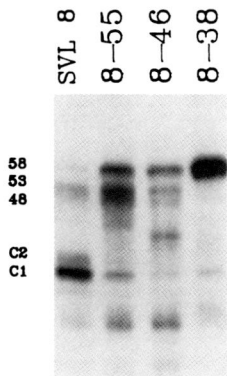

FIGURE 2. Crosslinking/label transfer analysis of 3' deletion derivatives of the SV40 late polyadenylation signal. A. Sequence of the downstream element region of the SV40 late poly(A) signal. B. UV crosslinking analysis was performed as described above using the RNAs indicated. Numbers represent the 5' and 3' ends of the RNAs employed and denote sequence positions downstream of the cleavage site.

with U,A or C residues, the 64kd polypeptide was detected by crosslinking analysis. Since the AAUAAA of the SV40 late polyadenylation signal is flanked by C residues, this result was not unexpected.

A second set of proteins, the hnRNP C proteins, was crosslinked efficiently to both transcripts (Fig. 1). Results obtained using deletion derivatives of several independent polyadenylation signals confirm that an intact downstream element is required for efficient crosslinking of these hnRNP proteins to polyadenylation substrate RNAs (18). When the 3' terminal 25 bases of the SV40 late polyadenylation signal are deleted (or substituted for by plasmid sequences), the efficiency of C protein crosslinking was greatly decreased (18).

A third set of proteins (48-53kd) was crosslinked to both SVL3 and SVL8 RNAs (Figure 1). These proteins appeared to be specific for the SV40 late polyadenylation signal as they were not crosslinked to any of the other polyadenylation signals tested (17,18). Transcripts containing radiolabel associated with G residues were exceptionally efficient label transfer substrates for these proteins. Partial purification of these proteins indicated that the 48-53kd band represents a protein doublet (data not shown).

Finally, three additional proteins are consistently observed by label transfer analysis on the SV40 late polyadenylation signal. The 58kd protein, crosslinked by both SVL3 and SVL8 RNAs, has been observed previously to be displaced from the RNA upon incubation of the extracts at $30^{\circ}C$ (17). This data suggests the 58kd protein is not involved in polyadenylation and may represent a non-specific protein-RNA interaction. An additional protein of approximately 63kd was identified using either transcript labeled at A or C residues. Preliminary evidence suggests that this protein may represent part of an hnRNP complex (data not shown). Finally, larger bands (>100kd) were crosslinked to both transcripts. The significance of these proteins is unclear, as they were also crosslinked to control transcripts lacking a polyadenylation signal (data not shown).

Mapping of Downstream Element Binding Proteins Using Oligonucleotide-derived RNAs

In order to further delineate sequences required for the downstream element specific protein interactions on the SV40 late polyadenylation signal, transcripts were produced using synthetic oligonucleotides, T7 polymerase and [^{32}P]labeled UTP as described (21). This approach provides the convenience of in vitro DNA template synthesis along with the removal of 5' polylinker sequences from the SVL RNAs. Because of bacteriophage promoter requirements, however, an extra 5' terminal G residue has been added to each RNA to allow efficient in vitro synthesis. Results obtained are shown in Figure 2. The numbers describing the RNAs correspond to nucleotide sequences they contain derived from the region downstream of the SV40 late cleavage site.

The 48 base 8/55 RNA crosslinked efficiently to the 48-53kd polypeptides but crosslinked relatively inefficiently to the hnRNP C proteins as compared to the SVL8 transcript. This result was unexpected because the two RNAs differ by only a 16 base polylinker derived sequence at the 5' end of SVL8. Several possible explanations exist, however, including a minimal size requirement for stable C protein binding, potential end effects, aberrant RNA folding and steric effects from other bound proteins. When RNAs containing small 3' deletions were tested (8/46, 8/38), crosslinking to the 48-53kd proteins was progressively inhibited, while crosslinking to the 58kd protein, which is normally displaced from the transcript, was enhanced.

These data indicate that the 48-53kd proteins require sequences between +38 and +55 relative to the cleavage site for interaction with the SV40 late polyadenylation signal. The highly efficient label transfer to the proteins when using transcripts with the radiolabel associated with G residues is consistent with this interpretation. A G-rich segment (13 out of 17 bases) lies within this segment and is only partially deleted in the 8/46 RNA. Finally, immunoprecipitation analyses indicate that the 48-53kd protein set does not include the La protein which has been implicated in in vitro polyadenylation (16; data not shown).

DISCUSSION

We have employed a UV crosslinking/label transfer technique to identify proteins associated with polyadenylation signal-containing RNAs in an *in vitro* polyadenylation system. A 64kd protein which required an intact AAUAAA sequence for interaction with RNA was identified as well as several proteins which required downstream element sequences for efficient crosslinking *in vitro*. While such an analysis alone does not demonstrate functional significance for these proteins, it does allow insight into the macromolecular associations of polyadenylation substrates *in vitro*.

Several lines of evidence suggest the 64kd protein may play an important role in polyadenylation. Six independent polyadenylation signal-containing RNAs crosslink to the protein *in vitro* (18). The AAUAAA motif, which is absolutely required for polyadenylation, it also required for efficient crosslinking to the 64kd protein. Deletion or mutation of this element severely affects the crosslinking of this protein to substrate RNAs (17). Biochemical fractionation indicates that this protein is not a poly(A) polymerase (17) and may, instead, represent a cleavage or specificity factor involved in 3' end processing (22).

The hnRNP C proteins, as well as the 48-53kd and 63kd polypeptides, require the downstream element for efficient crosslinking to the SV40 late polyadenylation signal. A requirement of downstream sequences for efficient C protein crosslinking was also seen with five additional polyadenylation signals (18). Deletion or substitution of sequences required for C protein interaction with the SV40 late polyadenylation signal correlates with a 3-4 fold reduced processing efficiency *in vitro* (18). This effect is similar to that reported by Sadofsky *et al.* for the SV40 late signal *in vivo* (4). These observations suggest that the function of the downstream element may be to support efficient hnRNP formation on polyadenylation substrates. The other proteins which require the downstream element for crosslinking may represent components of this complex, or, alternatively, may represent polyadenylation signal specific interactions. We are currently exploring these possibilities using biochemical and immunological approaches to dissect the polyadenylation signal recognition complex.

ACKNOWLEDGMENT

We wish to thank Elena Chiarchiaro for help in preparation of this manuscript.

REFERENCES

1. Birnstiel M L, Bussinger M and Strub K (1985). Transcription termination and 3' processing: the end is in site! Cell 41:349.
2. Proudfoot N J and Brownlee G G (1976). 3' non-coding region sequences in eukaryotic messenger RNA. Nature 263: 211.
3. Fitzgerald M and Shenk T (1981). The sequence 5'-AAUAAA-3' forms part of the recognition site for polyadenylation of late SV40 mRNAs. Cell 24:251.
4. Sadofsky M, Connelly S., Manley J L and Alwine J C (1985). Identification of a sequence element on the 3' side of AAUAAA which is necessary for simian virus 40 late mRNA 3'-end processing. Mol. Cell. Biol. 5:2713.
5. Conway L and Wickens M P (1985). A sequence downstream of AAUAAA is required for formation of simian virus 40 late mRNA 3' termini in frog oocytes. Proc. Natl. Acad. Sci. USA 82:3949.
6. McDevitt M A, Imperiale M J, Ali H, and Nevins J R (1984). Requirements of a downstream sequence for generation of a polyA addition site. Cell 37:993.
7. Zhang F, Denome R M and Cole C N (1986). Fine structure analysis of the processing and polyadenylation region of the herpes simplex virus type 1 thymidine kinase gene by using linker scanning, internal deletion and insertion mutations. Mol. Cell. Biol. 6:4611.
8. Mason P J, Elkington J A, Lloyd M M, Jones M B, and Williams J G (1986). Mutations downstream of the polyadenylation site of a Xenopus beta-globin mRNA affect the position but not the efficiency of 3' processing. Cell 46:263.

9. Darnell J E (1982). Variety in the level of gene control in eukaryotic cells. Nature 297:365.
10. Mather E L, Nelson K J, Haimovich J, and Perry R P (1984). Mode of regulation of immunoglobulin u and delta chain expression varies during β-lymphocyte maturation. Cell 36:329.
11. Humphrey T, Christofori G, Lucijanic V, and Keller W (1987). Cleavage and polyadenylation of mRNA precursors *in vitro* occurs within large and specific 3' processing complexes. EMBO J. 6:4159.
12. Skolnik-David H, Moore C L and Sharp P A (1987). Electrophoretic separation of polyadenylation-specific complexes. Genes Dev. 1:672.
13. Zarkower D and Wickens M (1987). Formation of mRNA 3' termini: stability and dissociation of a complex involving the AAUAAA sequence. EMBO J. 6:177.
14. Hashimoto C and Steitz J A (1986). A small nuclear ribonucleoprotein associates with the AAUAAA polyadenylation signal *in vitro*. Cell 45:581.
15. Moore C L, Skolnick-David H, and Sharp P A (1988). Sedimentation analysis of polyadenylation-specific complexes. Mol. Cell. Biol. 8:226.
16. Moore C L and Sharp P A (1984). Site-specific polyadenylation in a cell free reaction. Cell 36:581.
17. Wilusz J and Shenk T (1988). A 64kd nuclear protein binds to RNA segments that include the AAUAAA polyadenylation motif. Cell 52:221.
18. Wilusz J and Shenk T (1988). The C proteins of heterogenous nuclear ribonucleoprotein complexes interact with RNA sequences downstream of polyadenylation cleavage sites. Submitted for publication.
19. Melton D A, Krieg P A, Rebagliati M R, Maniatis T, Zinn K and Green M R (1984). Efficient *in vitro* synthesis of biologically active RNA and RNA hybridization probes from plasmids containing a bacteriophage SP6 promoter. Nucl. Acids Res. 12:7035.
20. Moore C L and Sharp P A (1985). Accurate cleavage and polyadenylation of exogenous RNA substrate. Cell 41:845.
21. Forster A C and Symons R H (1987). Self-cleavage of virusoid RNA is performed by the proposed 55-nucleotide active site. Cell 50:9.

22. Takagaki Y, Ryner L C and Manley J L (1988). Separation and characterization of a poly(A) polymerase and a cleavage/specificity factor required for pre-mRNA polyadenylation. Cell 52:731.

BUILDING THE RNA WORLD: EVOLUTION OF CATALYTIC RNA IN THE LABORATORY

Gerald F. Joyce

The Salk Institute for Biological Studies
La Jolla, California 92037

ABSTRACT We propose to construct an RNA-based evolving system in the laboratory, and use this system to explore the catalytic potential of RNA. The system requires the integration of three chemical processes: 1) residue-by-residue amplification of RNA; 2) introduction of random mutational errors; 3) selection of individuals in accordance with some imposed fitness criterion. Amplification is achieved by reverse transcribing from RNA to DNA and then using a high-turnover viral RNA polymerase to produce multiple copies of RNA. Mutations are introduced by replacing a segment of the plus strand DNA with a random sequence and then transcribing the partially mismatched template. Selection is performed by requiring each RNA molecule to carry out a transesterification reaction in order to become eligible for amplification. An RNA-based evolving system could be used to generate a phylogeny of catalytic RNAs in accordance with the imposed selection constraints. By altering the selection constraints it may be possible to modify the substrate specificity of an existing ribozyme or to develop ribozymes with novel catalytic function. In this way we hope to gain a better understanding of RNA's catalytic versatility and to assess its suitability for the role of primordial catalyst.

INTRODUCTION

The recent discovery of RNA molecules with catalytic activity (ribozymes) has forced a redefinition of the central dogma of biology [1-4]. It also suggests that an alternative life form could have existed at a very early stage in evolutionary history, prior to the development of protein synthesis and the translation apparatus. RNA is capable of serving as both genotype and phenotype, and thus could have provided the basis for an RNA organism (the so-called "RNA world" [5]). In the laboratory our goal is to combine amplification and mutation of an RNA genotype with selection of a corresponding RNA phenotype in order to construct an RNA-based evolving system. Such a system would constitute a working model of an RNA-based organism and would be a useful tool for exploring the catalytic potential of RNA.

There are two general approaches to the construction of an RNA-based evolving system. The "ground-up" approach focuses on the amplification of genetic information and assumes that adaptive phenotypic behaviors will emerge once natural selection has been established. This approach has led to the development of chemical systems in which a preformed RNA template is used to direct the synthesis of complementary oligomers (6,7). However, such systems are not general with respect to template sequence and, except under very limited circumstances, do not allow self-replication to occur (8,9). The "top-down" approach focuses on RNA phenotype and aims to solve the self-replication problem by designing an RNA enzyme with replicase activity. Several authors have suggested that it may be possible to construct an RNA replicase by modifying the core structure of an existing ribozyme (10-13). Unfortunately, there is no obvious route from modern ribozymes, which catalyze sequence-specific RNA cleavage/ligation reactions, to a general-purpose replicase, which would catalyze template-directed polymerization of activated mononucleotides.

A third approach, which is neither "ground-up" nor "top-down", is to rely on the activity of a protein enzyme in order to achieve amplification of RNA. This requires "stepping out" of the RNA world by using a biological catalyst that is not itself the product of RNA evolution. It may be worthwhile to introduce such an element of artificiality, so long as that element does not restrict the course of RNA evolution. The classic *in vitro* evolution experiments using Qß replicase and a variant form of Qß RNA, while providing a remarkable demonstration of evolutionary behavior in the test tube (14), are highly restricted due to the stringent substrate requirements of the replicase enzyme. All of the known RNA-dependent RNA polymerases (Qß replicase, MS2 replicase, etc.) are specific for RNA substrates whose sequence and secondary structure resembles that of the viral genome, and none are capable of amplifying any known catalytic RNA (15,16). What is needed is an amplification system that is blind to the sequence of the RNA that is being amplified.

We have devised a method based on a combination of standard molecular biology techniques that can be used to amplify a heterogeneous population of RNA molecules. The method requires the use of three polymerase enzymes, each of which is essentially indifferent to template sequence. Beginning with RNA, reverse transcriptase is used to generate a complementary DNA. This cDNA is converted to double-stranded DNA using a DNA-dependent DNA polymerase and then transcribed back to RNA using a DNA-dependent RNA polymerase. Amplification occurs at the level of transcription, relying on the high turnover of a viral RNA polymerase. T7 RNA polymerase, for example, generates 200-1200 moles of RNA transcript per mole of DNA template, depending on template length (17). Preliminary studies indicate that an entire amplification cycle, from RNA to DNA and back to RNA, can be carried out in a single reaction vessel in about 4 hours, with a net amplification factor of 100-200.

The construction of an RNA-based evolving system requires the integration of three chemical processes: 1) residue-by-residue amplification of RNA; 2) introduction of random mutational errors; 3) selection of individuals in accordance with some fitness criterion. Amplification maintains a constant total population size by producing copies of those individuals that survive the selection process. Mutation introduces variability through the formation of error copies during amplification. Selection occurs at the level of phenotype, acting to reduce variability by removing those individuals that do not conform to the imposed fitness criterion.

Through repeated cycles of amplification, mutation, and selection the population develops an organized structure consisting of one or more dominant individuals surrounded by a cluster of closely-related error copies (18). The dominant individuals are those that are most likely to survive the selection process, and therefore are most likely to replenish the population with additional copies of themselves. If the selection procedure is changed such that the dominant individuals are no longer the most likely to survive, then the population shifts in favor of some mutant species that is better suited to the new conditions. By changing the selection procedure in a gradual and progressive manner, it is possible to control the direction of evolution and to develop a population of RNA molecules whose behavioral characteristics are very different from those of individuals that were present in the initial population.

AMPLIFICATION OF RNA

As described above, it is possible to amplify RNA using a combination of three biological enzymes. RNA is reverse transcribed to cDNA, which is converted to double stranded DNA and then transcribed back to RNA (Fig. 1). Reverse transcription requires hybridization of a short oligodeoxynucleotide to the 3' end of an RNA template, followed by primer extension to produce a mixture of full-length and partial-length cDNAs. Typically, the yield of full-length products is 20-30% relative to the input of RNA template (19). For second-strand DNA synthesis a second oligodeoxynucleotide, which contains the complement of the T7 promoter sequence at its 5' end, is hybridized to the 3' end of the cDNA. Only full-length cDNA contains the hybridization site, so that partial-length products are excluded from this reaction step. The 5' overhang of both strands is converted to a complete duplex using the gap filling activity of *E. coli* DNA polymerase I (Klenow fragment). The conversion of cDNA to complete double-stranded DNA occurs with an efficiency of about 80-90% (20).

The crucial step in the amplification cycle is the *in vitro* transcription of RNA. The gene for T7 polymerase has been cloned (21), making it relatively easy to prepare very large amounts (2-10 MU) of the enzyme. This in turn allows high concentrations of polymerase to be used in the transcription reaction so as to maximize the yield of RNA. The polymerase binds to the promoter sequence and produces a run-off transcript that regenerates the original RNA. Premature transcripts do not contain the primer hybridization site for reverse transcription, and thus are unable to participate in subsequent rounds of amplification.

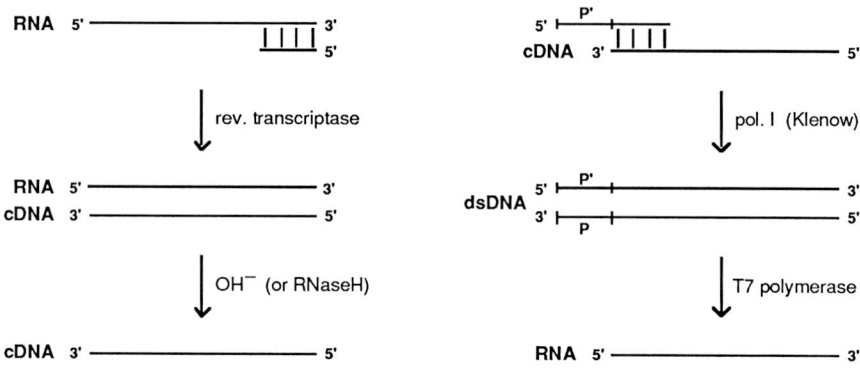

FIGURE 1. Procedure for *in vitro* amplification of RNA.

Amplification at the level of transcription restores the original population size, compensating for individuals excluded by selection or lost during the conversion of RNA to double-stranded DNA. Unless the mutation rate is very high or the selection regime very stringent, there will be more RNA at the end of a cycle of amplification/mutation/selection than there was at the beginning. From this excess an aliquot is removed to provide material for analytic purposes and to create an archive of the system's evolutionary history. If there is insufficient material due to strong mutational or selection pressure, the system can be put through a second round of amplification without performing mutation and selection.

There are two established techniques for carrying out further amplification at the level of double-stranded DNA. The first involves cloning the DNA into a suitable vector and then amplifying it within a bacterial host. A technique known as "shotgun cloning" is used to generate a population of chimeric vectors from a heterogeneous mixture of double-stranded DNA (22). This makes it possible to produce an essentially unlimited number of copies of the selected information. The major drawback to this technique is that it is very time consuming compared to *in vitro* amplification. In addition, clonal selection tends to occur during amplification within the bacterial host, thereby reducing population variability. A second technique for the amplification of double-stranded DNA is the "polymerase chain reaction" (23). This method utilizes two oligodeoxynucleotide primers (complementary to the 3' ends of the target DNA) and a thermostable DNA polymerase derived from the bacterium *Thermus aquaticus*. Amplification occurs through repeated cycles of heat denaturation, primer annealing, and polymerase-catalyzed primer extension, resulting in a 10^5-fold increase in the amount of DNA after about 2 hours (24). Both the shotgun cloning technique and the polymerase chain reaction can be used in conjunction with *in vitro* RNA amplification to produce very large amounts of material for analytic purposes.

INTRODUCTION OF MUTATIONS

Introduction of mutations during the amplification process results in a distribution of error copies clustered about the wild type. The distribution may include individuals whose behavior is more advantageous than that of the wild type with respect to some imposed selection constraint. In an evolving system, such individuals would be preferentially selected and then be used to generate a new distribution of error copies. Evolution is based on repetition of this cycle of events, leading to the development of molecules with adaptive behavioral properties.

The mutation process influences both the rate and direction of evolution. The rate of evolution is enhanced when the mutation rate is increased, subject to an upper limit determined by the relative fitness of the selected individuals (18). The direction of evolution is restricted by whatever sequence biases are introduced during mutagenesis. In the absence of *a priori* knowledge regarding what direction is likely to prove most advantageous, it is preferable to mutate in a random manner. Either one of two mutational strategies is adopted. The first attempts to scatter mutations throughout the molecule at a frequency of about $10^{-2} - 10^{-3}$. The second attempts to mutate with near certainty within a targeted region of interest by replacing a portion of the molecule with random nucleotides.

A number of techniques have been developed for *in vitro* mutagenesis of DNA. Chemical techniques, such as depurination (25), base modification (26,27), and the use of intercalating agents (28), have the disadvantage that each tends to produce only one type of mutational event. However, if one uses a combination of these techniques, it is possible to produce a variety of mutational lesions scattered throughout the molecule. The rate of mutation is enhanced by the use of a low-fidelity polymerase, such as AMV reverse transcriptase. This enzyme lacks the $3' \rightarrow 5'$ exonuclease (proofreading) activity found in DNA-dependent DNA polymerases (29), and can be used for both cDNA and second-strand synthesis (30). Its fidelity is further reduced by substituting Mn^{2+} for Mg^{2+} in the reaction mixture and/or by increasing the concentration of dNTP substrates (31,32).

The most widely used technique for site-directed mutagenesis involves hybridization of an oligodeoxynucleotide to single-stranded DNA, forming a partial duplex structure that contains a region of base mismatch. The oligomer strand is extended using a DNA-dependent DNA polymerase, and the resulting double-stranded DNA is used to transform bacterial cells (33). This technique is useful for producing a specific mutation at a defined location, but is awkward when one wishes to perform wholesale mutagenesis without taking the time to construct clones and harvest DNA from bacterial cells. Working in an entirely *in vitro* reaction system, we have modified the oligonucleotide-directed mutagenesis technique, enabling us to produce a population of RNAs that contain a short segment of random sequence (34). The modified technique relies on the ability of T7 RNA polymerase to generate many RNAs from a single DNA template (see above) and to operate efficiently on templates that are partially single-stranded (Fig. 2).

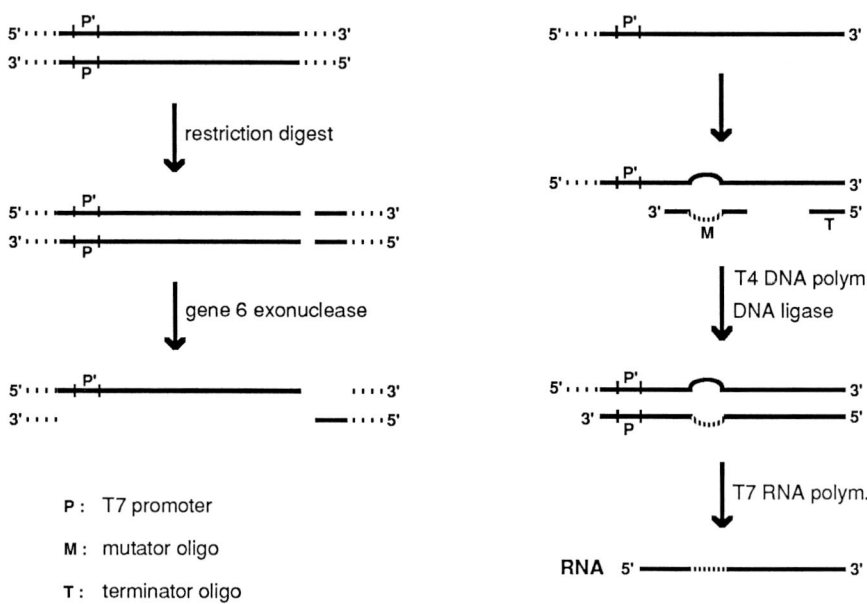

FIGURE 2. Procedure for *in vitro* mutagenesis.

Plasmid DNA, containing a T7 promoter and the gene for a catalytic RNA, is nicked by a restriction endonuclease at a site downstream from the target gene. The nicked plasmid is partially digested using a 5'→ 3' exonuclease (e.g. gene 6 exonuclease of T7 phage) to produce a stretch of single-stranded (minus strand) DNA. Two oligodeoxynucleotides are hybridized to this single-stranded segment. One, which we refer to as the "terminator oligo", forms a perfect duplex at the 3' end of the target gene. The other, which we refer to as the "mutator oligo", forms a partial duplex at a site of interest within the gene. The mutator oligo is designed such that it contains a central region of base mismatch flanked by two regions that form a perfect duplex. The mismatched region may be shorter or longer than the original complementary DNA, and may consist of a defined sequence or a mixture of random sequences. The two oligos are extended using T4 DNA polymerase and are ligated to form a template for transcription of the mutant RNA. T7 RNA polymerase binds to the double-stranded promoter and then moves along the template, transcribing RNA regardless of whether the DNA exists as a single or double strand (34).

SELECTION OF CATALYTIC RNA

There is a growing family of RNA molecules that have been found to exhibit catalytic activity *in vitro* (1,2,35,36). Probably the best studied of these molecules is the excised intron (IVS) from *Tetrahymena* pre-rRNA. The *Tetrahymena* ribozyme is able to catalyze sequence-specific RNA cleavage/ligation reactions, resulting in RNA products that have a 5'-phosphate and 3'-hydroxyl (37). The substrate requirements for this reaction have been well-characterized; the enzyme catalyzes nucleophilic attack by guanosine at a phosphodiester bond following a sequence of pyrimidines or nucleophilic attack by oligopyrimidine at a phosphodiester bond following guanosine (38). This activity provides a measure of RNA phenotype, and can be used to select functional ribozymes from a heterogeneous population of RNA molecules.

We have developed two techniques for the selection of catalytically active RNAs. Both involve the use of a truncated form of the *Tetrahymena* IVS that lacks the first 21 nucleotides at the 5' end and lacks the last 79 nucleotides at the 3' end. This molecule, which we refer to as L-21(Nhe), contains 314 nucleotides and has a 3'-terminal guanosine residue. It is able to catalyze transesterification reactions involving attack by the terminal G_{OH} at a phosphodiester bond following a sequence of pyrimidines.

The first selection technique measures the ability of the 3'-terminal G_{OH} to attack a phosphodiester bond near the 5' end of the catalyst itself to produce a circular molecule (Fig. 3a). An oligopyrimidine-terminated leader sequence is placed at the 5' end of L-21(Nhe), creating a target for nucleophilic attack. Circular products are easily distinguished from unreacted linear materials on the basis of characteristic differences in their electrophoretic mobility. Circular molecules migrate very slowly on polyacrylamide gels compared to linear molecules of the same size, and this difference can be enhanced by lowering the concentration of Tris/borate in the electrophoresis buffer (39). We have used this selection technique to harvest a number of alternative forms of the *Tetrahymena* ribozyme (40).

The second selection technique measures the ability of the 3'-terminal G_{OH} to attack a phosphodiester bond contained within an oligonucleotide substrate (Fig. 3b). The substrate is designed such that the product of the transesterification reaction creates a unique site for hybridization of an oligodeoxynucleotide primer. An oligodeoxynucleotide, which is complementary to 4-6 residues at the 3' end of the catalyst and 8-12 residues on the 3' side of the cleavage/ligation junction, binds preferentially to reacted materials. The bound oligomer is then extended using reverse transcriptase, producing a cDNA which carries the selected information. Selection is directly linked to amplification in that the production of cDNA serves as the first step in the amplification cycle.

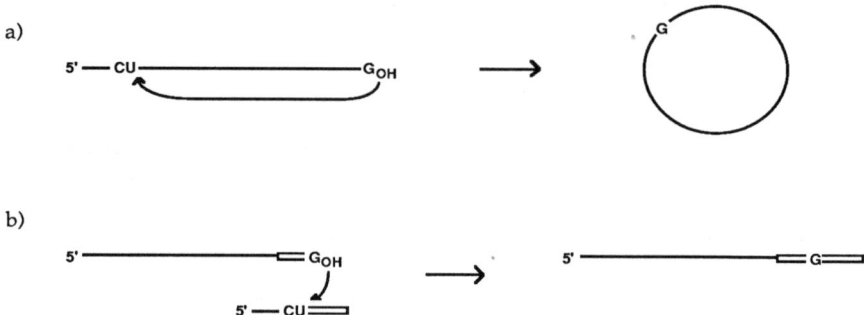

FIGURE 3. Two procedures for *in vitro* selection of catalytic RNA.

PROSPECTS FOR EVOLUTIONARY ENGINEERING

At the present time, the known catalytic repertoire of RNA enzymes is limited to hydrolysis and exchange reactions involving either an internal phosphodiester or terminal phosphomonoester of a nucleic acid substrate. This is not to say that ribozymes with some other function do not exist in modern biology or did not exist at an earlier stage in evolutionary history. But, given the small number of ribozymes that are currently available, how do we explore the possibility of alternative ribozyme function? An RNA-based evolving system offers one approach to this problem. Subjecting a population of RNA molecules to repeated cycles of amplification, mutation, and selection favors the emergence of individuals whose behavioral characteristics conform to the imposed selection constraints. By altering the selection constraints, it may be possible to develop RNA enzymes with novel catalytic activity. There are certainly some behaviors that lie beyond the capability of RNA and other behaviors that are so distant from the catalytic function of existing ribozymes that there is little hope of using evolution to bridge the gap. However, by selecting for behaviors that are chemically related to the behavior of existing ribozymes, it may be possible to extend the scope of RNA catalysis to include a broader range of phosphate ester reactions.

The selection constraints can be altered in two ways: either by changing the conditions under which a given RNA-catalyzed reaction is performed or by changing the nature of the reaction itself. With regard to changing reaction conditions, it is important to recognize that any *in vitro* selection system imposes constraints that are different from those which operate in the cellular environment. There is no reason to believe that the biological "wild type" is selectively advantageous compared to its mutant distribution when viewed outside of its native context. The wild type is further destabilized when more stringent reaction conditions are employed, favoring the emergence of mutant species that are better suited to the new conditions. By altering the reaction conditions in a gradual and progressive manner, one attempts to create an evolutionary pathway, leading through a succession of mutants to individuals that express the desired phenotype.

Changing the function of a catalytic RNA requires more than changing the conditions under which it operates; it becomes necessary to alter the substrate specificity of the reaction. This too should be done in a gradual and progressive manner in order to guide the evolving population toward the desired phenotype. One should provide a series of substrates that differ slightly in their chemical structure, each advancing one step closer to the ultimate goal. We refer to such as a series as a "molecular morphocline", in analogy to the evolutionary progression of morphologic characteristics that is observed at the organismal level. Beginning with a nucleoside 3'-phosphate or nucleotidyl-(3'→ 5')-nucleoside, one could imagine a molecular morphocline consisting of a series of nucleotide analogues that would lead to some other area of phosphate ester chemistry. The extent to which RNA enzymes can undergo functional modification by following such evolutionary pathways remains to be seen.

ACKNOWLEDGMENTS

The author wishes to thank Tan Inoue and Leslie Orgel for their helpful comments. G.F.J. is a Merck fellow of the Life Sciences Research Foundation.

REFERENCES

1. Kruger K, Grabowski PJ, Zaug AJ, Sands J, Gottschling DE, Cech TR (1982). Self-splicing RNA: autoexcision and autocyclization of the ribosomal RNA intervening sequence of *Tetrahymena*. Cell 31:147.
2. Guerrier-Takada C, Gardiner K, Marsh T, Pace N, Altman S (1983). The RNA moiety of ribonuclease P is the catalytic subunit of the enzyme. Cell 35:849.
3. Zaug AJ, Cech TR (1986). The intervening sequence RNA of *Tetrahymena* is an enzyme. Science 231:470.
4. Cech TR (1987). The chemistry of self-splicing RNA and RNA enzymes. Science 236:1532.
5. Gilbert W (1986). The RNA world. Nature 319:618.

6. Inoue T, Orgel LE (1983). A nonenzymatic RNA polymerase model. Science 219:859.
7. Joyce GF (1987). Nonenzymatic template-directed synthesis of informational macromolecules. Cold Spring Harbor Symp. Quant. Biol. 52:41.
8. von Kiedrowski G (1986). A self-replicating hexadeoxynucleotide. Angew. Chem. Int. Ed. Engl. 25:932.
9. Zielinski WS, Orgel LE (1987). Autocatalytic synthesis of a tetranucleotide analogue. Nature 327:346.
10. Sharp PA (1985). On the origin of RNA splicing and introns. Cell 42:397.
11. Pace NR, Marsh TL (1985). RNA catalysis and the origin of life. Orig. Life 16:97.
12. Cech TR (1986). A model for the RNA-catalyzed replication of RNA. Proc. Nat. Acad. Sci. USA 83:4360.
13. Szostak J (1986). Enzymatic activity of the conserved core of a group I self-splicing intron. Nature 322:83.
14. Mills DR, Peterson RL, Spiegelman S (1967). An extracellular Darwinian experiment with a self-duplicating nucleic acid molecule. Proc. Nat. Acad. Sci. USA 58:217.
15. Haruna I, Nozu K, Ohtaka Y, Spiegelman S (1963). An RNA "replicase" induced by and selective for a viral RNA: isolation and properties. Proc. Nat. Acad. Sci. USA 50:905.
16. Haruna I, Spiegelman S (1965). Specific template requirements of RNA replicases. Proc. Nat. Acad. Sci. USA 54:579.
17. Milligan JF, Groebe DR, Witherell GW, Uhlenbeck OC (1987). Oligoribonucleotide synthesis using T7 RNA polymerase and synthetic DNA templates. Nucl. Acids Res. 15:8783.
18. Eigen M (1971). Selforganization of matter and the evolution of biological macromolecules. Naturwiss. 58:465.
19. Berger SL, Wallace DM, Puskas RS, Eschenfeldt WH (1983). Reverse transcriptase and its associated ribonuclease H: interplay of two enzyme activities controls the yield of single-stranded complementary deoxyribonucleic acid. Biochemistry 22:2365.
20. Detera SD, Wilson SH (1982). Studies on the mechanism of *Escherichia coli* polymerase I large fragment: chain termination and modulation by polynucleotides. J. Biol. Chem. 257:9770.
21. Davanloo P, Rosenberg AH, Dunn JJ, Studier FW (1984). Cloning and expression of the gene for bacteriophage T7 RNA polymerase. Proc. Nat. Acad. Sci. USA 81:2035.
22. Messing J (1983). New M13 vectors for cloning. In Wu R, Grossman L, Moldave K (eds.): "Methods in Enzymology, Volume 101: Recombinant DNA, Part C," New York: Academic Press, p 43.
23. Scharf SJ, Horn GT, Erlich HA (1986). Direct cloning and sequence analysis of enzymatically amplified genomic sequences. Science 233:1076.
24. Saiki RK, Gelfand DH, Stoffel S, Scharf SJ, Higuchi R, Horn GT, Mullis KB, Erlich HA (1988). Primer-directed amplification of DNA with a

thermostable DNA polymerase. Science 239:487.
25. Kunkel TA, Schaaper RM, Loeb LA (1983). Depurination-induced infidelity of deoxyribonucleic acid synthesis with purified deoxyribonucleic acid replication proteins *in vitro*. Biochemistry 22:2378.
26. Shortle D, Koshland D, Weinstock GM, Botstein D (1980). Segment-directed mutagenesis: construction *in vitro* of point mutations limited to a small predetermined region of a circular DNA molecule. Proc. Nat. Acad. Sci. USA 77:5375.
27. Busby S, Dreyfus M (1983). Segment-specific mutagenesis of the regulatory region in the *Escherichia coli* galactose operon: isolation of mutations reducing the initiation of transcription and translation. Gene 21:121.
28. Shearman CW, Loeb LA (1983). On the fidelity of DNA replication: specificity of nucleotide substitution by intrcalating agents. J. Biol. Chem. 258:4477.
29. Battula N, Loeb LA (1976). On the fidelity of DNA replication: lack of exodeoxyribonuclease activity and error-correcting function in avian myeloblastosis virus DNA polymerase. J. Biol. Chem. 251:982.
30. Kacian DL, Myers JC (1976). Anticomplementary nature of small DNA produced during synthesis of extensive DNA copies of poliovirus RNA. Proc. Nat. Acad. Sci. USA 73:3408.
31. Sirover MA, Loeb LA (1977). On the fidelity of DNA replication: effect of metal activators during synthesis with avian myeloblastosis virus DNA polymerase. J. Biol. Chem. 252:3605.
32. Kunkel TA, Schaaper RM, Beckman RA, Loeb LA (1981). On the fidelity of DNA replication: effect of the next nucleotide on proofreading. J. Biol. Chem. 256:9883.
33. Gillam S, Smith M (1979). Site-specific mutagenesis using synthetic oligodeoxynucleotide primers: optimum conditions and minimum oligodeoxynucleotide length. Gene 8:81.
34. Joyce GF, Inoue T, unpublished results.
35. Peebles CL, Perlman PS, Mecklenberg KL, Petrillo ML, Jarrell KA, Cheng HL (1986). A self-splicing RNA excises an intron lariat. Cell 44:213.
36. Uhlenbeck OC (1987). A small catalytic oligoribonucleotide. Nature 328:596.
37. Zaug AJ, Kent JR, Cech TR (1984). A labile phosphodiester bond at the ligation junction in a circular intervening sequence RNA. Science 224:574.
38. Kay PS, Inoue T (1987). Catalysis of splicing-related reactions between dinucleotides by a ribozyme. Nature 327:343.
39. Grabowski PJ, Brehm SL, Zaug AJ, Kruger K, Cech TR (1983). Self-splicing of the ribosomal RNA precursor of *Tetrahymena*. In Hamer DH, Rosenburg MJ (eds.): "Gene Expression," New York: Alan R. Liss, p 327.
40. Joyce GF, Inoue T (1987). Structure of the catalytic core of the *Tetrahymena* ribozyme as indicated by reactive abbreviated forms of the molecule. Nucl. Acids Res. 15:9825.

A STEREOSELECTIVE BINDING SITE FOR ARGININE ON THE TETRAHYMENA SELF-SPLICING INTRON[1]

Michael Yarus

Department of Molecular, Cellular & Developmental Biology
University of Colorado at Boulder
Boulder, Colorado 80309-0347

L-arginine, uniquely among the standard set of modern amino acids, is an effective inhibitor of the self-splicing reaction carried out by a T7 RNA polymerase transcript of the Tetrahymena group I ribosomal intron (1). Inhibition is competitive with GTP, exhibiting a $K_i = 1.8$ mM at 5 mM Mg^{++}. By reincubating isolated splicing intermediates, it can be shown directly that splicing reactions after the attack of GTP on the precursor are unaffected by L-arginine, in agreement with the competitive kinetics.

This suggests that L-arginine can specifically occupy the guanosine site on the intron, and that the site distinguishes even similar amino acids such as L-lysine. The site is also stereoselective; inhibition with L-arginine occurs at lower concentrations than with D-arginine. Relative inhibition by varied analogues of L-arginine shows that there is considerable freedom for substitution of the alpha-amino and carboxyl groups of the amino acid. These observations are accommodated by a binding model which places arginine in the rG site by superposing the isoteric atoms of the guanidino group of arginine and the pyrimidine ring of GTP.

The existence of a specific binding site for an amino acid on a purified RNA suggests a possible route for the origin of the genetic code. Similar binding sites could also potentially serve to control the activities of allosteric RNA's by making them responsive to the concentrations of small molecules. Finally, if RNA's may specifically bind amino acids, then RNA enzymes can use that free

[1] This work was supported by NIH research grant R37 GM30881.

energy of binding to exert some order of catalysis on amino acid or peptide substrates, in addition to their action on currently known ribonucleotide substrates.

REFERENCES

1. Yarus M (1988). A specific amino acid binding site composed of RNA. Science, in press.

Index

AAUAAA sequence, polyadenylation signal, 3' end selection, protein interactions, 322, 323, 351–353, 355, 357
Adenylate kinase-like domain, ATP binding site, rho factor, *E. coli*, 330–331, 333
Alfalfa mosaic virus, treanslational enhancers, 238, 242, 243
Aminoacyl stem, RNase P, *B. subtilis*, catalysis of tRNA-like RNA of turnip yellow mosaic virus, 95
Aminoacyl-tRNA synthetase, 328
δ-Aminolevulinic acid (ALA) synthesis from glutamate, chloroplasts, 271–273, 276
Amplification of RNA in vitro, catalytic RNA, evolution in laboratory, 362–365
Amylase, wheat, 156
Anticodon nuclease (T4) (ACNase), phage T4 system, host (*E. coli*) tRNA reprocessing, 281, 282
  in vitro reconstitution, 283–284
Antidetermination by phage λ N protein, NusG cellular factor, 322
Antisense RNA, phage P22, gene expression regulation, 289–296
  immI region, 290–299
  ant gene, 290, 291
  negative regulation, 291, 296
  Pant, operon promoter, 290
  Psar promoter, 291
  RNA–RNA pairing model, 295–296
  small antisense regulatory (sar) RNA, 289
    pairing with *ant* mRNA, 293–296

RNase II, *E. coli*, 295, 296
  secondary structure, 291–293
  Shine-Dalgarno sequence of *ant*, 294, 296
Antisense RNA-mediated inhibition of MHV infection, 299–308
  micRNA-immune system, 300, 302, 304
  pathogen-derived resistance, 307
  schematic, MHV, 301
  Shine-Dalgarno sequence, 300
  tissue culture, 300
  transgenic mice, 300, 305–306
L-Arginine, stereoselective binding site, *Tetrahymena*, self-splicing intron, 373–374
ATP
  δ-aminolevulinic acid (ALA) synthesis from glutamate, chloroplasts, 271, 273
  binding site, adenylate kinase-like domain, rho factor structure–function correlations, *E. coli*, 330–331, 333
  and transcription initiation, vaccinia late mRNA poly(A) head generation, 202–204, 206
ATPase activity, rho factor, structure–function correlations, *E. coli*, 321, 326–328
AU content, intron recognition in plants, pre-mRNA splicing, 156, 159, 161–162
AUG initiation codon
  16S ribosomal RNA, UGA-specific translational suppressor *rrsB*, *E. coli*, 222

375

translational enhancers, eukaryotic viral (TMV) 5'-leader sequences (Ω) as, 238–239, 241, 244, 254
Autogenous regulation, S10 ribosomal protein operon, translational control, *E. coli*, 231
AUU codons, translational enhancers, eukaryotic viral (TMV) 5'-leader sequences (Ω) as, 238–239, 245

Bacillus subtilis, 34, 45
  glutamyl tRNAs, chloroplast, as cofactors in nonribosomal enzymatic reactions, 275–277
  phage SPOI, 59–60, 63–65
  see also Introns, group I, *B. subtilis* bacteriophage SPO1, self-splicing; RNase P, *B. subtilis* entries
Bacteriophage. *See* Phage *entries*
Base-paired stems of RNA, interactions with *Xenopus* transcription factor IIIa, specificity of 5S RNA binding activity, 129–130
Base pairing, mRNA–rRNA, and peptide chain termination, 16S ribosomal RNA, UGA-specific translational suppressor *rrsB*, *E. coli*, 224–227
Bean phaseolin, pre-mRNA splicing, 156
bla. See mRNA, *E. coli ompA* and *bla*, degradation initiation by 5' site-specific endonucleolytic cleavage
Branch points, intron recognition, 13–15, 17, 155, 156
Brome mosaic virus, translational enhancers, 238, 241–243
Bronze locus, maize, intron recognition in plants, pre-mRNA splicing, 156

C-202, rho factor modification, structure–function correlations, *E. coli*

chemical modification, 326–328, 332
  p-hydroxymercuribenzoate, 326–329, 332, 333
  NEM, 326–329, 332
  site-directed mutagenesis, 328–329, 332
C1054, 16S ribosomal RNA, UGA-specific translational suppressor *rrsB*, *E. coli*, 224–227
  A- and P-site cross-linking, 226
  S5 ribosomal protein effect, 224, 226–227
  sequence and secondary structure, 225
C1400 residue, 3OS ribosome function, role of 16S RNA, *E. coli*, 210, 211, 214
  insertion and deletion mutants, 215–217
  single base substitutions, 210, 214–216
Caenorhabditis elegans, U2 snRNP secondary structure, phylogenetic comparisons, 15–17
Calliphora erythrocephala fat bodies, splicing RNA in vivo, ultrastructure, 135–137
Cassette mutagenesis technique, 3OS ribosome function, role of 16S RNA, *E. coli*, 211
Catalysis, catalytic RNA. *See* Introns, group I *entries*; M1 RNA; RNase P RNA, *B. subtilis entries*; Structure of RNA, M1 RNA dimers, psoralen (HMT) cross-linking determination
Catalytic RNA, evolution in laboratory, 361–369
  amplification of RNA in vitro, 362–365
    polymerase chain reaction, 364–365
    shotgun cloning, 364
  evolutionary engineering, prospects, 368–369

mutagenesis in vitro, 365–367
  oligonucleotide-directed, 366
 RNA world, 361
 selection of RNA, 367–368
  3′ end, 367–368
Cauliflower mosaic virus, 35s promoter, 161–162
CCA, 3′-terminal, 80, 85, 86
 RNase P, *B. subtilis*, catalysis of tRNA-like RNA of turnip yellow mosaic virus, 95
Chlorophyll synthesis, rate-limiting step, δ-aminolevulinic acid (ALA) synthesis from glutamate, chloroplasts, 273
Chloroplasts, ALA synthesis, 271–273, 276; *see also* Glutamyl tRNAs, chloroplast, as cofactors in nonribosomal enzymatic reactions
Cis-acting sequences
 affect length, vaccinia late mRNA, poly(A) head generation at 5′ terminus, 204
 required for polyadenylation, 3′ end formation, pre-mRNA, polyadenylation and cleavage, 335, 336
Cleavage
 endonucleolytic. *See* mRNA, *E. coli* *ompA* and *bla*, degradation initiation by 5′ site-specific endonucleolytic cleavage
 hepatitis delta virus, human, self-cleaving RNAs, 102
 required sequences, 103–104
 cf. splicing RNA in vivo, ultrastructure, 134
Cleavage/specificity factor (CSF), 3′ end formation, pre-mRNA, polyadenylation and cleavage, 322–323, 336, 340–346
Conformational effects, ribosomal protein S4, RNA structure recognition, translational control, 120

Conformations, closed, open, and active, rho factor modification, structure–function correlations, *E. coli*, 332–333
Constitutive splicing factors, small nuclear U4 RNA, stage- and tissue-specific mRNA splicing, 174
Context dependence, 16S ribosomal RNA, UGA-specific translational suppressor *rrsB*, *E. coli*, 223–224
Crithidia fasiculata, 178, 179, 187, 188, 191, 193
cro mRNA and rho protein activity, RNA–protein interactions, 111
Cross-linking
 mapping procedures, structure of RNA, M1 RNA dimers, psoralen (HMT) cross-linking determination, 37–38
 30S ribosome function, role of 16S RNA, *E. coli*, 215–217
 structure of RNA, M1 RNA dimers, psoralen (HMT) cross-linking determination, 36
C-terminal and quaternary structure, rho factor modification, *E. coli*, 330, 331

Deletion tolerance at L6a-P6a, page T4 group I introns, self-splicing, structural requirements, 53–55
Delta antigen, hepatitis delta virus, human, self-cleaving RNAs, 104, 106
Dictyostelium, 258
DNA polymerase I of *E. coli*, cf. group I intron, *B. subtilis* bacteriophage SPO1, self-splicing, 63–64
Downstream protein-binding element, mapping, SV40 late polyadenylation signal, 3′ end selection, 356
Drosophila melanogaster
 embryos, splicing RNA in vivo, ultrastructure, 135, 138, 140, 141

Index

U2 snRNP, secondary structure, phylogenetic comparisons, 16–18

Early gene transcription, vaccinia, cf. late mRNA poly(A) head generation at 5' terminus, 200
Editing. See *Trypanosoma brucei*, mRNA editing of mitochondrial (kinetoplast) transcript
Elongation factors, 90
3' End formation, 321–323; see also mRNP assembly, 3'-poly(A) tracts in protein synthesis; SV40 late polyadenylation signal, 3' end selection, protein interactions
3' End formation, pre-mRNA, polyadenylation and cleavage, 322–323, 335–346
  cis-acting sequences required for polyadenylation, 335, 336
  cleavage/specificity factor (CSF), 322–323, 336, 340–346
  heat effect, 339
  HeLa cell nuclear extracts, 337, 342
  magnesium, 337–340, 342–345
  manganese, 340–345
  NEM effect, 339
  poly(A) polymerase, 322–323, 336, 340–346
    AAUAAA sequence, 332, 333, 336, 338, 341, 342, 345, 346
    bound cf. free, 345, 346
  reaction coupling, 345
  snRNP role, 323, 337
5' End, phosphorylation and M1 RNA dimer structure psoralen crosslinking, 36; see also mRNA, E. coli ompA and bla, degradation initiation by 5' site-specific endonucleolytic cleavage; Translational enhancers, eukaryotic viral (TMV) 5'-leader sequences (Ω) as; Vaccinia late mRNA, poly(A) head generation of 5' terminus
Enzymatic mapping, structure of RNA, pseudoknotted oligonucleotides, 26, 27
Error correction, glutamyl tRNAs, chloroplast, as cofactors in nonribosomal enzymatic reactions, 277–278
Evolution. See Catalytic RNA, evolution in laboratory

Finger structure, *Xenopus* transcription factor IIIa, specificity of 5S RNA binding activity, 123–124, 131
Folding, schematic, structure of RNA, pseudoknotted oligonucleotides, 26
Footprinting
  V1 nuclease, inaccuracy, 119–120
  Xenopus transcription factor IIIa, specificity of 5S RNA binding activity, 124–125, 128
Frog, U4X gene, 172–173

Gene expression. See Antisense RNA, phage P22, gene expression regulation; Transcription *entries*
Gene functionality, ribosomal protein S25 genes, *Tetrahymena*, 149
Genomic organization, small nuclear U4 RNA, stage- and tissue-specific mRNA splicing, 167, 168
α-Globin, human, 156, 161
β-Globin, human, 156, 159–161
  mRNA, 264
Glutamate and ALA synthesis, 271–273, 276
Glutaminyl tRNA synthesis, 274, 275
Glutamyl-tRNA reductase (GSA dehydrogenase), glutamyl tRNAs, chloroplast, as cofactors in nonribosomal enzymatic reactions, 273–274

Glutamyl tRNAs, chloroplast, as cofactors in nonribosomal enzymatic reactions, 271–278
  B. subtilis, 275–277
  E. coli, 275–277
  glutamyl-tRNA reductase (GSA dehydrogenase), 273–274
  GSA amino transferase, 273
  Hordeum vulgare, 272–275
  misaminoacylation by tRNA synthetase, 275–277
    error correction, 277–278
  mischarging of tRNA required for protein synthesis, 274–275
  tRNA^{Gln} species of glutamate isoacceptors, 274
  Synechocystis, 272–275, 277
GP1 pseudoknotted oligonucleotides, 26–31
GP4 pseudoknotted oligonucleotides, 27, 31
Growth hormone, human, 156
GSA aminotransferase, glutamyl tRNAs, chloroplast, as cofactors in nonribosomal enzymatic reactions, 273
GTP
  group I introns
    B. subtilis bacteriophage SPO1, self-splicing, 60, 61
    phage T4, self-splicing, structural requirements, 55, 56
  Tetrahymena self-splicing intron, stereoselective binding site for L-arginine, 373
GUS gene in good vs. bad context, translational enhancers, eukaryotic viral (TMV) 5'-leader sequences (Ω) as, 239–242, 246, 247, 254

Hairpins
  R17 coat protein recognition, 110–111

mRNA, E. coli ompA and bla, degradation initiation by 5' site-specific endonucleolytic cleavage, 312, 314, 315
S10 ribosomal protein operon, translational control, E. coli, 231–232, 234, 235
structure of RNA, M1 RNA dimers, psoralen (HMT) cross-linking determination, 39, 40, 42
see also Pseudoknots
Heat effect, 3' end formation, pre-mRNA, polyadenylation and cleavage, 339
HeLa cell nuclear extracts, 3' end formation, pre-mRNA, polyadenylation and cleavage, 337, 342
Helical parameters, structure of RNA, kinked [U(UA)$_6$A], X-ray crystallography, 2, 4, 10
Helicase activity, rho factor modification, structure–function correlations, E. coli, 321, 327–329
Hepadna virus, 99
Hepatitis B virus, 99
Hepatitis delta virus, human, self-cleaving RNAs, 99–106
  cleavage, required sequences, 103–104
  cleavage junction, 102
  delta antigen, 104, 106
  ligation, required sequences, 104
  cf. plant RNA viruses, 100
  reverse reaction (self-ligation), 100
  rolling circle replication method, 100, 104–106
  synthesized RNA
    in vitro, 101–102
    in vivo, 102
Hepatitis virus, mouse. See Antisense RNA-mediated inhibition of MHV infection
Hinge regions, Xenopus transcription factor IIIa, specificity of 5S RNA binding activity, 130

HIV-1 endcoded trans activator TAT, 322
Hordeum vulgare, glutamyl tRNAs, chloroplast, as cofactors in non-ribosomal enzymatic reactions, 272–275
p-Hydroxymercuribenzoate, rho factor modification, *E. coli*, 326–329, 332, 333

immI region antisense RNA, phage P22, gene expression regulation, 290–299
  ant gene, 290, 291
Internal control region, *Xenopus* transcription factor IIIa, specificity of 5S RNA binding activity, 123, 129, 130
Intron(s)
  branch point recognition region, U2 snRNP secondary structure, phylogenetic comparisons, 13–15, 17
  ribosomal protein S25 genes, *Tetrahymena*, 148–150
Intron recognition in plants, pre-mRNA splicing, 155–162
  AU content, 156, 159, 161–162
  branch point sequences, 155, 156
  human β-globin pre-mRNA in transfected plant protoplasts, 159–160
  sequences, cf. animal, and model construction, 159, 161–162
  soybean leghemoglobin
    in HeLa cell nuclear extract, 156–158
    pre-mRNA processing, transfected HeLa cells, 158
  splice site consensus, 155, 158
  cf. yeast, 156
Introns, group I, *B. subtilis* bacteriophage SPO1, self-splicing, 59–65

cf. DNA polymerse I of *E. coli*, 63–64
dot hybridization, 61–63
cf. eukaryotes, 65
GTP, 60, 61
open reading frames, 60, 63, 64
secondary structure model, 64
Introns, group I, phage T4, self-splicing, structural requirements, 49–57, 59–60, 63–65
  nrdB gene, 50, 51, 53
  cf. phage SPOI, *B. subtilis*, 59–60, 63–65
  sunY gene, 50, 51, 53
  td gene, 49–55, 60
    deletion tolerance, L6a-P6a, 53–55
    GTP, 55, 56
    model of secondary structure, 51
    open reading frames, 50, 51, 60
    P6 as catalytic cone structure, 50–53
    trans splicing, ribozyme cone, 55–57
    TS phenotype, 50, 54
Introns, group I, *Tetrahymena*, 367, 373, 374
Ionic conditions, *B. subtilis* RNase P, catalysis, 74, 76
  of tRNA-like RNA of turnip yellow mosaic virus, 92–94, 96
Irrelevant RNA, competition by, RNase P RNA, *B. subtilis*, site-directed mutagenesis and catalytic function, 72–73
IVS1, 2, and 3, 156–159, 161

Kinetoplast DNA, 177–178; see also *Trypanosoma brucei*, mRNA editing of mitochondrial (kinetoplast) transcript
Kinked RNA [U(UA)$_6$A], X-ray crystallography, 1–10

Index   381

L4 gene product, S10 ribosomal protein operon, translational control, *E. coli,* 231, 232, 234, 235

β-Lactamase. *See* mRNA, *E. coli ompA* and *bla*, degradation initiation by 5' site-specific endonucleolytic cleavage

La protein, SV40 late polyadenylation signal, 3' end selection, protein interactions, 356

Late mRNA. *See* Vaccinia late mRNA, poly(A) head generation at 5' terminus

Late promoter conserved sequence, vaccinia late mRNA, poly(A) head generation at 5' terminus, 200

5'-Leader sequences. *See* Translational enhancers, eukaryotic viral (TMV) 5'-leader sequences (Ω) as

Legumin, pea, pre-mRNA splicing, 156

Leishmania
   major, 192
   tarentolae, 178, 187, 191, 193

Ligation, required sequences, human hepatitis delta virus, self-cleaving RNAs, 104

Loop formation and removal, splicing RNA in vivo, ultrastructure, 134, 138–139, 141

Magnesium
   3' end formation, pre-mRNA, polyadenylation and cleavage, 337–340, 342–345
   Tetrahymena self-splicing intron, stereoselective binding site for L-arginine, 373

Maize bronze locus, intron recognition in plants, pre-mRNA splicing, 156

Manganese, 3' end formation, pre-mRNA, polyadenylation and cleavage, 340–345

Methylated bases, 3OS ribosome function, role of 16S RNA, *E. coli*, 216

Misaminoacylation by tRNA synthetase, glutamyl tRNAs, chloroplast, 275–277

Mischarging of tRNA required for protein synthesis, glutamyl tRNAs, chloroplast, 274–275

Mitochondria. *See Trypanosoma brucei*, mRNA editing of mitochondrial (kinetoplast) transcript

Molecular morphocline, 369

Monomers cf. M1 RNA dimers, psoralen (HMT) cross-linking determination of structure, 39–43

Morphocline, molecular, 369

Mutant ribosomes, in vitro generation, role of 16S RNA in 30S ribosome function, *E. coli*, 211–212

Mutational analysis of translational enhancer function, eukaryotic viral (TMV) 5'-leader sequences (Ω) as, 243–245, 253
   in prokaryotes (*E. coli*), 245–247

Mycoplasma capricolum, 226

NADH dehydrogenase, *Leishmania major*, 192

NADPH, δ-aminolevulinic acid (ALA) synthesis from glutamate, chloroplasts, 271, 276

Negative regulation, antisense RNA, phage P22, gene expression regulation, 291, 296

N-Ethylmaleimide (NEM), 326–329, 332, 339

Nicotiana tabacum, 159

NMR spectra, structure of RNA, pseudoknotted oligonucleotides, 28, 30, 31

nrdB gene, group I introns, phage T4, self-splicing, structural requirements, 50, 51, 53

Nuclear Overhauser effect spectroscopy, one-dimensional, structure of RNA, pseudoknotted oligonucleotides, 30
Nucleotidyl transferase, 90
NusG cellular factor, phage λ N protein transcription termination/antitermination, 322

Oat phytochrome type 3, intron recognition in plants, pre-mRNA splicing, 156
Oligonucleotide-directed mutagenesis, catalytic RNA, evolution in laboratory, 366
ompA. See mRNA, *E. coli ompA* and *bla*, degradation initiation by 5' site-specific endonucleolytic cleavage
Open reading frames, group I introns
 B. subtilis bacteriophage SPO1, self-splicing, 60, 63, 64
 td, phage T4, self-splicing, structural requirements, 50, 51, 60
Outer membrane protein A. See mRNA, *E. coli ompA* and *bla*, degradation initiation by 5' site-specific endonucleolytic cleavage
O. violaceus, 159

P6 as catalytic core structure, phage T4 group I introns, self-splicing, structural requirements, 50–53
Pant, operon promoter, phage 22 antisense RNA, gene expression regulation, 290
Pathogen-derived resistance, antisense RNA-mediated inhibition of MHV infection, 307
Pea legumin, pre-mRNA splicing, 156
Peptide-chain initiation, 30S ribosome function, role of 16S RNA, *E. coli*, 213, 217

Phage(s)
 λ N protein, transcription termination/antitermination, nusG cellular factor, 322
 SPOI, *B. subtilis*, cf. group I introns of phage T4, self-splicing, structural requirements, 59–60, 63–65
 see also Antisense RNA, phage P22, gene expression regulation; Introns, group I *entries*
Phage T4 promoter, 206
Phage T4 system, host (*E. coli*) tRNA reprocessing, 281–286
 anticodon nuclease (T4) (ACNase), 281, 282
 in vitro reconstitution, 283–284
 polynucleotide kinase (T4), 281, 282
 prr gene, host anticodon nuclease determinant, 282–285
 RNA ligase (T4), 281, 282
 RNase P, *E. coli*, and M1RNA, shared synthetic tRNA dimer precursor, 80
 stp, T4 anticodon nuclease determinant, 282–285
  amino acid sequence, 285
  p*stp* as catalytic component of ACNase, 284
Phaseolin, bean, intron recognition in plants, pre-mRNA splicing, 156
Phasianidae, 173
Phosphate–phosphate interaction, structure of RNA, kinked [U(UA)$_6$A], X-ray crystallography, 5–6
Phosphates, electrostatic repulsion, RNase P RNA, *B. subtilis*, site-directed mutagenesis and catalytic function, 73
Phosphorylation, 5' end, structure of M1 RNA dimers, psoralen (HMT) cross-linking determination, 36

partial photoreversal, 36, 38–41
Phylogenetic comparisons U2 snRNP, secondary structure, 13–21
Phytochrome type 3, oat, intron recognition in plants, pre-mRNA splicing, 156
Plant RNA viruses, cf. hepatitis delta virus, human, self-cleaving RNAs, 100; *see also* Intron recognition in plants, pre-mRNA splicing
poly(A)-binding protein, cytoplasmic, mRNP assembly, 3'-poly(A) tracts in protein synthesis, 258, 264, 267
Polyadenylation signals, ribosomal protein S25 genes, *Tetrahymena*, 152–153; *see also* 3' End formation, pre-mRNA, polyadenylation and cleavage; SV40 late polyadenylation signal, 3' end selection, protein interactions
poly(A) head. *See* Vaccinia late mRNA, poly(A) head generation at 5' terminus
poly(A) polymerase, 3' end formation, pre-mRNA, polyadenylation and cleavage, 322–323, 336, 340–346
  AAUAAA sequence, 332, 333, 336, 338, 341, 342, 345, 346
  bound cf. free, 345, 346
3'-poly(A) tracts. *See* mRNP assembly, 3'-poly(A) tracts in protein synthesis
poly(C), rho factor modification, structure–function correlations, *E. coli*, 321, 326–330, 332
poly(CAA) region, translational enhancers, eukaryotic viral (TMV) 5'-leader sequences (Ω) as, 245, 246
Polymerase chain reaction, catalytic RNA, evolution in laboratory, 364–365

Polynucleotide kinase (T4), phage T4 system, host (*E. coli*) tRNA reprocessing, 281, 282
Polypeptide synthesis, mRNA-dependent, 16S ribosomal RNA, UGA-specific translational suppressor *rrsB*, *E. coli*, 222
Polysomes, mRNP assembly, 3'-poly(A) tracts in protein synthesis, 260–264
P protein, *B. subtilis* RNase P, catalysis of tRNA-like RNA of turnip yellow mosaic virus, 96
Protein synthesis. *See* mRNP assembly, 3'-poly(A) tracts in protein synthesis
prr gene, host anticodon nuclease determinant, phage T4 system, host (*E. coli*) tRNA reprocessing, 282–285
Psar promoter, antisense RNA, phage P22, gene expression regulation, 291
Pseudoknots
  GP1, 26–31
  α-mRNA structure recognition, ribosomal protein S4, translational control, 114–116
  RNase P, *B. subtilis*, catalysis of tRNA-like RNA of turnip yellow mosaic virus, 90
  structure, 25–32
  U2 snRNP, secondary structure, phylogenetic comparisons, 14, 17–19
P site binding of tRNA, 30S ribosome function, role of 16S RNA, *E. coli*, 210, 213, 215, 217
Psoralen (HMT) cross-linking. *See* Structure of RNA, M1 RNA dimers, psoralen (HMT) cross-linking determination

Quail, U4X gene, 172–173

R17 coat protein recognition of hairpin site, 110–111
Reaction coupling, 3' end formation, pre-mRNA, polyadenylation and cleavage, 345
Recognition. *See* Intron recognition in plants, pre-mRNA splicing
Relaxed scanning model, translational enhancers, eukaryotic viral (TMV) 5'-leader sequences (Ω) as, 238
Release factor 2 mRNA, 16S ribosomal RNA, UGA-specific translational suppressor *rrsB*, *E. coli*, 222, 225–226
Reverse reaction (self-ligation), hepatitis delta virus, human, self-cleaving RNAs, 100
Reverse transcriptase-dideoxy sequencing, *B. subtilis* RNase P, catalysis of tRNA-like RNA of turnip yellow mosaic virus, 91
rho factor modification, structure–function correlations, *E. coli*, 321, 325–334
   ATPase activity, 321, 326–328
   ATP binding site, adenylate kinase-like domain, 330–331, 333
   conformations, closed, open, and active, 332–333
   C-terminal and quaternary structure, 330, 331
   cysteine-202, chemical modification, 326–328, 332
      p-hydroxymercuribenzoate, 326–329, 332, 333
      NEM, 326–329, 332
   cysteine-202, site-directed mutagenesis, 328–329, 332
   helicase activity, 321, 327–329
   poly(C), 321, 326–330, 332
   RNA- and ATP-binding sites, interaction at ATP hydrolysis, 321, 332–333

   RNA binding specificity, 321, 326, 329–333
rho protein and cro mRNA, 111
Ribonucleoprotein particle. *See* RNP entries
S4 Ribosomal protein, RNA structure recognition, translational control, 110, 113–120
   α-mRNA cf. 16S rRNA binding sites, 117–120
   α-mRNA pseudoknot recognition, 114–116
   secondary structure, 114, 115
   Shine-Dalgarno sequence, 115
   2 16S rRNA non-contiguous domains, 117
   5' domain, 118, 119
   V1 nuclease footprinting, inaccurate mapping, 119–120
   α operon, autoregulation, 113–114
S8 Ribosomal protein and spc RNA, 111
S10 Ribosomal protein operon, translational control, *E. coli*, 231–235
   autogenous regulation, 231
   hairpin, 231–232, 234, 235
   L4 gene product, 231, 232, 234, 235
   secondary structure, 232, 233
   translation initiation complex, 232, 235
S25 Ribosomal protein genes, *Tetrahymena*, 145–153
   gene functionality, 149
   introns compared, 148–150
      pre-mRNA splicing reaction, 150
   nucleotide and deduced amino acid sequences, 148–149
   polyadenylation signals, 152–153
   T. pigmentosa, 146–153
      TAA codon for Gln, 149
   transcription initiation regions, 150–151
   T. thermophila, 146–153
      cf. r proteins L1–3, 151, 153
30S Ribosome function, role of 16S RNA, *E. coli*, 209–217

A site binding, 216
C1400 residue, 210, 211, 214
   insertion and deletion mutants, 215–217
   single base substitutions, 210, 214–216
   cross-linking, 215–217
   3' end of 16S RNA, 210, 211, 213, 216
   methylated bases, 216
   mutant ribosomes, in vitro generation, 211–212
      cassette mutagenesis technique, 211
      functional activity, 213–214
   peptide-drain initiation, 213, 217
   P site binding of tRNA, 210, 213, 215, 217
Ribosome–mRNA binding, translational enhancers, eukaryotic viral (TMV) 5'-leader sequences (Ω) as, 238, 239, 244, 245, 249–251, 253, 254
Ribosome protection, *E. coli ompA* and *bla* mRNA, degradation initiation by 5' site-specific endonucleolytic cleavage, 316
Ribozyme. *See* Catalytic RNA, evolution in laboratory
RNA binding
   sites
      and ATP binding sites, interaction at ATP hydrolysis, rho factor structure–function correlations, *E. coli*, 321, 332–333
      ribosomal protein S4, 16S rRNA structure recognition, translational control, 117–120
   specificity, rho factor modification, structure–function correlations, *E. coli*, 321, 326, 329–333
RNA editing. *See Trypanosoma brucei*, mRNA editing of mitochondrial (kinetoplast) transcript

RNA fiber studies, structure, cf. kinked [U(UA)$_6$A], X-ray crystallography, 2, 4–6
RNA ligase T4, 80–81, 102, 281, 282
RNA polymerase, T7, 27, 63, 91
RNA–protein interactions, 109–111
   cro mRNA and rho protein activity, 111
   R17 coat protein recognition of hairpin site, phosphate residues and, 110–111
   hnRNA A1 protein, 111
   spc RNA and ribosomal protein S8, 111
   see also S4 Ribosomal protein, RNA structure recognition, translational control; *Xenopus* transcription factor IIIa, specificity of 5S RNA binding activity
RNA–RNA pairing model, antisense RNA, phage P22, gene expression regulation, 295–296
RNA structure. *See* Structure of RNA entries
RNA T$_1$ digestion, translational enhancers, eukaryotic viral (TMV) 5'-leader sequences (Ω) as, 239
RNA world, catalytic RNA, evolution in laboratory, 361
hnRNA A1 protein, RNA–protein interactions, 111
mRNA
   β-globin, 264
   splicing. *See* Small nuclear U4 RNA, stage- and tissue-specific mRNA splicing
   see also mRNP assembly, 3'-poly(A) tracts in protein synthesis; *Trypanosoma brucei*, mRNA editing of mitochondrial (kinetoplast) transcript
mRNA, *E. coli ompA* and *bla*, degradation initiation by 5' site-specific endonucleolytic cleavage, 311–319

Index

3' hairpin, 312, 314, 315
physical gene maps, 314, 317
ribosome protection, 316
5' site-specific endonucleolytic cleavages, 316–319
stability at different growth rates, 313
half-lives, 313
stem-loop structure, 312
5' terminus, 312, 314–315
translational efficiency and premature stop codons, effects on stability, 315–316
α-mRNA, cf. 16S rRNA binding sites, ribosomal protein S4, RNA structure recognition, translational control, 117–120
α-mRNA pseudoknot recognition, ribosomal protein S4, translational control, 114–116
secondary structure, 114, 115
Shine-Dalgarno sequence, 115
M1 RNA. *See* RNase P, *E. coli*, and M1 RNA, shared synthetic tRNA dimer precursor; Structure of RNA, M1 RNA dimers, psoralen (HMT) cross-linking determination
micRNA, 300; *see also* Antisense RNA entries
micRNA-immune system, antisense RNA-mediated inhibition of MHV infection, 300, 302, 304
pre-mRNA splicing reaction, ribosomal protein S25 genes, *Tetrahymena*, 150; *see also* 3' End formation, pre-mRNA, polyadenylation and cleavage; Intron recognition in plants, pre-mRNA splicing
5S RNA, *Escherichia coli*, 222; *see also* *Xenopus* transcription factor IIIa, specificity of 5S RNA binding activity
16S rRNA non-contiguous domains, ribosomal protein S4, RNA structure recognition, translational control, 117
5' domain, 118, 119
see also 30S Ribosome function, role of 16S RNA, *E. coli*
16S rRNA, UGA-specific translational suppressor rrsB, *E. coli*, 221–227
C1054, 224–227
A- and P-site cross-linking, 226
S5 ribosomal protein effect, 224, 226–227
sequence and secondary structure, 225
context dependence, 223–224
polypeptide synthesis, mRNA-dependent, 222
AUG initiation codon, 222
base pairing, mRNA-rRNA, and peptide chain termination, 224–227
release factor 2 mRNA, 222, 225–226
site-directed mutagenesis, 227
suppressor isolation, 222–223
23S RNA, *Escherichia coli*, 210, 222
snRNA. *See* Small nuclear U4 RNA, stage- and tissue-specific mRNA splicing
spc RNA and ribosomal protein S8, RNA-protein interactions, 111
tRNA
glutaminyl, 274, 275
protein-dependent processing, RNase P RNA, *B. subtilis*, site-directed mutagenesis and catalytic function, 69–72, 75, 76
yeast (Asp), cf. kinked [U(UA)$_6$A], X-ray crystallography, 2, 4–6
see also Glutamyl tRNAs, chloroplast, as cofactors in nonribosomal enzymatic reactions; RNase P, *B. subtilis*; RNase P, *E. coli*, and M1 RNA, shared synthetic tRNA dimer precursor

tRNA^{Gln} species of glutamate isoacceptors, glutamyl tRNAs, chloroplast, as cofactors in nonribosomal enzymatic reactions, 274
tRNA^{Glu}, δ-aminolevulinic acid (ALA) synthesis from glutamate, chloroplasts, 276
tRNA^{phe}, yeast, RNase P, *E. coli,* and M1 RNA, shared synthetic tRNA dimer precursor, 80–86
RNase II, *E. coli,* phage P22 antisense RNA, gene expression regulation, 295, 296
RNase A$_1$, structure of RNA, M1 RNA dimers, psoralen (HMT) crosslinking determination, 41
RNase P, *B. subtilis*
  catalysis of tRNA-like RNA of turnip yellow mosaic virus, 89–96
  ionic conditions and, 74, 76, 92–94, 96
  RNA, site-directed mutagenesis and catalytic function, 67–76
    irrelevant RNA, competition by, 72–73
    tRNA processing, protein-dependent, 69–72, 75, 76
    secondary structure model, 68
  structure of M1 RNA dimers, psoralen (HMT) cross-linking determination, 34, 45
RNase P, *E. coli,* and M1 RNA, shared synthetic tRNA dimer precursor, 79–86
  cf. phage T4, 80
  ^{32}P labeling, 83, 84
  tRNA^{phe}, yeast, 80–86
  3'-terminal CCA, 80, 85, 86
  T4 RNA ligase, 80–81
RNase P, *E. coli,* and M1 RNA dimers, psoralen (MHT) cross-linking determination, 41, 45
RNase T$_1$, structure of RNA, M1 RNA dimers, psoralen (HMT) cross-linking determination, 41

RNase V, structure of RNA, pseudoknotted oligonucleotides, 28, 29, 31
RNP fibril map construction, splicing RNA in vivo, ultrastructure, 135, 138, 140
RNP, splicing RNA in vivo, ultrastructure, 133–134, 136
hnRNP C protein, SV40 late polyadenylation signal, 3' end selection, protein interactions, 323, 355–357
mRNP assembly, 3'-poly(A) tracts in protein synthesis, 257–267
  length of tract (tail), 258–265, 267
    capped vs. uncapped, 260, 262–264, 267
    polysomes, 260–264
    protein content, 265
    translational efficiency, 260, 264, 266
  model, 268
  poly(A)-binding protein, cytoplasmic, 258, 264, 267
snRNP
  role in 3' end formation, pre-mRNA, polyadenylation and cleavage, 323, 337
  U2, phylogenetic comparisons, 13–21
Rolling circle replication method, hepatitis delta virus, human, self-cleaving RNAs, 100, 104–106
Rous sarcoma virus, translational enhancers, cf. TMV, 238, 241–243
rrsB. See 16S rRNA, UGA-specific translational suppressor *rrsB,* *E. coli*
Rubisco small subunit, intron recognition in plants, pre-mRNA splicing, 156

Saccharomyces cerevisiae, U2 snRNP, secondary structure, phylogenetic comparisons, 14–21

Sarcoma virus, Rous, translational enhancers, cf. TMV, 238, 241–243
Satellite tobacco necrosis virus, 285
Secondary structure
  antisense RNA, phage P22, gene expression regulation, 291–293
  model, *B. subtilis*
    group I intron, bacteriophage SPO1, self-splicing, 64
    RNase P RNA, *B. subtilis*, 68
    S10 ribosomal protein operon, translational control, *E. coli*, 232, 233
Self-cleaving RNAs. *See* Hepatitis delta virus, human, self-cleaving RNAs
Self-splicing. *See* Introns, group I entries
Shine-Dalgarno sequence, 214, 222, 232, 234
  ant antisense RNA, phage P22, gene expression regulation, 294, 296
  antisense RNA-mediated inhibition of MHV infection, 300
  ribosomal protein S4, RNA structure recognition, translational control, 115
  cf. translational enhancers, eukaryotic viral (TMV) 5′-leader sequences (Ω) as, 238, 239, 247, 248, 253, 254
Shotgun cloning, catalytic RNA, evolution in laboratory, 364
Single-stranded residues, highly conserved, *Xenopus* transcription factor IIIa, specificity of 5S RNA binding activity, 127–129
Single-stranded-specific S nuclease, pseudoknotted oligonucleotides, 28, 29, 31
Site-directed mutagenesis
  rho factor modification, *E. coli*, 328–329, 332

16S ribosomal RNA, UGA-specific translational suppressor *rrsB*, *E. coli*, 227
  see also under RNase P RNA
Size exclusion, ribosomal protein S4, RNA structure recognition, translational control, 119
Slippage mechanism, vaccinia late mRNA, poly(A) head generation at 5′ terminus, 199, 205–206
Small antisense regulatory (sar) RNA, antisense RNA, phage P22, gene expression regulation, 289
Small nuclear U4 RNA, stage- and tissue-specific mRNA splicing, 165–174
  adult, 167, 169–170
  constitutive splicing factors, 174
  developing chick, 170–171
  genomic organization, 167, 168
  splice site, 173–174
  SP6 RNA polymerase, 167, 169, 172
  U4B gene, 165–171, 173
  U4X gene, 165–168, 170, 171
  phylogenetic distribution, 171–173
Sm site, anti-Sm antibodies, U2 snRNP secondary structure, phylogenetic comparisons, 16–17, 20
Soybean leghemoglobin, intron recognition in plants, pre-mRNA splicing, 156–158
Sparsomycin, 238
Spectroscopy, NOE, pseudoknotted oligonucleotides, 30
Spliceosome, 13
  in vivo, ultrastructure, 134, 139, 141
Splice site consensus, intron recognition in plants, pre-mRNA splicing, 155, 158
Splicing RNA in vivo, ultrastructure, 133–141
  Calliphora erythrocephala fat bodies, 135–137
  cf. cleaving, 134

Index

Drosophila melanogaster embryos, 135, 138, 140, 141
  loop formation and removal, 134, 138–139, 141
  RNP, 133–134, 136
  RNP fibril map construction, 135, 138, 140
  spliceosome, 134, 139, 141
  splice site selection, first-come first-served principle, 141
  see also Small nuclear U4 RNA, stage- and tissue-specific mRNA splicing
SP6 RNA polymerase, small nuclear U4 RNA, stage- and tissue-specific mRNA splicing, 167, 169, 172
Stage-specificity. *See* Small nuclear U4 RNA, stage- and tissue-specific mRNA splicing
Stem-loop structure, mRNA, *E. coli ompA* and *bla*, degradation initiation by 5′ site-specific endonucleolytic cleavage, 312; *see also* Structure of RNA, U2 snRNP secondary structure, phylogenetic comparisons
Stereoselective binding site for L-arginine, *Tetrahymena*, self-splicing intron, 373–374
stp, T4 anticodon nuclease determinant, phage T4 system, host (*E. coli*) tRNA reprocessing, 282–285
  amino acid sequence, 285
  p*stp* as catalytic component of ACNase, 284
Structure–function correlations. *See* rho factor modification, structure–function correlations, *E. coli*
Structure of RNA, kinked [U(UA)$_6$A], X-ray crystallography, 1–10
  cf. DNA, 2, 3, 4, 8
  helical parameters, 2, 4, 10
  phosphate–phosphate interaction, 5–6
  cf. RNA fiber studies, 2, 4–6

cf. tRNA, yeast (Asp), 2, 4–6
  stabilization of kinks, 8–10
  H bonds, 8
  tetradecamer sequence, 5
  torsion angles, 6–8, 10
  twist angle variation, 7
Structure of RNA, M1 RNA dimers, psoralen (HMT) cross-linking determination, 33–46
  cross-linking, 36
  mapping procedures, 37–38
  hairpins, 39, 40, 42
  long-range interactions, 40, 42–43
  cf. monomers, 39–43
  5′ phosphorylation of, 36
  photoreversal, partial, 36, 38–41
  ^{32}P-M1-RNA (unmodified) probe, 36–41
  synthesis of M1 RNA, 35–36
Structure of RNA, pseudoknotted oligonucleotides, 25–32
  enzymatic mapping, 26, 27
  folding, schematic, 26
  GP1 oligonucleotides, 26–31
  GP4, 27, 31
  NMR spectra, 28, 30, 31
  nuclear Overhauser effect spectroscopy, one-dimensional, 30
  turnip yellow mosaic (RNA) virus, 26
Structure of RNA, U2 snRNP secondary structure, phylogenetic comparisons, 13–21
  alignment, 14–17
  Caenorhabditis elegans, 15–17
  chick, 16–17
  Drosophila melanogaster, 16–18
  human, 14–17, 19
  intron branch point recognition region, 13–15, 17
  mouse, 16–17
  positions 1–6, 15
  pseudoknots, 14, 17–19
  rat, 16–17

Saccharomyces cerevisiae, 14–21
Sm site, anti-Sm antibodies, 16–17, 20
    stem I, 15
    stem II, 16–17, 19, 20
    stem IV, 20–21
    stem-loop, 18
    trypanosome (*T. brucei*), 14–21
    Vicia faba, 16–17
    Xenopus laevis, 16–17
sunY gene, group I introns, phage T4, self-splicing, structural requirements, 50, 51, 53
Suppressor isolation, 16S ribosomal RNA, UGA-specific translational suppressor *rrsB*, *E. coli*, 222–223
SV40 late polyadenylation signal, 3' end selection, protein interactions, 322, 323, 351–358
    AAUAAA sequence, 322, 323, 351–353, 355, 357
    downstream protein-binding element, mapping, 356
    La protein, 356
    hnRNP C protein, 323, 355–357
    UV crosslinking/label transfer analysis, 352–354, 357
Synechocystis, glutamyl tRNAs, chloroplast, as cofactors in nonribosomal enzymatic reactions, 272–275, 277

TAA codon for Gln, ribosomal protein S25 genes, *Tetrahymena*, 149
Td gene, phage T4 self-splicing group I introns, structural requirements, 49–55, 60
3'-Terminal CCA, *E. coli* RNase P and M1 RNA shared synthetic tRNA dimer precursor, 80, 85,86
Tetradecamer sequence, structure of RNA, kinked [U(UA)$_6$A], X-ray crystallography, 5

Tetrahymena
    group I introns, 367, 373, 374
    self-splicing intron, stereoselective binding site for L-arginine, 373–374
    see also S25 Ribosomal protein genes, *Tetrahymena*
Tetrapyrroles, 271
Thermus aquaticus, 365
α-Thio-NTP, 110
Tissue specificity. *See* Small nuclear U4 RNA, stage- and tissue-specific mRNA splicing
Tobacco mosaic virus. *See* Translational enhancers, eukaryotic viral (TMV) 5'-leader sequences (Ω) as
Tobacco necrosis virus, satellite, 285
Tobacco protoplasts, translational enhancers, eukaryotic viral (TMV) 5'-leader sequences (Ω) as, 240–242, 244
Torsion angles, kinked [U(UA)$_6$A] RNA, X-ray crystallography, 6–8, 10
Trans activator TAT, hIV-1 endcoded, 322
Transcription factor. *See Xenopus* transcription factor IIIa, specificity of 5S RNA binding activity
Transcription initiation
    coupling to, vaccinia late mRNA poly(A) head generation at 5' terminus, 202–204
    ATP, 202–204, 206
    regions, ribosomal protein S25 genes, *Tetrahymena*, 150–151
Transcription termination, 321–323
    antitermination by phage λ N protein, NusG cellular factor, 322
    see also rho factor modification, structure–function correlations, *E. coli*
Transgenic mice, antisense RNA-mediated inhibition of MHV infection, 300, 305–306

Translational control. *See* S4 Ribosomal protein, RNA structure recognition, translational control; S10 Ribosomal protein operon, translational control, *E. coli*
Translational efficiency
and premature stop codons, effects on stability of *E. coli ompA* and *bla* mRNA, 315–316
mRNP assembly, 3'-poly(A) tracts in protein synthesis, 260, 264, 266
Translational enhancers, eukaryotic viral (TMV) 5'-leader sequences (Ω) as, 237–254
AUG initiation codon, 238–239, 241, 244, 254
AUU codons, 238–239, 245
capped vs. uncapped, 241, 242
GUS gene in good vs. bad context, 239–242, 246, 247, 254
mechanism, in vitro analysis, 247, 249–252
mutational analysis of enhancement function, 243–245, 253
in prokaryotes (*E. coli*), 245–247
poly(CAA) region, 245, 246
relaxed scanning model, 238
ribosome mRNA binding, 238, 239, 244, 245, 249–251, 253, 254
RNA T_1 digestion, 239
cf. Shine-Dalgarno sequence, 238, 239, 247, 248, 253, 254
tobacco protoplasts, 240–242, 244
T4-RNA ligase, 239
Xenopus laevis oocytes, 243, 244, 246, 253
Translational suppressor. *See* 16S rRNA, UGA-specific translational suppressor *rrsB, E. coli*
Translation initiation complex, S10 ribosomal protein operon, translational control, *E. coli*, 232, 235

Trans splicing, ribozyme cone, group I introns, phage T4, self-splicing, structural requirements, 55–57
Trypanosoma brucei, mRNA editing of mitochondrial (kinetoplast) transcript, 177–184, 187–195
bloodstream, 183, 184, 192, 193
COII gene, 188, 189, 191, 192
COIII, 189, 193–195
co- vs. posttranscriptional occurrence, 195
CYb gene, 188–193
frame shift corrections, 191
initiation codons, creation, 189–191
AUG, 178, 179
MURF-2, 188–190
poly (AU) tail, 183, 184, 188, 191, 192, 195
sequences, cf. DNA, 190, 191
stage-specific, 177–184, 192–193
TURF2 gene, 178–184
3' end of transcript, 180, 181, 183, 184, 188, 190–192, 195
unassigned reading frames (URF), 177–178
maxicircle, 178, 181, 188, 189
uridines, 187–189, 191
Trypanosoma brucei, U2 snRNP secondary structure, phylogenetic comparisons, 14–21
TS phenotype, phage T4 self-splicing group I introns, structural requirements, 50, 54
Turnip yellow mosaic virus
pseudoknotted oligonucleotides, 26
translational enhancers, cf. TMV, 238, 241, 242, 243
see also under RNase P, *B. subtilis*
Twist angle variation, structure of RNA, kinked [U(UA)$_6$A], X-ray crystallography, 7

U2 snRNP, secondary structure phylogenetic comparisons, 13–21

U4 RNA. *See* Small nuclear U4 RNA, stage- and tissue-specific mRNA splicing

U4B gene, small nuclear U4 RNA, stage- and tissue-specific mRNA splicing, 165–171, 173

U4X gene, small nuclear U4 RNA, stage- and tissue-specific mRNA splicing, 165–168, 170, 171
  phylogenetic distribution, 171–173

UGA. *See* 16S rRNA, UGA-specific translational suppressor *rrsB*, *E. coli*

Unassigned reading frames (URFs). *See under Trypanosoma brucei*, mRNA editing of mitochondrial (kinetoplast) transcript

UV crosslinking/label transfer analysis, SV40 late polyadenylation signal, 3' end selection, protein interactions, 352–354, 357

V1 nuclease footprinting, inaccurate mapping, ribosomal protein S4, RNA structure recognition, translational control, 119–120

Vaccinia late mRNA, poly(A) head generation at 5' terminus, 199–206
  cell-free transcription system, in vitro, 200–203, 205, 206
  cis-acting sequences affect length, 204
  cf. early gene transcription, 200
  late promoter conserved sequence, 200
  slippage mechanism, 199, 205–206
  transcription initiation, coupling to, 202–204
  ATP, 202–204, 206

Valyl-tRNA synthetase, 90

Vicia faba, structure of RNA, U2 snRNP, secondary structure, phylogenetic comparisons, 16–17

Wheat amylase, intron recognition in plants, pre-mRNA splicing, 156

Xenopus laevis
  oocytes, eukaryotic viral (TMV) 5'-leader sequences (Ω) as translational enhancers, 243, 244, 246, 253
  U2 snRNP secondary structure, phylogenetic comparisons, 16–17
  U4X gene, 172–173

Xenopus transcription factor IIIa, specificity of 5S RNA binding activity, 109–110, 123–131
  base-paired stems of RNA, interactions with, 129–130
  hinge regions, 130
  borders of binding site on 5S RNA, 125–126
  finger structure, 123–124, 131
  footprinting, 124–125, 128
  internal control region, 123, 129, 130
  single-stranded residues, highly conserved, 127–129
  somatic vs. oocyte RNA affinity for TFIIIa, 126–127
  consensus sequence, 127

X-ray crystallography, kinked RNA [U(UA)$_6$A] structure, 1–10

Yeast
  intron recognition, cf. plants, pre-mRNA splicing, 156
  tRNA, structure, cf. kinked RNA, 2, 4–6
  tRNA^{phe}, 80–86
  S. cerevisiae U2 snRNP secondary structure, 14–21